DEVELOPMENTAL AND CELL BIOLOGY SERIES

EDITORS
P.W. BARLOW   P.B. GREEN   C.C. WYLIE

EMBRYOGENESIS IN ANGIOSPERMS

EMBRYOGENESIS IN ANGIOSPERMS

A Developmental and Experimental Study

V. RAGHAVAN
Department of Botany
The Ohio State University

FOREWORD BY
M.S. SWAMINATHAN
Director-General
International Rice Research Institute

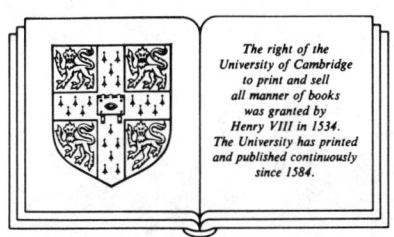

CAMBRIDGE UNIVERSITY PRESS
CAMBRIDGE
LONDON   NEW YORK   NEW ROCHELLE
MELBOURNE   SYDNEY

Published by the Press Syndicate of the University of Cambridge
The Pitt Building, Trumpington Street, Cambridge CB2 1RP
32 East 57th Street, New York, NY 10022, USA
10 Stamford Road, Oakleigh, Melbourne 3166, Australia

© Cambridge University Press 1986

First published 1986

Printed in the United States of America

Library of Congress Cataloging in Publication Data
Raghavan, V. (Valayamghat), 1931–
Embryogenesis in angiosperms.
(Developmental and cell biology series; 17)
Bibliography; p.
Includes index.
1. Botany–Embryology. 2. Angiosperms. 3. Botany, Experimental. I. Title. II. Series.
QK665.R328 1986 582.13'04333 85-11361

British Library Cataloguing in Publication Data
Raghavan, V.
Embryogenesis in angiosperms: a developmental
and experimental study.
1.Angiosperms 2.Botany–Embryology
I.Title
582.13'0433 QK495.A1
ISBN 0 521 26771 4

On many an idle day have I grieved over lost time. But it is never lost, my lord. Thou hast taken every moment of my life in thine own hands.

    Hidden in the heart of things thou art nourishing seeds into sprouts, buds into blossoms, ripening flowers into fruitfulness.

    I was tired and sleeping on my idle bed and imagined all work had ceased. In the morning I woke up and found my garden full with wonders of flowers.

Rabindranath Tagore
Gitanjali (1916)

Contents

	Foreword by M. S. Swaminathan	*page* ix
	Preface	xi
	Abbreviations	xiii
1	**Embryogenesis in angiosperms–the changing scene**	1
	How the scene has changed	2
	Methods of studying embryogenesis	4
	Concluding comments	11
2	**Developmental embryogenesis**	13
	Embryo development	15
	Endosperm development	36
	Formation of accessory embryos	41
	Concluding comments	44
3	**Cellular and biochemical aspects of embryogenesis**	46
	The egg, zygote and early division-phase embryos	46
	The suspensor	58
	Storage protein synthesis	72
	Loss of viability of embryos	80
	Concluding comments	82
4	**Experimental embryogenesis**	84
	Seed embryo culture	84
	Proembryo culture	97
	Precocious germination	103
	Genetics of embryogenesis	111
	Concluding comments	113
5	**Somatic embryogenesis**	115
	Totipotency and somatic embryogenesis	116
	Survey of somatic embryogenesis in angiosperms	120
	Cytology of somatic embryogenesis	134
	Physiology of somatic embryogenesis	136

Contents

    Genetics of somatic embryogenesis   147
    Concluding comments   150

6 Pollen embryogenesis   152
    Survey of pollen embryogenesis in angiosperms   153
    Embryogenesis in isolated pollen grains   162
    Pathways of pollen embryogensis   165
    Cytology of pollen embryogenesis   170
    Physiology of pollen embryogenesis   177
    Embryogenic competence of pollen grains   184
    Genetics of pollen embryogenesis   186
    Concluding comments   188

7 Regulation of gene activity during embryogenesis   190
    Zygotic embryogenesis   191
    Somatic embryogenesis   197
    Pollen embryogenesis   202
    Regulation of gene activity in embryo mutants   205
    Concluding comments   206

8 Applied aspects of embryogenesis   207
    Embryo rescue in inviable crosses   208
    Clonal multiplication of plants   221
    Haploids for breeding and selection   223
    Preservation of the germ plasm   225
    Concluding comments   228

References   230
Author index   282
Subject index   294

Foreword

Studies on the growth and development of embryos in flowering plants have attracted interest since ancient times. The advent of the microscope has led to rapid progress in our understanding of embryogenesis, hence the availability of several books and monographs on different aspects of embryogenesis. Professor Raghavan has brought together for the first time in synoptic form, and in a style characterized by clarity of expression and authority of understanding, the most recent knowledge on the theoretical, developmental and experimental facets of embryogenesis in angiosperms. He has analyzed the complexity of the different aspects of embryo development using a systems approach based on data from morphology, anatomy, genetics and biochemistry.

A unique feature of this book is its comparative description of *in vivo* and *in vitro* processes concerned with embryogenesis. The vast quantity of data available in this field has been presented in such a way that the knowledge becomes directly useful for those interested in producing embryoids from somatic cells as well as from pollen grains. Consequently, this book will be of help to all scholars and scientists interested in the control of somatic and pollen embryogenesis. Those working in tissue-culture research and genetic engineering in flowering plants will find this book particularly useful.

The chapter on somatic embryogenesis helps us to understand the mechanisms controlling morphogenetic changes in differentiated cells. The author has stressed the need for further research on overcoming the difficulty of triggering somatic embryo production, particularly in monocots. For those interested in the commercial exploitation of hybrid vigor in self-pollinated plants, like rice, the potential offered by somatic embryogenesis for the rapid multiplication of F_1 seeds will be particularly attractive. China already has more than 7 million hectares under hybrid rice, and a major determinant of the commercial viability of this technology in other countries will be the cost of the F_1 seed.

One area of research discussed in the book that is of considerable theoretical and practical significance is direct embryogenesis from pollen. Normally, in cereal crops, the pathway used is the production of callus in anther culture followed by subsequent plant regeneration. However, there appears to exist the possibility of producing embryoids directly from pollen grains, as in tobacco. If such techniques can be standardized in major economic plants, this will be the method of choice for producing haploids from F_1 hybrids, as it would totally eliminate the occurrence of chimeras and mixoploids. Such haploids could then become sources of homozygous diploids.

Another area that the author has dealt with in a superb manner is the technique of embryo rescue in hybrid combinations in which the embryo normally aborts. In many crop plants, we urgently need new genes for pest resistance, grain quality, adaptation to adverse soils and environmental conditions. Genes for these characters are often available in wild relatives. Embryo rescue helps transfer such genes, whereas normally hybrid seeds cannot be obtained because of embryo abortion. This monograph also deals with several other applied aspects of embryogenesis, such as the preservation of germ plasm.

We owe a deep debt of gratitude to Professor Raghavan for his labor of love resulting in this outstanding book on a topic of supreme relevance to the goal of working for a world where no human being goes to bed hungry.

Manila, Philippines   *M. S. Swaminathan*

Preface

The idea of writing a small book on plant embryogenesis that will deal with the normal development of the embryo as well as with the apparently similar processes that form embryo-like structures from somatic cells and pollen grains had been brewing in my mind for some time. The objective was to present an integrated version of the facts about the morphology, ontogeny, biochemistry and genetics of different modes of embryogenesis encountered in angiosperms in a book that would be suitable for a one-semester course on plant embryogenesis. As for the relatively small group of botanists who specialize in angiosperm embryogenesis, it was my hope that the book will serve as a review of the current perspectives in the field. I was happy, therefore, to accept an invitation extended by Dr. Paul Green on behalf of the editors of the Developmental and Cell Biology Series of Cambridge University Press to write such a book.

The view of angiosperm embryogenesis presented in this book owes as much to the research performed during the first half of this century as it does to the work done during the past two decades. Traditionally, embryogenesis has been of enduring interest to the botanist as a means to study the complex series of events leading from the fertilized egg to the formation of a full-grown embryo. Later, tissue culture methodology opened up new avenues of investigation that allowed us to demonstrate the development of facsimiles of embryos from single cells and pollen grains. Much of the recent progress in our understanding of embryogenic processes has resulted from the application of new and innovative biochemical and molecular techniques. The study of angiosperm embryogenesis is therefore a subject that finds itself at the crossroads of a multidisciplinary approach and one that will stimulate the reader to draw on the disciplines of tissue culture, biochemistry, cell biology and genetics. It is my view that only through such an approach can we fully comprehend the problems underlying embryogenic development in plants.

In writing this book, I have attempted to draw a broad picture of the different processes involved in embryogenic development from the fertilized egg, the somatic cell and the pollen grain and to point out the directions in which important advances are made. This account occupies the first six chapters of the book. Chapter 7 discusses the mechanism of programming of developmental information during embryogenic episodes. An inevitable source of disappointment in this account is that, thus far, so little information about gene activation during embryogenesis has come to light. The concluding chapter focuses on the possible practical applications of the new knowledge gained from the developmental and experimental study of angiosperm embryogenesis in enhancing crop productivity in the field. It is hoped that this approach will prove useful to the student as well as to those engaged in research in developmental embryogenesis.

Although the proportion of original research articles to reviews cited in the book is very high, no attempt has been made to include all the relevant literature. The appearance of ever-increasing numbers of reviews on somatic embryogenesis and pollen embryogenesis and the extended treatment given to some of the topics covered here in other publications led to the arbitrary decision to concentrate mainly, but not exclusively, on papers published since 1975. Even then, the fact that more than 1,000 original references have been cited in the book testifies to the vigorous condition of research in this field. The references cited are intended to provide access to the mainstream of research upon which the account in each chapter is developed; in those cases in which the reference is for the same phenomenon discovered in another plant, the intent is to convey the wide basis for the discovery reported.

For a book of this kind, the number of illustrations has been kept to a minimum. A few figures were prepared by me specially for this book, but most were borrowed from published works. I am grateful to authors and publishers who have granted me permission to reproduce these figures. I also appreciate the courtesy of colleagues who have generously provided original prints from their works for my use in the book.

The entire manuscript was reviewed by Paul Green and Peter Barlow, two editors of the Developmental and Cell Biology Series. I am deeply indebted to them for the many useful suggestions for improvements of the style and scientific content of the book. Finally, it is a pleasure to thank my wife, Lakshmi, and my daughter, Anita, for their help and devotion and, above all, for enduring with equanimity the long hours I spent in the preparation of this work.

V. Raghavan

Abbreviations

ABA	abscisic acid
ATP	adenosine triphosphate
BA	benzyladenine
BAP	benzylaminopurine
BUDR	5-bromo-2'-deoxyuridine
DMSO	dimethylsulfoxide
DNA	deoxyribonucleic acid
cDNA	complementary DNA
dT	deoxythymidine
2,4-D	2,4-dichlorophenoxyacetic acid
EMS	ethyl methanesulfonate
ER	endoplasmic reticulum
EDTA	ethylenediaminetetraacetic acid
FeEDTA	ferric ethylenediaminetetraacetic acid
GA	gibberellic acid
IAA	indoleacetic acid
IBA	indolebutyric acid
2iP	$N^6(\Delta^2$-isopentenyl) adenine
NAA	naphthaleneacetic acid
NOA	naphthoxyacetic acid
RNA	ribonucleic acid
mRNA	messenger RNA
rRNA	ribosomal RNA
tRNA	transfer RNA
SDS	sodium dodecyl sulfate
2,4,5-T	2,4,5-trichlorophenoxyacetic acid
2,4,6-T	2,4,6-trichlorophenoxyacetic acid

1
Embryogenesis in angiosperms – the changing scene

This book is devoted to an analysis of a single important phase in the life cycle of the most highly evolved class of plants, known as Angiospermae (angiosperms) or flowering plants (Division: Anthophyta). As a class, the angiosperms have perfected a specialized reproductive structure, the flower, which counts as one of its major functions the production of eggs and sperm. The egg is formed within a privileged location in the female gametophyte, which itself is concealed in the ovule with its multiple layers of integuments and nucellus. Sperm are generated from pollen grains born within the anther locule. Following fertilization, as the ovule is transformed into a seed, the enclosed zygote enters a pathway of cell division and differentiation to give rise to an embryo. A full-grown embryo consists of a hypocotyl–root axis that bears the root meristem at one end and at the other, one or two cotyledons and the meristem of the shoot. The knowledge that the number of cotyledons subtending the shoot apex in the seed embryo is a diagnostic character paved the way to separate angiosperms into two major subclasses: Dicotyledoneae (dicotyledons, dicots), having two cotyledons, and Monocotyledoneae (monocotyledons, monocots), having only one cotyledon.

Embryogenesis is the study of development of the embryo beginning with its single-celled progenitor, the zygote, and ending with the formation of a mature embryo. In angiosperms, maturity of the embryo is reckoned in terms of its ability to give rise to a normal seedling plant upon germination of the seed. Broadly speaking, embryogenesis involves an array of developmental episodes occurring in an ordered sequence whereby the structural and functional organization of the adult plant becomes progressively expressed. The fundamental phenomena operating during embryogenesis are fabrication of the different chemical and structural components of the cell and the organization of cells into patterns distinctive of tissues and organs. The course of embryogenesis is described in terms of (1) changes in morphology, structure and cellular patterns of the developing embryo, (2) how individual characteristics are inherited within an overall plan and (3) how biochemical and molecular changes combine to store, replicate and retrieve developmental information. A primary aim of this book is to analyze, at different levels of complexity, the processes and reactions taking place in embryos of angiosperms that cause them to develop the way they do and produce structures we see appearing within a given time frame.

How the scene has changed

The logical outcome of advances in plant microtechnique, plant physiology, genetics, tissue culture and cell and molecular biology over the years has been the development of several specialized subfields in angiosperm embryogenesis. Using cytological and histological methods, work done up to about 1940 established the basic cell lineage involved in the transformation of the zygote into an embryo. These studies described with precision the diversity of form and development of embryos in a number of angiosperms and traced the origin, position and histogenetic role of specific cells or cell tiers of early embryos. During this period in the history of the study of embryogenesis, an understanding of embryo development in different species took center stage, and the determination of cell lineages led to the view that they were programmed by a blueprint characteristic of each species. Descriptive embryology is the preferred term used to characterize this type of work, which continues to make important contributions even now. These studies have been reviewed by Maheshwari (1950, 1963), Davis (1966) and Johri (1984).

With the successful isolation and culture of plant organs and tissues under aseptic conditions in the mid-1930s, methods were at hand for the experimental analysis of embryo growth and its dependency on the tissues of the ovule. Although immature angiosperm embryos were first grown to maturity in culture by Hannig as early as 1904, it was the systematic analysis of growth requirements of continuously growing callus and isolated root cultures by Gautheret, White and others that provided the main impetus for the study of growth and differentiation in culture of small embryos. The range of possible experiments on the control of growth of cultured embryos was greatly extended by the isolation of the growth hormones, indoleacetic acid (IAA) in 1934, gibberellic acid (GA) in 1939 and kinetin in 1955. Great as has been the impact of tissue culture methodology on the study of embryogenesis in angiosperms, its impact on the work of descriptive embryologists has been even greater, for many of them began experiments ostensibly designed to determine the controlling mechanisms in the orderly growth of embryos *in vivo* by isolation and culture of progressively small embryos *in vitro* in defined media. Despite the fact that in some cases the connection between culture conditions and embryo morphogenesis was absent or discouragingly obscure, a compendium of nutritional requirements for the successful culture of embryos of a variety of plants resulted from these studies. The outcome of these investigations was surveyed by

Wardlaw (1955) in a book entitled, *Embryogenesis in Plants,* covering the period up to the early 1950s.

A major influence on the analysis of angiosperm embryogenesis, beginning in the early 1960s, has been the use of the transmission electron microscope to monitor the fine structural changes that take place during the transformation of the egg into an embryo. This examination resulted in the accumulation of a considerable body of knowledge of the complexity and structure of the subcellular organization of embryos of different stages of development. The structural studies provided new insights into the metabolic status of the egg before and after fertilization, the ultrastructural basis for polarity of the egg and zygote and the functional differentiation of cells formed from the first division of the zygote. The electron microscope has also been one of the most powerful tools for understanding of the function of cells of the female gametophyte and of the subtending part of the embryo known as the suspensor.

Recent advances in biochemistry, genetics and molecular biology are beginning to have a marked influence on our understanding of the control of gene activity during embryogenesis in angiosperms. Most questions about gene expression during embryogenesis relate to the synthesis and accumulation of storage proteins in the embryonic cotyledons. In many plants storage protein accumulation encompasses a major part of embryogenic development. Some progress has been made in isolating messenger ribonucleic acid (mRNA) coding for storage proteins, in monitoring the fate of this mRNA during embryogenesis and in cloning and sequencing the genes for storage proteins. We can therefore now begin to account for the synthesis of the most abundant embryogenic proteins in terms of activation of specific genes instead of in physiological or cytological terms. To understand the course of embryogenesis, mRNAs coding for certain structural proteins of the embryo have also been cloned and their expression followed in embryos of different ages; however, molecular biology is yet to make a distinct contribution to our understanding of gene expression during the early division phase of the zygote.

Two spectacular achievements that enlivened the field of angiosperm embryogenesis came about from unexpected quarters during the early 1960s. Paradoxically, research that led to these discoveries had nothing to do with embryos at all, but was the outcome of experiments on the growth and differentiation of plant cells, tissues and organs cultured under controlled nutritional and environmental conditions. One such achievement was the demonstration that when single cells and clusters of cells sloughed off from a rapidly proliferating callus were nurtured in a nutrient broth, they gave rise to organized growth through somatic facsimiles of embryos. From these structures normal plants could be regenerated. The other was the discovery that when anthers of certain plants at an appropriate stage of development were cultured in a relatively simple medium, a small number of the enclosed pollen grains dedifferentiated and developed into embryo-like structures and plantlets with the haploid or gametic number of chromosomes. These results acknowledged the principle that the angiosperm zygote does not hold an exclusive patent to produce embryos, as apparently similar structures can also be generated from somatic cells and germ cells in tissue cultures. However, the chief interest of students of embryogenesis was centered on the factors regulating the marvellous transformation of a single somatic cell or a pollen grain into an embryo-like structure. Since these discoveries, reports of the transformation of somatic cells and pollen grains into embryo-like structures by tissue culture methods have continued to the present day. In fact, the remarkable progress made in these two areas has all but overwhelmed the traditional field of angiosperm embryogenesis. The manipulative advantages afforded by somatic cells and pollen grains, as contrasted with the zygote, are making an investigation into the biochemical and molecular changes during the early phase of transformation of a cell into an embryo a reality instead of a pious hope. Developments in the field of embryogenesis from somatic cells and pollen grains up to the mid-1970s have been reviewed in *Experimental Embryogenesis in Vascular Plants* (Raghavan, 1976b) and in a book of nearly the same title edited by Johri (1982).

Discussions about the currently active areas in angiosperm embryogenesis referred to in the previous pages will resurface in Chapters 2 to 7. Chapter 8 is devoted to a survey of applications resulting from the study of embryogenesis in angiosperms in an effort to improve our agriculture.

Methods of studying embryogenesis

The changing emphasis in the analysis of angiosperm embryogenesis has led to the use of increasingly sophisticated methods to study various aspects of embryo development. Although methodological details are largely assumed as familiar background, we will nevertheless briefly review some of these methods. This list is neither complete nor the descriptions adequate for an untrained person to try a hand at the game. Our purpose is merely to introduce the reader to the basic principles of the more important methods employed, since the results accruing from their use have contributed to the account presented in the following chapters. Further details on these techniques may be found in reference works dealing with microtomy, histochemistry, tissue culture and molecular biology.

Microscopy

What is known about the pathways of embryogenesis in angiosperms is based almost exclusively on light microscopic observations of fixed, sectioned and stained tissues and organs. Fixation in formalin or in mixtures containing formalin, glacial acetic acid and ethanol in a specific proportion arrests the developmental processes of cells without greatly altering their internal structure. Following dehydration of materials by transfer through a series of alcohols of increasing molecular weight, they are embedded in paraffin wax. Serial sections of the paraffin-embedded material cut on a rotary microtome and suitably stained provide a three-dimensional view of the internal structure of the organ. In order to follow the entire sequence of embryogenesis, it is common practice to fix ovules of different ages over a span of time. With most paraffins, it is difficult to obtain sections thinner than 5 μm. However, electron microscopy has introduced the use of improved fixatives such as glutaraldehyde and hard epoxy plastics such as Epon. This has made possible sectioning of ovules at 0.5- to 1.0-μm thickness suitable for light microscopy.

Histochemistry, cytochemistry and autoradiography

Histochemistry is the method of choice for the localization of specific chemical substances or sites of chemical activity in cells, tissues and organs. Histochemical methods are especially useful when heterogeneity of the tissues precludes the use of conventional biochemical techniques. Because fixatives might alter the chemical nature of the substances intended for study, histochemical tests are generally done on fresh tissues. Typically, the material is rapidly cooled in liquid nitrogen and sectioned on a special low-temperature microtome known as the cryostat. The sections are subsequently treated on slides with reagents specific for the detection of the chemical substance. The result of the reaction is the deposition of a colored product at the site of activity. Histochemical methods have been used extensively to localize proteins, nucleic acids and carbohydrates during embryogenesis in a number of plants. These methods remain a primary source of information on the biosynthetic activities of the egg and early division phase embryos of angiosperms. With appropriate controls, many histochemical methods can be quantified to yield data on the concentration of a given chemical substance in the cell. The well-known Feulgen reaction for deoxyribonucleic acid (DNA) is a classic example of quantitative cytophotometry.

Another method used to localize biochemical events at the cellular level is autoradiography, which permits detection, by methods similar to those used in photography, of a radioactive isotope introduced into cells or tissues. In a typical experiment designed to follow the synthesis of nucleic acids or proteins, the material is incubated in a medium containing a radioactive precursor of DNA (^{3}H-thymidine), RNA (^{3}H-uridine) or proteins (^{3}H-amino acids) for a short period of time. After incubation in the isotope, the material is fixed and sectioned as usual. The sections attached to slides are covered with a sensitive photographic emulsion and exposed in the dark for a period of time ranging from a few days to several weeks. During the period of exposure, radiation emitted by the isotope exposes the photographic emulsion at the site of incorporation. When the slide is developed in much the same manner as a photographic film, sites of incorporation of the isotope in the tissue are revealed by the presence of silver grains on the overlying emulsion. Thus, autoradiography permits localization of the sites of synthesis of specific macromolecules in cells at a given time. Within certain limits, the rate of synthesis of the macromolecules can be determined by silver grain counts per unit area of the section or from cells of comparable size.

Tissue culture

The term tissue culture, as it is now used in plant sciences, employs a family of techniques involving aseptic culture of plant organs, tissues, cells and protoplasts. Each type of culture requires slightly different methods, but the principle of culture is the same. For embryo culture, surface sterilization of the embryos as such is not necessary; instead, entire ovules, seeds or capsules containing ovules are sterilized and embryos excised aseptically and transferred to a nutrient medium. In a similar way, in anther culture investigations, unopened flower buds are sterilized and individual anthers subsequently excised and cultured. A basic medium consisting of simple materials such as sucrose, inorganic salts and vitamins will support continued growth of isolated seed embryos cultured *in vitro* or will induce the dedifferentiation of pollen grains of cultured anthers into plantlets.

Specific tissues of the plant, such as the secondary phloem of *Daucus carota* (carrot), may also be excised and grown in sterile culture; in addition to sugars, mineral salts and vitamins, most tissue cultures require one or more plant hormones such as 2,4-dichlorophenoxyacetic acid (2,4-D) or kinetin to foster growth as a continuously proliferating mass of callus. If the callus is transferred to a liquid medium of the same composition and agitated gently on a shaker, the individual cells and cell groups at the surface are often broken away, floating free in the medium to give a suspension culture. By mechanical filtration of this culture through sieves of different pore sizes, a suspension consisting mostly of single cells can be prepared as an inoculum for a new batch of cell cultures. However, propagation of free plant cell cultures requires a more elaborate medium than does a callus culture.

From the culture of single cells and cell suspensions, it is one step further to strip cells of their walls and culture isolated protoplasts. Protoplasts have been isolated mostly from leaves, although cell cultures have also been used occasionally as starting material for protoplast isolation. In a common protocol to isolate protoplasts, leaves are sterilized, cut into small pieces and floated in an enzyme mixture containing cellulase, hemicellulase and pectinase as well as an osmotic stabilizer. The enzymes will hydrolyze the cellulosic cell wall and middle lamella, whereas the osmoticum will prevent bursting of the protoplasts liberated from the confines of the wall. After the protoplasts are released into the incubation medium, they are purified by repeated centrifugation and cultured on a complex medium. Within a few days after culture, a new cell wall is re-formed on the isolated protoplast that now behaves like an isolated cell.

Biochemical techniques

The biochemical techniques used to study angiosperm embryogenesis generally involve those that measure the rates of synthesis of DNA, RNA and proteins or that isolate and characterize mRNA and fractionate nucleic acids and proteins. Most protocols employed to follow the rate of macromolecule synthesis use a radioactive precursor of the macromolecule to be measured. After incubation of the material in the isotope for a short period of time, the material is homogenized and the homogenate treated with a solution that precipitates the macromolecule with the incorporated label. The radioactivity in the precipitate as determined by scintillation counting is generally taken as a measure of the added label incorporated into the macromolecule during the period of incubation. In computing the actual rate of incorporation of the isotope, consideration must be given to the changing precursor uptake pattern in the different samples and to the size of the precursor pool (the amount of precursor actually available for synthesis at a given time). The latter consideration is crucial in investigations dealing with the rate of DNA and RNA synthesis because the direct precursors of nucleic acid synthesis are the nucleotide triphosphates and not the radioactive nucleosides that are generally administered exogenously. Thus, to determine the final dilution of the exogenous nucleoside precursor added initially, it is necessary to measure the specific activity of the direct precursor to the macromolecule. In a few studies undertaken to determine the rates of RNA synthesis during embryogenesis, a sensitive method developed by Emerson and Humphreys (1971) utilizing ^{3}H-adenosine as a precursor has permitted accurate measurement of adenosine triphosphate (ATP) pool size, specific activity of ATP and the actual rates of RNA synthesis.

mRNA separation. Several methods are currently used to isolate mRNA from cells. One technique involves separating mRNA from total RNA by chromatography on an oligo deoxythymidine-(dT) cellulose column. Since most mRNAs carry a polyadenylic acid [poly(A)] sequence that binds to oligo-(dT) cellulose, this protocol permits isolation of mRNA without contamination from other RNA species. mRNA is subsequently eluted from the column with a suitable buffer. Alternatively, poly(A)-containing-RNA [poly(A)RNA] is isolated from polysomes treated with ethylenediaminetetraacetic acid (EDTA) to dissociate and release mRNA. Poly(A)RNA is further selected from the mixture by oligo-(dT) cellulose chromatography.

Cell-free protein synthesis. The ability of mRNAs to code for proteins is verified by *in vitro* or cell-free translation. In this method, the isolated mRNA is mixed with preparations from wheat germ or rabbit reticulocytes that contain all the essential cytoplasmic ingredients necessary for protein synthesis. A labeled amino acid is added to the cell-free system to detect the amount of proteins synthesized. These proteins can be further characterized by polyacrylamide gel electrophoresis.

Gel electrophoresis. Although several fractionation methods are used in biochemistry to separate nucleic acids and proteins, one of the best yet developed is polyacrylamide gel electrophoresis. The polyacrylamide is a support medium that functions as a molecular sieve to separate individual molecules of nucleic acids and proteins by differentially restraining their movement when placed in an electrical field. Since the rate of migration of molecules is dependent on their charge and molecular weight, small molecules will move faster than larger ones, which have trouble making their way through a column of acrylamide gel. After separation, the gel is stained to locate the bands or autoradiographed against a sheet of X-ray film to locate radioactively labeled macromolecules. When polyacrylamide gel electrophoresis is performed in the presence of the ionic detergent sodium dodecyl sulfate (SDS), the method can be used to determine the molecular weight of migrating proteins. This is based on the fact that when a protein mixture is treated with SDS, the detergent complexes with the proteins in proportion to their molecular weight. The result is the production of protein species of uniform shape and equivalent charge per molecular weight. During electrophoresis on an acrylamide gel, the migration of SDS-complexed proteins will be determined solely by their molecular weights.

One of the most powerful of electrophoretic techniques to fractionate mixtures of proteins is two-dimensional gel electrophoresis. As the name implies, the technique involves fractionation of proteins in two dimensions based on two independent properties, such as the isoelectric point and ability to complex with SDS. In this method, the proteins are initially separated in one direction according to their isoelectric point using a substance that produces a stable pH gradient such as ampholyte. This enables the different proteins to settle in different positions in the electrical field when they encounter a pH value equal to their isoelectric point. The proteins are subsequently separated in a second direction by virtue of their ability to migrate according to their molecular weight when complexed with SDS. A staggering number of proteins can be separated by this method.

Molecular techniques

An amazing property of the unique sequence of bases that constitute a strand of DNA or RNA is that they can match up with complementary sets of bases and form stable double-stranded molecules. This principle of nucleic acid hybridization has become a versatile tool for the study of gene activity. Using this method, it is possible to assay quantitatively for the immediate products of gene expression by RNA–DNA hybridization and to determine the gene sequence complexity by DNA–DNA hybridization. Hybridization techniques have been extended to the cytological level (*in situ* hybridization) to localize gene sequences on chromosomes by DNA–DNA hybridization or by hybridization with cloned genes and to probe for mRNA by hybridization of cellular poly(A) with polyuridylic acid [poly(U)].

Molecular hybridization. Experiments are generally carried out under conditions in which one population of a single-stranded nucleic acid is radioactively labeled and present at a low concentration, while the other population is unlabeled and present at a very high concentration. After hybridization, the hybrids are separated from the nonhybridized fraction by passage over a column of hydroxyapatite (a calcium phosphate complex). Depending on the phosphate concentration of the elution buffer, hydroxyapatite has the peculiar property of releasing or binding single-stranded or double-stranded molecules. By determining the radioactivity in the fractions thus separated, the percentage of hybrids formed is determined. In some cases, an enzyme S1 nuclease, which selectively degrades single-stranded molecules but not duplexes, has also been used to determine the percentages of hybrids formed. Many variations have been introduced into nucleic acid hybridization techniques during the short span of their use, and it has become an exceedingly complex subject. The following account therefore focuses on a few such experiments described in this book.

In order to determine changes in mRNA sequences coding for a specific protein during embryogenesis, mRNA is prepared from embryos during the period in which synthesis of this protein is the dominant activity of the cell. The mRNA isolated is used as a template to prepare a highly labeled complementary DNA (cDNA) by the enzyme reverse transcriptase. Samples of labeled cDNA are hybridized with an excess of unlabeled mRNA isolated from embryos of different ages. On the basis of measurements of the reassociation kinetics and the percentage of hybrids formed, the changes in the relative concentration of mRNA sequences are determined.

The question of whether particular messages are transcribed on repetitive or single-copy DNA sequences is explored by comparison of the reassociation kinetics between labeled cDNA prepared from the mRNA in question and excess of unlabeled DNA, and labeled single-copy DNA and excess of unlabeled DNA. Single-copy DNA is what is implied by its name – a unique sequence – and is isolated in a pure state uncontaminated by the presence of repeated sequences. For this purpose, denatured fragments of total DNA are reannealed long enough for all repetitive sequences to form duplexes, leaving the single-copy population made up of single-stranded fragments. Hydroxyapatite fractionation of the mixture at this stage will separate single-stranded fragments from the double-stranded hybrids. An additional reannealing cycle will yield a relatively pure fraction of single-copy DNA. If the course of reaction kinetics in hybridization experiments between labeled cDNA and excess of unlabeled DNA is similar to that between single-copy DNA and excess DNA, this indicates the presence of repetitive sequences. The rate of hybridization to single-copy DNA of mRNA populations can also provide information on the sequence complexity of mRNA. For example, a hybridization curve with a single rate constant will result if all mRNA sequences are present in equal concentrations, while RNAs of different sequence complexities will introduce corresponding changes in the curve.

In situ *hybridization*. A hybridization method using ^{3}H-poly(U) as a probe has been developed to follow the cellular accumulation of poly(A)RNA during embryogenesis. The protocol followed involves the application of aliquots of the radioactive probe to paraffin or plastic sections on slides. The slides are then sealed with a coverslip and the preparation incubated in a moist chamber. After annealing of the probe to sections, the uncomplexed probe is removed by digestion with pancreatic ribonuclease (RNase) A, and sites of mRNA accumulation are detected autoradiographically.

Cloning of cDNA. Since cDNA represents a number of mRNA sequences, recombinant DNA technology is used to isolate individual mRNA sequences. In this method, known as cloning, a piece of cDNA is introduced into a bacterial cell, where its replicates autonomously from that of the host chromosomes. Clones of cDNA thus made are used to hybridize mRNA preparations from embryos in search of specific sequences used as templates. In order to clone cDNA, the single-stranded cDNA is converted into a double-stranded molecule by the use of DNA polymerase I.

Next, the double-stranded DNA fragments are covalently linked to the DNA of a plasmid by enzymatic recombination. Essentially, in this technique a restriction endonuclease makes a staggered cut in the circular plasmid DNA at a specific site, and the cDNA fragment is inserted into the gap, where it fits snugly with the sticky complementary ends. The circular plasmid so re-formed is taken up by bacterial cells (transformation), where it replicates. The small number of bacterial cells transformed by the recombinant plasmid are selected from the nontransformed cells by using plasmids that carry resistance to certain antibiotics. If these plasmids carrying recombinant DNA are added to bacterial cultures containing the specific antibiotic, only those bacteria that have taken up the plasmids will survive. Because different bacterial cells take up different recombinant plasmids, the next step in the procedure is to isolate those bacterial colonies that contain cDNA inserts of interest. This is done by colony hybridization, in which DNA from different colonies of bacteria with recombinant plasmids is immobilized on a nitrocellulose filter and subjected to *in situ* hybridization with a labeled probe carrying a sequence complementary to that being sought. The location of the labeled hybrids is detected by autoradiography. Bacterial cultures containing representatives of the labeled hybrids are allowed to multiply and the plasmids subsequently extracted and purified. Treatment of the recombinant plasmids with the same restriction enzyme used in their formation will release cDNA intact for use in molecular hybridization with mRNA.

Cloned DNA fragments are widely used in molecular hybridization experiments with complementary mRNA sequences. Two techniques that verify that a cloned DNA is in fact complementary to a specific gene are hybrid-arrested translation and hybrid-selected translation. The former technique depends on the fact that complementary mRNA sequences hybridized with cloned DNA are unavailable for translation in a cell-free system in which noncomplementary mRNAs are readily translated. Dissociation of the hybrid by heat treatment restores the translational ability of mRNA and permits characterization of proteins coded by the gene. In the hybrid-selected translation method, cloned DNA immobilized on a filter is hybridized with a mixture of different mRNA sequences. As only complementary mRNA sequences will hybridize with the cloned DNA, the hybrids formed can be recovered from the mixture. The mRNA sequences are subsequently dissociated from the hybrids and subjected to *in vitro* translation to characterize the polypeptides formed.

Concluding comments

The progress made during the past 50 years in the study of embryogenesis in angiosperms is remarkable. It appears that this cannot be dissociated from the technical advances that have taken place during this period as well as the increasing use of embryos as an experimental system. The convergence of the old and new studies has made the subject of angiosperm embryogenesis a large one, with many ramifications and profound implications for the developmental biologist. Although it is hazardous to predict how this field is likely to develop in the future, it seems certain that during the next few years our knowledge of embryogenesis through pathways involving the zygote, somatic cell and pollen grain will be greatly deepened by the advancing frontiers of molecular biology.

2
Developmental embryogenesis

The overriding fact about the life cycle of angiosperms is that it is set in the context of an alternation of generations between a diploid sporophyte and a haploid gametophyte. Repeated divisions of the fertilized egg, the zygote, give rise to the embryo from which the plant body or the sporophyte is formed. Certain specialized cells within the reproductive organs of the adult sporophyte undergo meiosis to form the gametophyte. As the name implies, the function of the gametophyte is to produce gametes; when that is accomplished with adequate provision to ensure the union of the egg and the sperm, a zygote is formed. At this point, not only is diploidy restored and the life cycle completed, but the zygote presents an opportunity to capitalize on the attributes of a new genetic combination. Our account of embryogenesis in angiosperms begins with a brief consideration of the events that initiate gametophyte development, so that we can later set the cellular and biochemical changes in proper perspective.

The gametophytes of angiosperms exhibit little phenotypic organization beyond the basic competence to house and nurture the egg and the sperm. However, the female gametophyte as the embryo sac and the male gametophyte as the pollen grain are profoundly differentiated from each other by a sweeping series of structural transformations during their ontogeny, resulting in strikingly different final organization. The classic review of this subject remains P. Maheshwari's (1950) textbook, *An Introduction to the Embryology of Angiosperms;* the ontogeny of the different types of embryo sacs collated and described by Maheshwari are still relevant. The development of the female gametophyte begins with the delimitation of a single hypodermal cell in the nucellus of the incipient ovule. This cell, called the archesporial cell, can be recognized in histological preparations by its large size and densely staining cytoplasm as well as by certain histochemical characteristics such as high RNA and protein content. Although in most angiosperms investigated the archesporial cell functions directly as the megaspore mother cell (megasporocyte), in some there is a deviation initiated by a periclinal division. Following this division, the inner cell functions as the megasporocyte. Irrespective of its origin, the megasporocyte goes through a meiotic division to generate a cluster of four haploid megaspores, which are generally arranged as a linear tetrad. Of the four megaspores, three toward the micropylar pole of the ovule usually abort, leaving the surviving one at the chalazal pole to develop into the embryo sac. This cell enlarges at the same time that its nucleus undergoes three mitotic divisions to produce eight haploid nuclei floating in the common cytoplasm of a sac-like supercell, the embryo sac. These nuclei soon rearrange themselves in a characteristic manner by the migration of three nuclei as distinct membrane-enclosed cells to each pole of the embryo sac. The three cells at the micropylar pole organize as the egg apparatus, consisting of an egg cell flanked on either side by a synergid. At the opposite chalazal end are three antipodal cells. The main body of the embryo sac remaining after the definition of the egg apparatus and antipodals is occupied by the central cell containing two polar nuclei. Before fertilization, the two polar nuclei become enclosed within a single membrane to form a diploid fusion nucleus. The mature embryo sac derived from a single megaspore as described above is known as the monosporic type and is a seven-celled haploid structure. The typical monosporic eight-nucleate embryo sac was first described in *Polygonum divaricatum* and, according to convention, is known as the Polygonum type.

Not every species of angiosperm investigated possesses a seven-celled embryo sac, but there is always an egg cell and a fusion nucleus. Deviations from the modal pattern concern the number of megaspores participating in the development of the embryo sac and consequent changes in the composition of the egg apparatus and ploidy level of the fusion nucleus. In one category, following meiosis, only two megaspores abort; the surviving two divide by two mitotic divisions to yield an eight-nucleate bisporic embryo sac. These nuclei organize as a typical seven-celled haploid structure, like the monosporic type. Some very bizarre types of embryo sacs are produced when all four megaspores participate in gametogenesis to form a tetrasporic embryo sac. The four megaspore nuclei may undergo a single postmeiotic division to produce an eight-nucleate embryo sac or undergo two postmeiotic divisions to form a 16-nucleate embryo sac; alternatively, the four nuclei may fuse in a complex manner before division. Regarding the organization of the egg apparatus in the tetrasporic type of embryo sac, in one variation known as Peperomia type, there is only one synergid whereas the Plumbago and Plumbagella types have no synergids. The central cell may contain four haploid nuclei (Plumbago and Penaea types), eight haploid nuclei (Peperomia type) or one haploid and one triploid nucleus (Fritillaria and Plumbagella types). These nuclei apparently fuse before fertilization to form fusion nuclei with more than the usual diploid chromosome complement. The different types of embryo sacs also show a trend toward reduction in the number of antipodals, ranging from 11 in the Drusa type to none at all in the Oenothera and Plumbago types. This finding has led to the suggestion that the major types of angiosperm embryo sacs have probably evolved from a gymnosperm ancestry by progressive specialization of a structure resembling an archegonium (Favre-Duchartre, 1978). Recent reports of enzymatic isolation of megasporocytes (Yang and Zhou, 1984) and intact embryo sacs from ovules (Zhou and Yang, 1982, 1984) offer promise that megasporogenesis in angiosperms can be monitored under controlled conditions.

The corresponding cell in the incipient anther primordium that generates the pollen grain is the microspore mother cell (microsporocyte). This cell undergoes a meiotic division to form a tetrad of four microspores. The microspores released from behind the barrier of the wall of the tetrad mature into pollen grains. During gametogenesis, the uninucleate pollen grain divides asymmetrically into a large vegetative cell and a small generative cell (first haploid mitosis). Under favorable conditions of culture in the laboratory, or after transfer to the stigmatic surface of a flower in nature, the pollen grain begins to germinate. Upon germination of the pollen grain, the generative cell devoid of its cell wall and the nucleus of the vegetative cell, which is now loose from its cytoplasm, migrate into the emerging pollen tube. During the terminal phase of differentiation of the male gametophyte, the generative cell divides to form two sperm involved in the act of double fertilization in the embryo sac. In this event, which is unique to angiosperms, one of the sperm fuses with the egg to form the zygote, while the second sperm fuses with the fusion nucleus of the central cell to form the primary endosperm nucleus; as the name implies, this nucleus is the forerunner of the endosperm.

The following sections concentrate on the fate of the zygote and of the endosperm nucleus. The title of this chapter, which refers to developmental embryogenesis, therefore indicates only the main thread of the discussion, a thread that is woven into the fabric of postfertilization development of the ovule.

Embryo development

One of the consequences of fusion of the egg with the sperm is initiation of divisions in the fertilized egg. The first division may not occur immediately after fertilization, since the zygote remains quiescent for a period of time ranging from a few hours to several months before it divides. The product of the first few rounds of division of the zygote is generally designated as the proembryo. There is some disagreement, however, as to how few the first few rounds of divisions are and as to where the dividing line between the proembryo and embryo proper falls. Whereas Maheshwari (1950) used the term without defining it, Johansen (1950) regarded the proembryo as a filament of cells constituting the early phase of embryogenesis, the upper limit being the completion of four rounds of divisions or the formation of

16 cells. According to Souèges (1936), the proembryo possesses the characteristic axial symmetry of the egg and includes stages of embryogenesis prior to the appearance of the primordia of cotyledons. Although the initiation of cotyledon primordia varies with respect to time and the number of cells in embryos of different angiosperms examined, the essential elements of this transition appear to be quite similar. For this reason, the stage of initiation of cotyledon primordia is considered a good cutoff point when proembryo stage ends.

Classification of embryo ontogenesis

The distinguishing feature of the first division of the zygote is that it is nearly always transverse to the long axis of the cell. Although in most cases this division has invariably been described as asymmetrical, producing a large basal cell and a small apical cell, a significant number of cases suggest that the reverse may be true or that both cells may be of the same size (Sivaramakrishna, 1978). A consistent unequal division is morphogenetically significant, because it points to some underlying cytoplasmic program that will determine the position of the mitotic spindle and orientation of the cell wall. Following this division, the cell formed toward the cavity of the embryo sac and facing the chalazal end of the ovule is known as the apical or terminal cell; the cell attached to the embryo sac wall at the micropylar end is the basal cell. The embryo is derived by subsequent divisions of these two cells, but there is considerable variation in the extent to which the basal cell contributes to the formation of the organogenetic part of the embryo and the subtending suspensor. On the basis of these variations and the plane of division of the terminal cell, five different types of embryo ontogenesis have been described (Maheshwari, 1950).

Conceptually, there are two main pathways by which embryogenesis is accomplished in both dicotyledons and monocotyledons – either by a longitudinal or by a transverse division of the terminal cell of the two-celled proembryo. Let us first consider the pathway in which the division of the terminal cell is in the longitudinal plane. In plants belonging to Ranunculaceae, Anonaceae, Onagraceae, Cruciferae, Pedaliaceae and Scrophulariaceae, in which a longitudinal division of the terminal cell is the rule, the basal cell contributes very little or nothing at all, by way of cell derivatives, to the formation of the organogenetic part of the embryo. Here the basal cell is destined to form the suspensor. As this type of embryo ontogenesis was first described in a member of the Cruciferae, it is conventionally known as the Crucifer type. By contrast, in plants belonging to Compositae, Balsaminaceae, Vitaceae, Rosaceae and Violaceae, in which the terminal cell is also partitioned longitudinally, the strategy is for derivatives of both terminal and basal cells to blend indistinguishably to form the em-

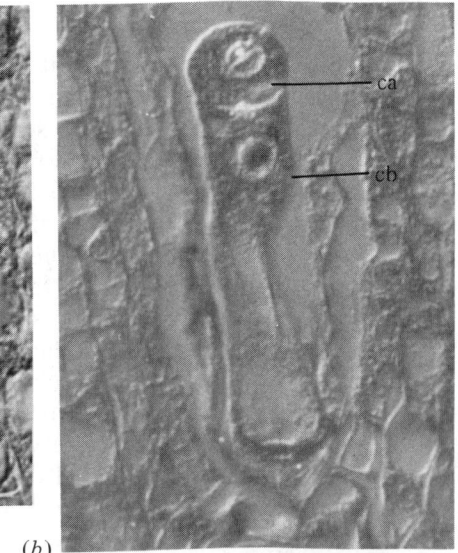

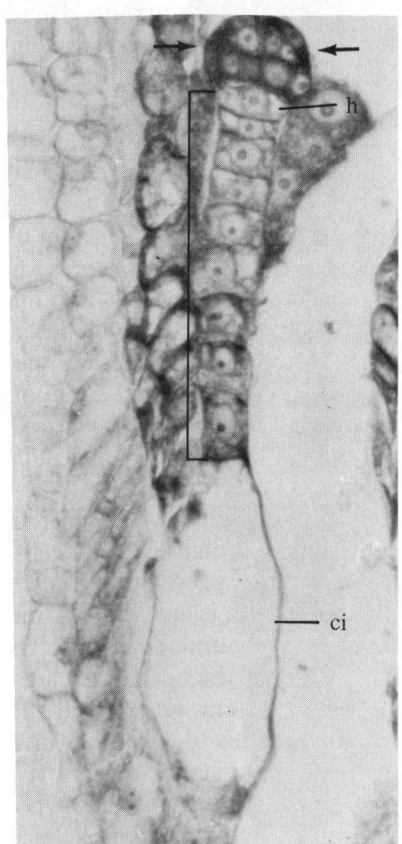

bryo. However, not all descendants of the basal cell are incorporated into the embryo proper; a small number of cells assume the characteristics of a suspensor. This is known as the Asterad type.

The Chenopodiad type described in members of Chenopodiaceae and Boraginaceae is basically similar to the Asterad type except that here the first division of the terminal cell is in the transverse plane. Two other types of embryo ontogenesis in which the first division of the terminal cell is transverse are the Solanad type and the Caryophyllad type. In the former type, represented in Solanaceae, Campanulaceae, Theaceae, Linaceae and Rubiaceae, the basal cell forms a well-defined suspensor, with most of the organogenetic part of the embryo derived from the terminal cell. The Caryophyllad type differs fundamentally from the Solanad type in the absence of divisions in the basal cell and the formation of the suspensor from the derivatives of the terminal cell. This type of embryo ontogeny described in members of Caryophyllaceae, Crassulaceae and Haloragaceae has also been documented in *Saxifraga* (Saxifragaceae), *Androsaemum* (=*Hypericum;* Guttiferae) and *Drosera* (Droseraceae). In these three genera, it is not unusual for the basal cell to divide once or twice and in fact become part of the suspensor formed mainly from the derivatives of the terminal cell (Maheshwari, 1950; Bhojwani and Bhatnagar, 1978). In this system of embryo classification we must be prepared, then, to discover that some subjective judgment is inevitable in the assignment of embryo development to particular classes.

A more complicated classification of embryo ontogenesis has been developed by Souèges (1937) (see also Crété, 1963). Although the pattern of early segmentation of the embryo forms the backbone of this system, it looks upon angiosperm embryos as adhering to certain laws concerned with the origin of cells at any particular stage, the number of cells produced in each generation, the role of cells formed and the deterministic fates of cells produced during the early division phase of embryos. A simplified version of Souèges' system of classification of embryo types adopted by Mestre (1967) has focused attention on the fate of the two cells formed from the first division of the zygote. Periasamy (1977) developed the thesis that the increasing exposure of newly formed cells of the zygote to the internal environment of the future embryo, rather than to the external environment of the endosperm, makes the segmentation of the first internal cell an important morphogenetic event. This has led to a new approach in the classification of embryo types based on the number of cell tiers in the proembryo when the first series of cells whose surfaces are not exposed to the exterior is segmented. The relative location of the tiers with varying cell numbers and the terminal or subterminal location of the tier in which formation of the first internal cells occurs are employed as additional criteria with which to assign embryos to different groups in the proposed classification.

Developmental embryogenesis

The early division sequences of embryos of a large number of plants investigated seem to be remarkably precise. In many instances the cell lineages during embryogenesis also appear to be regular, making it possible to anticipate the destiny of particular cells in the four- or eight-celled proembryo as giving rise to specific organs in the mature embryo. However, an increasing number of investigations have found the crucial first and second divisions of the zygote to be highly variable; in other studies, embryo ontogenesis cannot be referred to recognized types, and still others point out irregularities in the method of origin of embryo parts. An extreme case of variation is the occurrence of a vertical or longitudinally oriented oblique wall, rather than a transverse wall, initiating the first division of the zygote; this mode of embryo origin designated Piperad type, is however observed only in a few obscure plants (Natesh and Rau, 1984). Stemming from these and other reports, there is some question as to whether a reliable system of classifying angiosperm embryo ontogenesis can be formulated on the basis of successive steps in the division of the zygote and a dogmatic emphasis on the destination of cells formed. Histological studies of early embryogenesis have produced such a wealth of information on the phenomenon relative to other aspects of embryogenesis that one must consider whether the importance of this information may have been exaggerated.

From proembryo to mature embryo

The discussion thus far has dealt only with the development of the earliest stages of the embryo. The student of embryogenesis is faced with the need to understand how an adult embryo is produced from this limited number of cells. Toward this end, the successive steps in the embryogenesis of representative species from both dicots and monocots will be described. Needless to say, this includes only a fraction of the number of species whose embryo ontogeny has been followed; descriptions of embryo development in numerous other species can be found in such sources as Johansen (1950), Maheshwari (1950) and Davis (1966) as well as in the review of Crété (1963). Of some practical concern and interest for the light it sheds on the problems of analyzing embryogenesis in angiosperms is the fact that in tracing the development of the embryo from its single-celled origin we still depend on reconstruction of serial sections of fixed and processed materials.

Capsella bursa-pastoris (Shepherd's purse), a member of the Cruciferae, is a common weed that does not evoke much general attention but that is noted for its regular pattern of embryo development. For this reason, *Capsella* has been widely used as an instructional model to illustrate the course of development of a dicot embryo. As we shall see in later chapters, bryo. However, not all descendants of the basal cell are incorporated into the embryo proper; a small number of cells assume the characteristics of a suspensor. This is known as the Asterad type.

The Chenopodiad type described in members of Chenopodiaceae and Boraginaceae is basically similar to the Asterad type except that here the first division of the terminal cell is in the transverse plane. Two other types of embryo ontogenesis in which the first division of the terminal cell is transverse are the Solanad type and the Caryophyllad type. In the former type, represented in Solanaceae, Campanulaceae, Theaceae, Linaceae and Rubiaceae, the basal cell forms a well-defined suspensor, with most of the organogenetic part of the embryo derived from the terminal cell. The Caryophyllad type differs fundamentally from the Solanad type in the absence of divisions in the basal cell and the formation of the suspensor from the derivatives of the terminal cell. This type of embryo ontogeny described in members of Caryophyllaceae, Crassulaceae and Haloragaceae has also been documented in *Saxifraga* (Saxifragaceae), *Androsaemum* (=*Hypericum;* Guttiferae) and *Drosera* (Droseraceae). In these three genera, it is not unusual for the basal cell to divide once or twice and in fact become part of the suspensor formed mainly from the derivatives of the terminal cell (Maheshwari, 1950; Bhojwani and Bhatnagar, 1978). In this system of embryo classification we must be prepared, then, to discover that some subjective judgment is inevitable in the assignment of embryo development to particular classes.

A more complicated classification of embryo ontogenesis has been developed by Souèges (1937) (see also Crété, 1963). Although the pattern of early segmentation of the embryo forms the backbone of this system, it looks upon angiosperm embryos as adhering to certain laws concerned with the origin of cells at any particular stage, the number of cells produced in each generation, the role of cells formed and the deterministic fates of cells produced during the early division phase of embryos. A simplified version of Souèges' system of classification of embryo types adopted by Mestre (1967) has focused attention on the fate of the two cells formed from the first division of the zygote. Periasamy (1977) developed the thesis that the increasing exposure of newly formed cells of the zygote to the internal environment of the future embryo, rather than to the external environment of the endosperm, makes the segmentation of the first internal cell an important morphogenetic event. This has led to a new approach in the classification of embryo types based on the number of cell tiers in the proembryo when the first series of cells whose surfaces are not exposed to the exterior is segmented. The relative location of the tiers with varying cell numbers and the terminal or subterminal location of the tier in which formation of the first internal cells occurs are employed as additional criteria with which to assign embryos to different groups in the proposed classification.

18 *Developmental embryogenesis*

The early division sequences of embryos of a large number of plants investigated seem to be remarkably precise. In many instances the cell lineages during embryogenesis also appear to be regular, making it possible to anticipate the destiny of particular cells in the four- or eight-celled proembryo as giving rise to specific organs in the mature embryo. However, an increasing number of investigations have found the crucial first and second divisions of the zygote to be highly variable; in other studies, embryo ontogenesis cannot be referred to recognized types, and still others point out irregularities in the method of origin of embryo parts. An extreme case of variation is the occurrence of a vertical or longitudinally oriented oblique wall, rather than a transverse wall, initiating the first division of the zygote; this mode of embryo origin designated Piperad type, is however observed only in a few obscure plants (Natesh and Rau, 1984). Stemming from these and other reports, there is some question as to whether a reliable system of classifying angiosperm embryo ontogenesis can be formulated on the basis of successive steps in the division of the zygote and a dogmatic emphasis on the destination of cells formed. Histological studies of early embryogenesis have produced such a wealth of information on the phenomenon relative to other aspects of embryogenesis that one must consider whether the importance of this information may have been exaggerated.

From proembryo to mature embryo

The discussion thus far has dealt only with the development of the earliest stages of the embryo. The student of embryogenesis is faced with the need to understand how an adult embryo is produced from this limited number of cells. Toward this end, the successive steps in the embryogenesis of representative species from both dicots and monocots will be described. Needless to say, this includes only a fraction of the number of species whose embryo ontogeny has been followed; descriptions of embryo development in numerous other species can be found in such sources as Johansen (1950), Maheshwari (1950) and Davis (1966) as well as in the review of Crété (1963). Of some practical concern and interest for the light it sheds on the problems of analyzing embryogenesis in angiosperms is the fact that in tracing the development of the embryo from its single-celled origin we still depend on reconstruction of serial sections of fixed and processed materials.

Capsella bursa-pastoris (Shepherd's purse), a member of the Cruciferae, is a common weed that does not evoke much general attention but that is noted for its regular pattern of embryo development. For this reason, *Capsella* has been widely used as an instructional model to illustrate the course of development of a dicot embryo. As we shall see in later chapters,

Capsella has also served as a favorite material for analysis of the ultrastructural aspects of embryo development as well as for investigations in embryo culture. Figure 2.1 shows a series of stages in the development of *Capsella* embryos. These stages do not cover the whole spectrum of embryogenesis but trace it far enough to show the main events and thus indicate the kinds of problems confronting plant embryologists. Our description of embryogenesis in *Capsella,* based on the work of Souèges (1919), involves some cell lineage terminology. As seen in Figure 2.1, the first division of the zygote produces two cells, ca and cb. Next, the basal cell cb divides transversely to produce cells ci and cm. Characteristic of the Crucifer type of embryo development, a longitudinal division follows in the terminal cell (ca). As a result of these divisions, a four-celled embryo is formed as a three-tiered structure with a terminal tier composed of two cells. Each of the two terminal cells is now partitioned by vertical walls to yield a quadrant. Quadrant is quite a transient stage, as it is immediately partitioned by transverse walls and phased out into the octant stage, consisting of eight cells arranged in two tiers. In the adult embryo, the stem tip and cotyledons are formed from the terminal tier, and the hypocotyl is generated from the basal tier. At the octant stage of the embryo, cells ci and cm move into action to form the suspensor. Generally, the most basal cell ci does not divide, but rather increases in size to form a large vesicular cell. During expansion, there may be a severalfold increase in the volume of the cell, largely attributable to the development of a central vacuole. Meanwhile, the middle cell cm of the three-celled embryo forms a row of six to ten cells held together as a filament and constituting the main body of the suspensor. A curious and interesting cell of the suspensor is the hypophysis h. This cell, which lies next to the embryo, divides transversely to form two cells, h' and h" and the descendants of these cells contribute to the root cortex, root cap and root epidermis of the mature embryo.

The first histogenetic event in the organ-forming part of the embryo is a periclinally oriented division in each of the octant cells. The result of this process is the formation of a 16-celled globular mass (globular embryo), consisting of eight external protoderm cells destined to form the embryonic epidermis and eight internal cells that gradually differentiate the primary tissues of the procambium and ground meristem of the hypocotyl and cotyledons. As is well known, the primary vascular tissues and the entire complex of ground tissues (i.e., tissues other than the epidermis and vascular tissues) of the adult plant are differentiated from the procambium and ground meristem, respectively. The globular embryo has a radial symmetry; it continues to enlarge by divisions within the sphere. Toward the end of the globular stage, the embryo expands laterally as a result of periclinal divisions in the terminal lateral poles, giving rise to a pair of cotyledons. At the onset of cotyledon initiation, the embryo assumes a more or less cor-

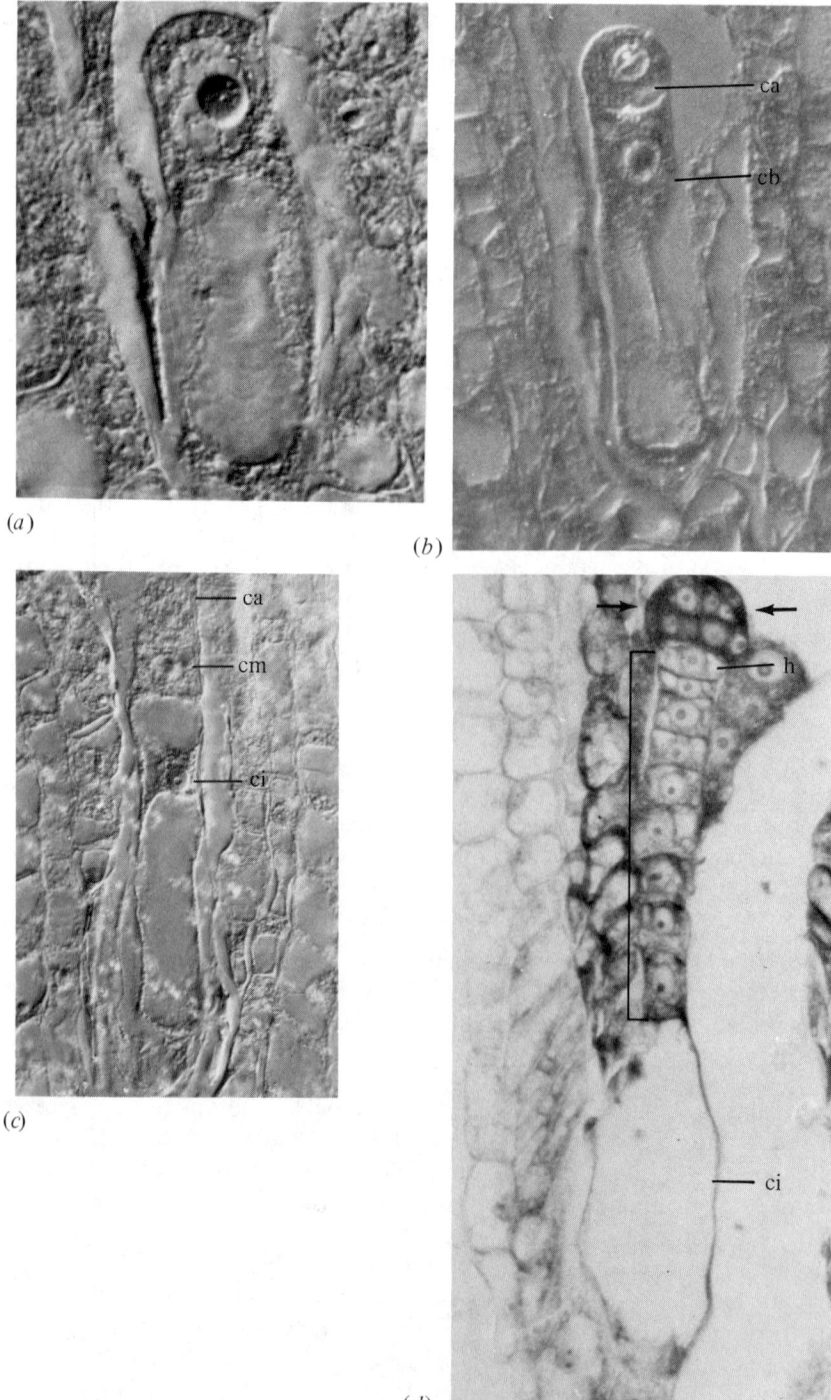

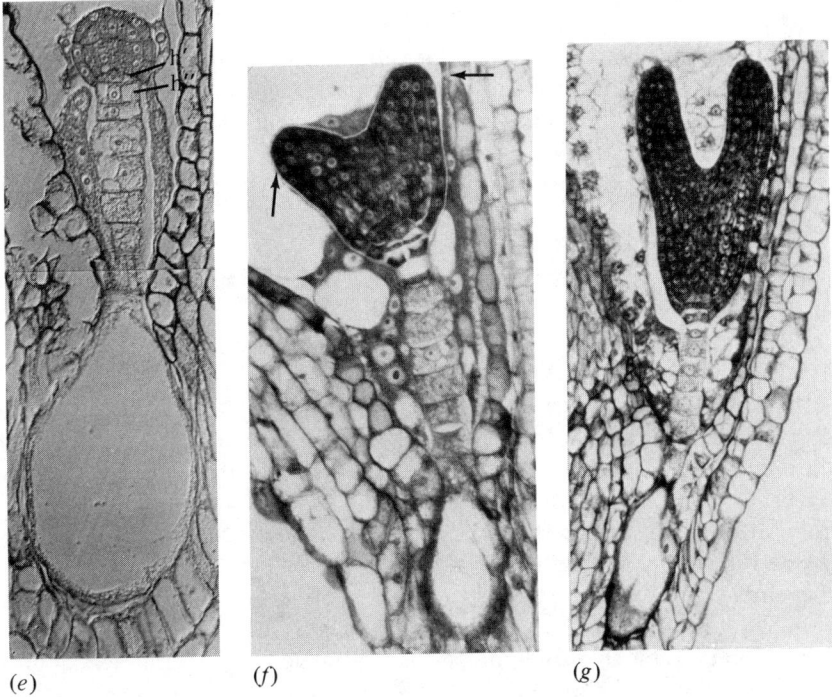

(e) (f) (g)

FIGURE 2.1. Embryogenesis of *Capsella bursa-pastoris;* in all cases, the micropylar part of the ovule is toward the bottom of the page. (a) A zygote about to divide. (b) First division of the zygote to form cells ca and cb. (c) Division of cell cb to form cells ci and cm, all photographed with Nomarski optics. (d) A globular embryo (*arrows*); cells of the suspensor enclosed in the square bracket are formed by the division of cell cm. Cell h is the hypophysis. (e) Periclinal division in the globular embryo. Cells h' and h" are formed by the division of cell h; photographed with Nomarski optics. (f) A heart-shaped embryo, showing cotyledons (*arrows*). (g) A torpedo-shaped embryo.

date shape (heart-shaped stage). At the same time that the cotyledons are initiated, division and differentiation of cells in the basal tier of the embryo give rise to the hypocotyl. The processes of cell differentiation in the cotyledons and hypocotyl are usually accompanied by considerable elongation of these organs, resulting in the formation of the torpedo-shaped embryo. During cotyledon initiation, a mound of rapidly dividing cells that constitute the future shoot apex of the embryo is organized in the depression between the cotyledons. The root system of the future embryo is delimited in the form of a root apex at the torpedo-shaped stage of the embryo, increasing in prominence with further growth of the embryo.

During this period, the continued elongation of the cotyledons and hypocotyl compounded by spatial restrictions within the ovule causes the embryo to become curved at the tip, ultimately assuming the shape of a horseshoe. Although the inception of a basic organization characteristic of the adult plant is unmistakably clear in the torpedo-shaped and older embryos, subtle physiological and biochemical changes continue to occur in the embryo and in the surrounding tissues of the ovule before they become mature.

Like *Capsella*, embryos of *Phaseolus vulgaris* (French bean) have also been the subject of extensive experimental investigations, meriting a brief account here. In *P. vulgaris*, the basal cell of the two-celled embryo divides transversely, followed by a longitudinal division in the terminal cell to form a four-celled embryo. Whereas the embryo proper is formed from the terminal cell by a series of divisions very similar to those described in *Capsella*, the behavior of the two cells derived from the basal cell is quite different here. The cell next to the most basal cell divides by transverse and longitudinal walls to form four pairs of cells of which the one closest to the embryo proper gives rise to the hypophysis. The other three pairs of cells divide repeatedly and form part of the suspensor. During further growth of the embryo, the basal cell also divides transversely and longitudinally, adding to the bulk of the suspensor (Souèges, 1950). The result is the formation of a massive columnar suspensor made up of several tiers of four cells each. Since the descendants of the basal cell contribute only a limited number of cells to the development of the embryo proper, embryogenesis in *P. vulgaris* can be assigned to the Crucifer type.

The early divisions of the zygote in monocots involve the same orderly sequence of steps described for representatives of dicots. In *Halophila ovata* (Swamy and Lakshmanan, 1962a) and *Najas lacerata* (Swamy and Lakshmanan, 1962b), after the first division of the zygote in the transverse plane the basal cell enlarges modestly into a vesicular structure, and the entire embryo including the suspensor is derived from the terminal cell. By contrast, in *Luzula forsteri* (Souèges, 1923), *Muscari comosum* (Souèges, 1932) and *Sagittaria sagittifolia* (Swamy, 1980) part of the embryonic root and the entire suspensor are contributed by the descendants of the basal cell. During the late phase of embryogenesis, the single cotyledon appears terminally on the cylindrical embryo, and the shoot apex arises in a lateral depression at the base of the cotyledon.

The apparent lateral position of the shoot apex in relationship to the terminal cotyledon has led to a lively debate on the origin of these structures on the monocot proembryo. Souèges (1954) forcefully championed the view that the cotyledon is derived from the terminal cell of a three-celled proembryo and that the shoot apex arises from the derivatives of the subterminal cell. On the opposite side of the debate, Swamy (1979),

Swamy and Lakshmanan (1962a, b), Guignard (1975, 1984) and Ba, Cavé, Henry and Guignard (1978) have claimed that the cotyledon and shoot apex arise from the same terminal cell of the proembryo while derivatives of the subterminal cell are incorporated into the hypocotyl. According to Swamy and Guignard and their co-workers, what is apparently perceived as a lateral shoot apex is a shoot apex displaced from its terminal position by the aggressive growth of the cotyledon. This is very nicely illustrated in a scanning electron microscopic study of embryogenesis in *Potomogeton lucens* (Ba et al., 1978).

Despite the ontogenetic similarity between dicot and monocot embryos up to the octant stage, the form of the mature embryos is dramatically dissimilar in the two subclasses. From a morphogenetic point of view, Swamy (1962) advanced the hypothesis that the terminal octants of both dicot and monocot proembryos transcribe more or less identical developmental programs to initiate two lateral cotyledons and a terminal shoot apex. The program is faithfully played back in the dicots as the cotyledonary initials are established before the relatively quiescent shoot apex springs into action. The striking feature of the monocot embryo is the simultaneous growth of cotyledon initial cells and the shoot apex, the derivatives of which are consolidated into one cotyledon. This view implies that because the single cotyledon incorporates the primordia of both cotyledons and the original shoot apex, a new functional shoot apex arises laterally from the subterminal tier of cells of the proembryo. Interesting as this hypothesis is, it is only fair to remark that the apparent promise that this might lead to a morphogenetic investigation of the origin of the shoot apex and cotyledons in monocot embryos has not been fulfilled.

In their mature structure and their ontogeny, embryos of Gramineae have much less in common with other monocots. The main feature of the grass embryo, as exemplified by *Triticum aestivum* (wheat) (Fig. 2.2) is the development of an absorptive organ known as the scutellum. Other special features of the embryo are the presence of a cap-like tissue covering the root (coleorhiza) and one that covers the plumule (coleoptile). On one side of the coleorhiza there is also a small flap-like outgrowth called the epiblast.

In wheat, the first division of the zygote is oblique, and both cells formed divide further by oblique walls to form a four-celled proembryo. Derivatives of these cells contribute to the formation of the mature embryo. During its ontogeny, the embryo begins to take shape at the 16- to 32-celled stage. In the club-shaped embryo, a vaguely outlined elevation due to accelerated cell division is initiated in the apical–lateral region, marking the appearance of the scutellum. Although the scutellum is the first structure of the mature embryo to be initiated, it does not attain its actual bulk until much later. Soon after initiation of the scutellum, the

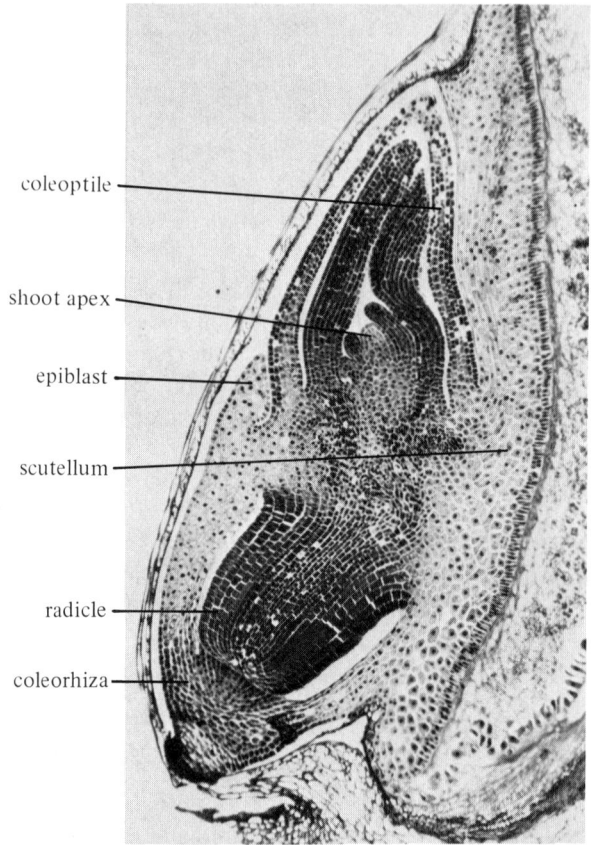

FIGURE 2.2. Section of the embryo of *Triticum aestivum*.

opposite side of the embryo begins to enlarge, foreshadowing the differentiation of the shoot apex and leaf primordia. The final major event of embryogenesis is the differentiation of the radicle, which is formed endogenously in the central zone of the embryo (Batygina, 1969). In certain other members of the Gramineae there is no regular pattern of cell division in the proembryo and, as seen in *Zea mays* (maize), divisions become chaotic even beginning the second round (Randolph, 1936).

This discussion concludes with an account of the remarkable type of embryogenesis described in *Paeonia anomala*, *P. moutan* and *P. wittmanniana* (peony) by Yakovlev and Yoffe (1957). The unique feature of embryogenesis in the peony is that the first division of the zygote as well as several subsequent divisions are unaccompanied by wall formation, thus giving rise to a free-nuclear proembryo. During divisions, the nuclei tend to move toward the periphery of the cluster, leaving a large vacuole in the center. Within a few days after anthesis, the proembryo has sufficient free nuclei with which to begin the task of organizing a multicellular proembryo. Evidently, multicellularity in the proembryo brings with it new properties conducive to morphogenesis. This becomes apparent as certain marginal cells of the proembryo begin to divide and emerge as embryo primordia (Fig. 2.3). Although numerous developmentally equivalent embryo primordia are formed on a proembryo, only one outlives the others and differentiates into a typical dicot embryo. Independent investigations by other workers, notably Cave, Arnott and Cook (1961) and Carniel (1967), on different species of *Paeonia* have confirmed the essential features of the unusual embryogenesis in this genus. Free nuclear division of the zygote observed in peony appears to be the angiosperm equivalent of the situation so widely documented in gymnosperms.

Organization of tissue systems and meristems

The organization of the three primary meristems – the protoderm, ground meristem and procambium – as well as the apical meristems of the shoot and root during embryogenesis has been followed in studies that have coordinated observations on gross morphological changes and histogen differentiation. As these tissues do not appear simultaneously, but differentiate during progressive growth of the embryo, an increase in cell number may be an important factor determining the type of meristem that differentiates. A basic plan pervades the organization of tissues and meristems in embryos of angiosperms examined. As noted in *Capsella,* the elements of the protoderm, which is generally the first meristem to differentiate, are initiated by a periclinal division of cells of the octant embryo. After they are cut off, cells of the embryonic protoderm continue to divide anticlinally to keep pace with the increasing volume of the embryo, merging at the basal end with the root apical meristem. The ground meristem and the procambium are derived from the inner core of cells of the globular embryo bounded by a superficial protoderm. The differentiation of the ground meristem is signaled by the appearance of cells with decreased stainability and increased vacuolation. The cells of the ground meristem may differentiate into a cortex or into both cortex and pith. The former situation prevails in embryos of *Phlox drummondii* (annual phlox) (Miller and Wetmore, 1945), *Dianthus chinensis* (rainbow pink) (Buell, 1952), *Sphenoclea zeylanica* (Swamy and Padmanabhan, 1961), *Downingia bacigalupii* and *D. pulchella* (Kaplan, 1969) and *Stellaria media* (chickweed) (Ramji, 1975). The peripheral layer of cells of the central meristematic core of the globular embryo are the progenitors of the cortex of the mature embryo. During elongation of the embryo, these cells divide periclinally to

26 *Developmental embryogenesis*

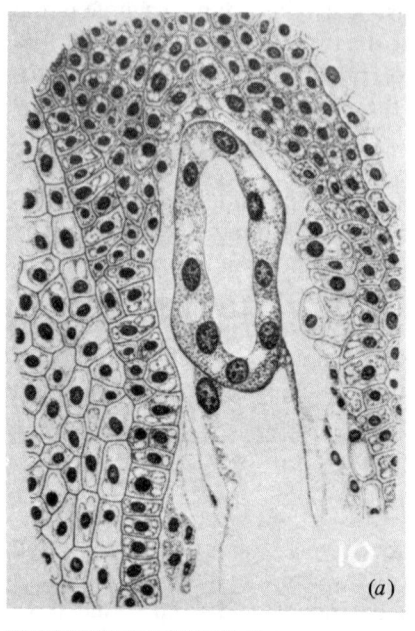

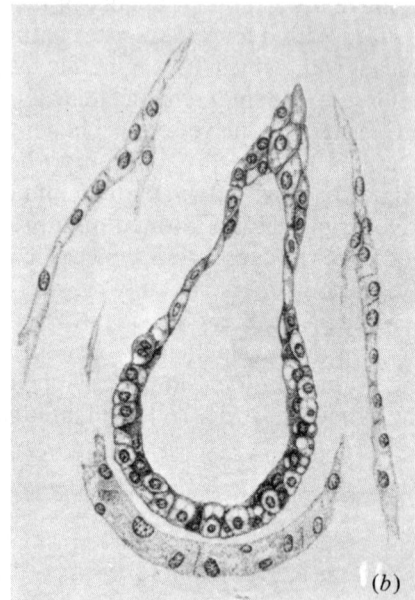

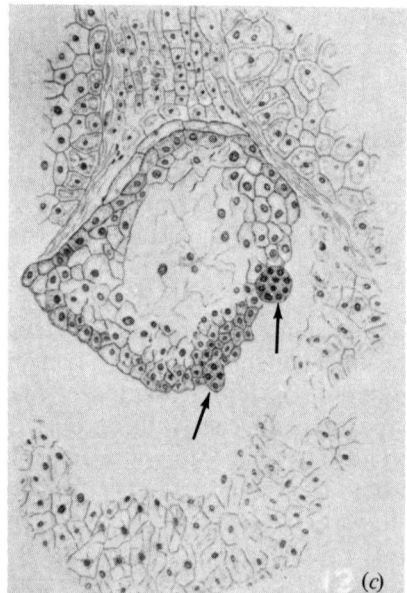

FIGURE 2.3. Embryogenesis of *Paeonia*. (a) Free-nuclear proembryo in *P. wittmanniana*. (b) Organization of multicellular proembryo in *P. anomala*. (c) Multicellular proembryo of *P. anomala* showing embryo primordia (*arrows*). (From Yakovlev and Yoffe, 1957.)

form three to four concentric layers of cells of the embryonic cortex. In *Juglans regia* (English walnut) (Nast, 1941), *Pisum sativum* (pea) (Reeve, 1948), *Nerium oleander* (Mahlberg, 1960) and *Sesamum indicum* (sesame) (Hanawa, 1960), a pith is also derived from the ground meristem. First evidence of the pith is the presence of enlarged light-staining cells in the central region of the globular embryo; by repeated divisions these cells continue as the pith of the hypocotyl part of the mature embryo.

Unlike the cells of the ground meristem, procambial cells stain somewhat intensely, indicating a dense protoplasmic content, and are characteristically narrow and elongate in shape. Procambium initiation is a critical feature of histogenesis of the embryo and occurs in the cells of the cortex or pith of globular stage embryos before the primordia of cotyledons are formed (Miller and Wetmore, 1945; Balfour, 1957; Mahlberg, 1960; Hanawa, 1960; Kaplan, 1969). According to Mahlberg (1960), in *Nerium oleander* a procambium is first seen in the innermost part of the embryonic cortex of the early heart-shaped embryo at the level of the presumptive cotyledons and is discernible by its small transectional area. As more initials are laid down on either side of the original trace, a complete ring of procambium is established. Presumably, after the procambium is developed in the hypocotyl axis of the embryo, it extends acropetally into the cotyledons. The procambial connection to the shoot and root apical meristems are also formed later. Thus, by the time the embryo attains the torpedo-shaped stage, it has a solid procambium cylinder extending from the root apex to the tip of the cotyledon.

Embryos of two members of the Rhizophoraceae that were recently investigated, namely, *Cassipourea elliptica* (Juncosa, 1984a) and *Bruguiera exaristata* (Juncosa, 1984b), appear to differ from those of most other angiosperms in one important respect: here, the procambium is not distinguishable until after the initiation of cotyledons. In *Rhizophora mangle* embryo, it has been claimed that the procambium connecting the shoot and root apical meristems is discontinuous for a brief period until it is joined to the hypocotyl procambium cylinder by cotyledonary strands (Juncosa, 1982). This might indicate independent origins of the procambium in the different organs of the developing embryo. This view is reinforced in a study of the histogenic pattern in embryos of *Sphenoclea zeylanica* (Swamy and Padmanabhan, 1961), in which procambium initiation occurs in the cotyledons soon after they are formed but is delayed in the hypocotyl until seed germination. These reports indicate that a reevaluation of the generally accepted views on the procambium interrelationships between cotyledons, hypocotyl and shoot and root meristems of developing embryos is necessary to reconcile the apparent contradictions.

Another important aspect of procambial differentiation in embryos that deserves comment is the maturation of the first vascular elements. Do the xylem and the phloem elements follow a course of maturation coincident with that of the procambium? Is there a consistent pattern with regard to the type of vascular element that matures first in the embryo? As the degree of morphological differentiation attained by the embryo in seeds varies widely, the available information presents a confusing story. According to Dauphiné and Rivière (1940), sieve tubes are the first vascular elements to differentiate in mature embryos of *Lupinus albus* (lupine) and are first discernible in the hypocotyl–root axis, from which they extend acropetally into the cotyledons and basipetally into the vicinity of the root apex. These investigators also found that sieve elements are the first to differentiate in embryos of *Helianthus annuus* (sunflower), *Ricinus communis* (castor bean) and *Mirabilis jalapa* (four-o'clock plant). In *Juglans regia* embryos, phloem, which precedes xylem, first appears in the root and the first xylem elements are seen at the base of the cotyledonary node (Nast, 1941). First xylem elements are seen in the cotyledons of embryos of *Phlox drummondii* (Miller and Wetmore, 1945), *Cassipourea elliptica* (Juncosa, 1984a) and *Bruguiera exaristata* (Juncosa, 1984b), where they appear ahead of the phloem and generally in more than one locus. Despite the versatility of developing embryos as an experimental system, it is clear that vascular tissue maturation during embryogenesis has not been comprehensively studied. The variability in the pattern of vascular tissue maturation observed in embryos thus far investigated justifies the need for more comparative studies before generalizations can be attempted.

Origin of shoot and root apices. Careful ontogenetic studies have established that the potential shoot and root apical meristems become recognizable in the globular embryo at about the same time that the primary tissues are blocked out. In *Nerium oleander,* the earliest indication of the shoot and root apices is the appearance of zones of partially differentiated meristematic cells at either end of the embryo. These zones are interpreted as residual groups of undifferentiated cells of the globular embryo, implying that they are the direct descendants of the octant-stage embryo rather than new cells generated as a result of tissue organization in the embryo (Mahlberg, 1960). Considering their importance as determinative centers of the adult plant, it appears that shoot and root meristems have an undistinguished origin.

In evaluating the gross aspects of differentiation of the shoot apex in the globular embryo we can say that it is augured by a change from the random distribution of cell divisions to a concentration of mitoses in the apical half and by cytohistological differentiation into a peripheral and a central zone (Mahlberg, 1960; Kaplan, 1969). Since these changes become evident before the initiation of cotyledons, the latter are considered the first formative organs produced by the shoot apex. The visible differentiation of the shoot apex is a reflection of continued meristematic activity in the protoderm and of the cells immediately below the protoderm in the apical part of the embryo as it changes shape from globular to broadly ovoid to heart-shaped. At a still later stage of embryogenesis, a further cytological differentiation of three to four layers of cells below the protoderm occurs, foreshadowing the formation of the central zone of the embryonic shoot apex. Subsequently, the apex becomes well defined as a rounded dome and lapses into a state of quiescence or dormancy in the mature embryo.

The embryonic root consists of the root meristem and the root cap. Depending on the type of embryo ontogenesis, these tissues are formed from the derivatives of the terminal or the basal cell of the two-celled proembryo. In *Downingia,* in which the root is derived from the terminal cell, at the time of root differentiation the proembryo has four superposed tiers of cells. During organization of the root apex, cells of the most basal tier divide periclinally to give rise to the root cap, which merges with the protodermal cells formed from the cells of the next tier toward the apical part of the embryo. This latter tier also generates cells of the root meristem (Kaplan, 1969). In *Capsella* in which the embryonic root is derived from the two cells of the hypophysis, the root cap is formed from the cell of the hypophysis close to the suspensor by periclinal divisions, while the hypophyseal cell close to the embryo proper by anticlinal and periclinal divisions gives rise to the root meristem. The cells of the root meristem remain densely protoplasmic throughout embryogenesis and are thus distinguished from the light-staining cells of the pith and the cortex.

Origin of the quiescent center. It is now well established that the apical meristem of the angiosperm root contains a region known as the quiescent center, in which cells seldom synthesize DNA and hence are relatively inactive mitotically. From the point of view of origin of the quiescent center, an important question is whether it is present during the development of the root meristem in the embryo. At what stage of development does the embryo acquire a quiescent center? What is the fate of the quiescent center during maturation of the embryo? Based on the tempo of mitotic activity and staining reactions, some early studies showed the existence in the root meristem of developing embryos of a group of cells analogous to the quiescent center. For example, Sterling (1955) identified a zone of mitotically inactive cells surrounded by actively dividing cells in the radicle meristem of maturing *Phaseolus lunatus* (lima bean) embryos; these cells could correspond to the quiescent center of mature roots. Rondet (1962) found the embryonic root meristem of *Alyssum maritimum* to contain a region that stains lightly with methyl-green pyronin and indicated that it is comparable to the quiescent center of seedling roots. On the basis of cell lineage studies, Vallade (1972) showed that a group of four cells comparable to the quiescent center is present throughout embryogenesis in

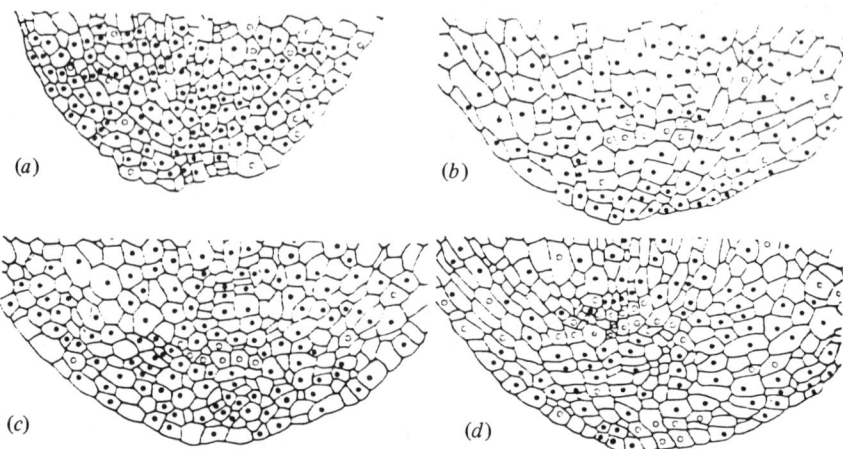

FIGURE 2.4. Development of the quiescent center in young embryos of *Pisum sativum* incubated in ^{3}H-thymidine; labeled nuclei are represented by closed circles and unlabeled nuclei by open circles. In (*b*), (*c*) and (*d*) the quiescent center is indicated by regions in the root meristem where the open circles are concentrated. Lengths of embryo axes are (*a*) 320 μm, (*b*) 640 μm, (*c*) 470 μm, (*d*) 760 μm. (From Clowes, 1978b.)

Petunia hybrida, beginning with the globular stage. In other studies, a protocol involving administration of ^{3}H-thymidine to ovules or incubating excised embryos in the radioactive isotope combined with autoradiography was employed to monitor DNA synthetic activity in the root meristem and to identify the quiescent center. Analysis of the distribution of ^{3}H-thymidine-labeled nuclei and mitotic figures has led to the delimitation of a quiescent center in devloping embryos of pea (Jones, 1977; Clowes, 1978b) and maize (Clowes, 1978a) with well-developed root cap initials or a primitive root cap. As a model for the earliest seen quiescent center, in the pea embryo it is represented by a band of no more than four to seven cells (Fig. 2.4). As the embryo matures, decay occurs in the meristematic activity throughout the root meristem, followed by the disappearance of the quiescent center. These results indicate that the quiescent center is a product of embryogenic development that originates as soon as the root meristem is organized in the early-stage embryos.

Morphology of the suspensor

As seen in *Capsella,* a filamentous suspensor attached to the organogenetic part of the embryo is one of the most familiar images of embryogenesis in angiosperms. Early histological studies of embryos of numerous plants demonstrated that the suspensor is a short-lived organ that varies greatly in size and shape. The suspensor attains full development in the proembryo; it then begins to degenerate and is completely obliterated by the growing embryo. From a mechanistic standpoint, the suspensor is thought to push the growing embryo into the cavity of the embryo sac, close to the source of cellular nutrients. Research on the ultrastructure of the suspensor over the past 20 years (reviewed in Chapter 3) has steadily accumulated evidence in favor of its role in absorbing and transmitting nutrient substances to the embryo. Here we shall consider the general morphology of the suspensor in representative species demonstrating a whole new panorama of structural modifications.

The family Orchidaceae is a good example of the remarkable diversity exhibited by the suspensor (Swamy, 1949). It ranges from species of *Cypripedium*, *Spiranthes*, *Listera* and *Neottia*, which do not possess a suspensor, to *Geodorum*, *Eulophia* and *Cymbidium*, in which the suspensor cells begin to elongate into tubular structures immediately after they are cut off. Intermediate forms are found mostly in *Epipactis*, *Gastrodia* and *Coelogyne*, in which the suspensor initial cell does not divide and persists throughout embryogeny without much elongation. In *Ophrys*, *Orchis* and *Habenaria*, a five- to ten-celled filamentous suspensor is formed initially; later the filament elongates out of the micropyle, penetrating the placenta by means of tubular structures. Another modification of suspensor morphology is found in certain species of *Epidendrum*, in which the initial cell divides transversely and longitudinally to form a multiseriate swollen structure. In *Vanda*, *Phlaenopsis* and others, the suspensor initial produces eight cells that grow toward the embryo and envelop it.

Leguminosae is known to display a range of suspensor morphology matching that of Orchidaceae. According to Lersten (1983), who has collated much of the information on suspensors in this family, the form of the suspensor may be (1) spheroid, accompanied by various degrees of swelling of the cells (*Lotus*); (2) long, slender, biseriate and filamentous (*Cicer*); (3) uniseriate, with grossly inflated cells (*Ononis*) or (4) massive, elongate and two to six cells wide (*Phaseolus*). In the subfamily Genisteae, two strikingly different forms of the suspensor have evolved; in *Genista monosperma*, it appears as an enormous spherical structure virtually engulfing the small proembryo, whereas in the different species of *Lupinus* it ranges from a uniseriate filament to a biseriate column of short, broad cells (Fig. 2.5).

There are probably few other families in which the suspensor exhibits such variations as in Orchidaceae and Leguminosae. Modified suspensors are also found in plants belonging to Fumariaceae, Crassulaceae, Rubiaceae, Podostemaceae, Trapaceae and Tropaeolaceae. In *Dicraea stylosa* (Podostemaceae), the basal cells of the filamentous suspensor enlarge and

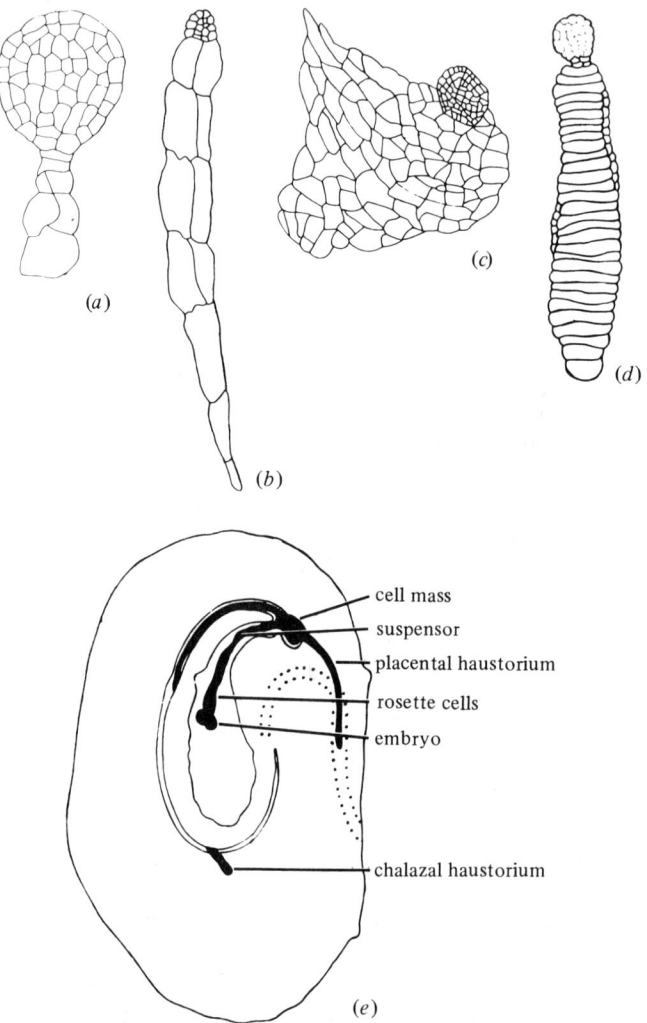

FIGURE 2.5. Variations in suspensor morphology. (a) *Lotus corniculatus*. (b) *Cicer soongaricum*. (c) *Genista monosperma*. (d) *Lupinus pilosus*. (From Lersten, 1983.) (e) Longitudinal section of the carpel of *Tropaeolum majus* showing the growth of suspensor haustoria. (From Walker, 1947.)

put out several thin-walled densely cytoplasmic tubes that invade the integument (Mukkada, 1962). In *Trapa bispinosa* (Trapaceae), the suspensor initial divides repeatedly by transverse and longitudinal divisions to produce a long, coiled, multiseriate structure that pushes the embryo deep into the embryo sac cavity. In addition, a collar-like layer of cells originating from the suspensor covers the embryo on the side facing the bulk of the nucellus (Ram, 1956). A most interesting modification of the suspensor is found in *Tropaeolum majus* (Tropaeolaceae), in which it attains several millimeters in length at the globular stage of the embryo (Fig. 2.5). At one end the suspensor is attached to the embryo by a rosette of elongate cells, and at the micropylar end it forms a cellular mass from which two multicellular branches arise. One penetrates the integument near the micropyle and grows around the ovule in the cells of the carpel (chalazal haustorium). The other branch (placental haustorium) traverses through the integument and funiculus to reach its final destination in the vascular bundle of the placenta (Walker, 1947).

In the above account, the focus of morphological variations in the suspensor has been on the extension growth of cells reaching the extraembryonal tissues of the ovule and even the carpel. In the embryology literature these outgrowths have been frequently referred to as haustoria, meaning that they are specialized structural adaptations for the absorption of nutrient substances. Complex suspensors described in Orchidaceae, Podostemaceae and Trapaceae are correlated with the absence of endosperm or the presence of an extremely reduced endosperm in the ovules. This indicates that nutrients absorbed by the suspensor haustoria may be an important source of energy for the growth of embryos. However, direct evidence has been difficult to obtain, as most observations are based on the close association of the haustoria with the extraembryonal tissues and do not bear directly on the question of actual transfer of nutrients or on the ultrastructural modifications of the haustoria for nutrient uptake. Some evidence along these lines is available, however, for the cells of the suspensor itself (see Chapter 3).

Timetable for embryogenesis

A timetable for embryogenesis identifies the times at which stable and reproducible changes are initiated in the embryo. Using embryos of different ages as a frame of reference, attempts have been made to relate physiological age or developmental stage of embryos to changes in their mitotic activity, length, volume or fresh and dry weights. A generally accepted notion is that cell division predominates during the early period of embryogenesis and that cell enlargement prevails during the later period. However, not many quantitative studies have been undertaken to support this view. In an investigation in which actual counts of nuclei in division during embryogenesis in *Capsella* and *Gossypium hirsutum* (cotton) were made, increase in cell number was found to be interrupted by a lag during cotyledon initiation, resulting in the formation of a conventional S-shaped curve for cell number from zygote to globular embryo and another S-shaped curve from heart-shaped embryo to mature embryo (Pollock and Jensen, 1964). In the globular embryo of *Hordeum vulgare* (barley), *Oryza sativa* (rice) and wheat an exponential increase in cell number is maintained only for a fleeting period – up to the stage of 20 to 40 cells – and thereafter the number of cells formed shows a rhythmic variation. The occurrence of subtle changes in cell metabolism preparatory to form change in the embryo is suggested by these findings (Nagato, 1978). On the basis of qualitative observations in developing *Glycine max* (soybean) embryos, the period up to the initiation of cotyledons (8 to 10 days after anthesis) has been identified with cell division activity, followed by a period of cell enlargement lasting 7 to 10 days (Meinke, Chen and Beachy, 1981). Cell doubling times during the exponential phase of embryo growth have been reported to vary from 20 to 22 hours in cotton (Pollock and Jensen, 1964), 8.0 to 19.2 hours in wheat (Bennett, Rao, Smith and Bayliss, 1973), 9.2 to 12.9 hours in barley and 15.7 to 22.7 hours in *Secale cereale* (rye) (Forster and Dale, 1983a); in carrot embryos, the cell doubling time is 42 hours (Gray, Ward and Steckel, 1984).

Although linear measurements that yield S-shaped curves continue to be used to construct timetables for embryogenesis (Tykarska, 1980; Rogers and Quatrano, 1983), it has become clear that fresh and dry weight changes are the most accurate indicators of embryo development in time. These approaches have shown that during the period following fertilization when morphogenetic changes leading to the establishment of embryonic organs take place, the embryo acquires only a fraction of its fresh or dry weight at maturity. This is followed by considerable increases in fresh and dry weights of embryos during the next several days, most of this increase coinciding with the expansion of cotyledons, accumulation of storage reserves and attainment of maturity. While dry weight of the embryo continues to increase at a reduced rate through its mature stage to the period of quiescence or dormancy, fresh weight registers a peak a few days before maturity and decreases thereafter. These findings are less surprising when we consider changes in the embryo water content. There is a dramatic increase in the water content during the first half of embryogenesis and an equally dramatic decrease during the second half leading to a loss of up to 90 percent of water previously present in the cells (Walbot, Clutter and Sussex, 1972; Walbot, Harris and Dure, 1975; Crouch and Sussex, 1981). The results plotted in Figure 2.6 illustrate these changes separately for cotyledons and embryo axes of *Phaseolus vulgaris,* along with length and dry weight changes of pods and seeds. In practical terms, the physiological and developmental changes in the embryos are correlated with changes in the dimensions of ovules and ovaries; thus, from the latter data generated from plants grown under constant light and temperature regimens, it is possible to determine the stage of development of the enclosed embryo for use in experimental investigations.

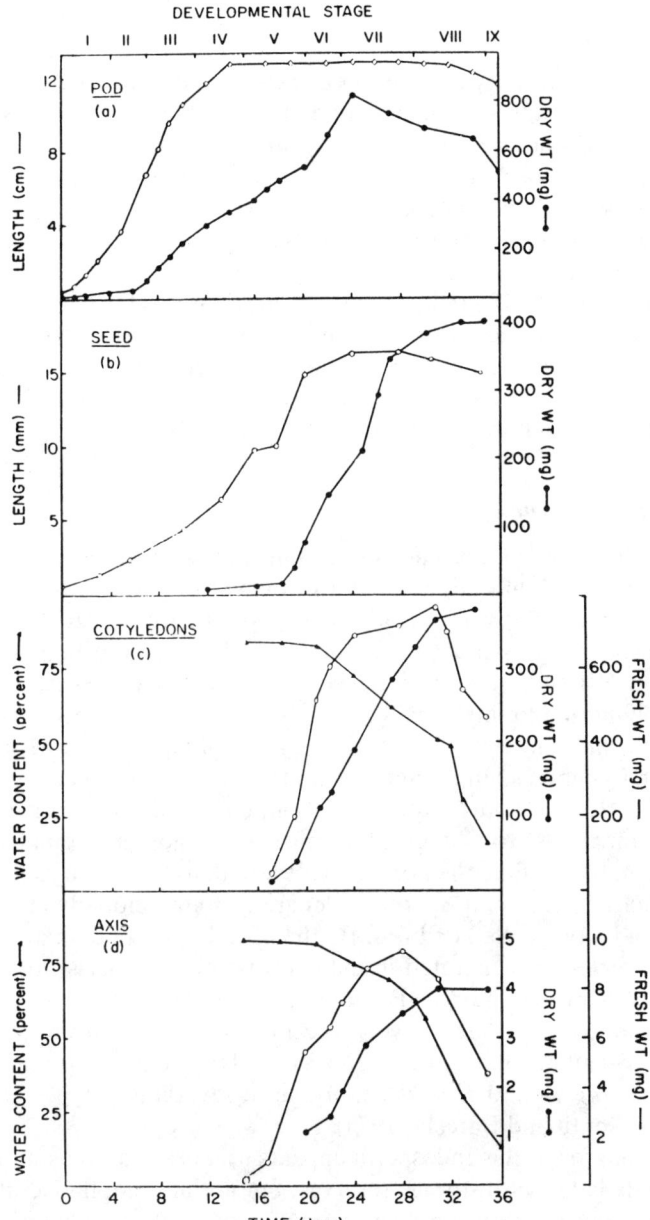

FIGURE 2.6. Postfertilization changes in length, fresh weight, dry weight and water content of pods, seeds, cotyledons and embryo axes of *Phaseolus vulgaris*. (From Walbot et al., 1972.)

Endosperm development

The fate of the primary endosperm nucleus in the central cell is a question very much related to our analysis of embryogenesis. A tissue known as the endosperm is generated by the mitotic activity of this nucleus, and as the endosperm is formed it surrounds the embryo and fills the entire central cell. During its development, the endosperm also begins to accumulate enormous quantities of storage products such as starch, proteins and fats in the cells. From the point of view of production of cells with triploid or higher levels of ploidy and piling up of nutrient products in the cells, the utilization of genomic information in the primary endosperm nucleus has important implications in embryogenesis and seed formation. The pattern of development of the endosperm and its influence on embryogenesis constitute the theme of this section.

Endosperm ontogenesis

The development of the endosperm has now been followed at the light microscopic level in a large number of plants. In general, the endosperm nucleus begins to divide ahead of the zygote; the extent to which the endosperm develops prior to the division of the zygote ranges from the formation of a few isolated nuclei to a fully developed tissue. A comparison involving the timing of mitoses in the primary endosperm nucleus and the zygote in *Triticum aestivum* shows the striking shift in the balance of competing genomes. In plants grown at 20°C, the endosperm nucleus divides as early as 6 hours after pollination, but the division of the zygote is delayed until 22 hours. At least four nuclear divisions are completed in the endosperm by the time the zygote divides, so that by 24 hours after pollination at least 16 free endosperm nuclei are drifting randomly in the central cell around a two-celled embryo. By the fifth day after pollination as many as 5,000 nuclei are generated in the endosperm, in contrast to little more than 96 cells in the embryo (Bennett et al., 1973). Differences of more or less the same magnitude between the rates of mitosis in the early endosperm and embryo have also been recorded in other species of *Triticum*, *Hordeum vulgare* and *H. bulbosum* and in several genotypes of *Triticale* (Bennett, Smith and Barclay, 1975).

In its final form, the endosperm appears as a tissue or as a multinucleate cell heavily laden with storage reserves, but nothing emphasizes the differences in endosperm morphology more effectively than its ontogenesis. Of the three principal modes of endosperm development that have been recognized, the nuclear endosperm is of widespread occurrence. Here, the first several rounds of division of the primary endosperm nucleus are unaccompanied by cytokinesis and result in the formation of a few thousand free nuclei in the central cell. The initial step in cellularization is the migration of the nuclei through the cytoplasm to more peripheral regions of the central cell, from which cytokinesis spreads centripetally until the entire tissue becomes cellular. Following this step, the endosperm continues to expand by the production of more cells to generate a massive tissue. In a variation of this pattern, cellularization may involve only the free nuclei around the embryo or in the micropylar region of the embryo sac, or the endosperm may remain as a multinucleate mass of protoplasm throughout its life. Another type known as the cellular endosperm is ontogenetically distinguished from the nuclear type by the absence of a free nuclear phase and the occurrence of a regular cycle of mitosis and cytokinesis in the primary endosperm nucleus and its division products. An endosperm type infrequently encountered in certain monocots is the helobial type, in which, following the first division of the primary endosperm nucleus, the central cell is partitioned asymmetrically into a large micropylar cell and a small chalazal cell. As regards the subsequent fate of these cells, the micropylar cell becomes multicellular following an initial free nuclear phase, while the chalazal cell remains undivided or undergoes a limited round of divisions to produce a cluster of free nuclei. These three types of endosperm ontogenesis have been described in representative plants by Maheshwari (1950); although descriptions of endosperm development in an increasing number of plants continue to appear in the botanical literature, they have not introduced any new concepts on the structure and morphology of this tissue.

One of the unique modifications of the endosperm that has been described in exquisite detail in a number of angiosperms is the formation of outgrowths known as endosperm haustoria. Whereas serial sections of fertilized ovules have revealed the patterns of endosperm ontogenesis, the structure and morphology of the haustoria have been analyzed by means of whole mounts of the entire endosperm, facilitating the isolation of the parts intact. The haustoria arise either at the micropylar or the chalazal end or at both ends of the developing endosperm and appear as branched tubular processes driven into the adjacent parts of the ovule, such as the chalaza and integument. Histochemical observations seem to suggest that starch, proteins and other metabolites pass into the endosperm through the haustoria (Bhatnagar and Kallarackal, 1980; Vijayaraghavan and Prabhakar, 1984). For descriptive accounts of the haustorial outgrowths in representative species, reference is made to Maheshwari (1950).

Endosperm and embryo nutrition

From the time embryologists first began to understand the relationship between the embryo and the endosperm, there were two functions the endosperm was expected to fulfill: (1) to nurture the embryo during its heterotrophic phase of growth and (2) to provide combustible sources of energy during seed germination. Although the evidence in favor of the role of the endosperm as a nurse tissue for the developing embryo is circumstantial rather than experimental, the idea has gained great credibility over the years. The absence of the endosperm in mature seeds of certain plants known to produce this tissue at early stages of ovule development seems to indicate that the endosperm is an ephemeral substrate apparently consumed by the growing embryo. In considering the possible dependency of the embryo on the endosperm nutrients, one puzzling problem is the stage at which the embryo switches on to endosperm nutrition. Light and electron microscopic studies have suggested that the young embryo depends on nutrients derived from the suspensor and that only after the suspensor has degenerated does direct absorption of endosperm nutrients commence (Schulz and Jensen, 1969; Erdelská, 1980). In *Linum usitatissimum* (flax), the switchover to endosperm nutrition takes place in the torpedo-shaped embryo, and utilization of the endosperm is indicated by the appearance of a conspicuous zone of destroyed endosperm cells around the cotyledons (Erdelská, 1980). In wheat, endosperm nutrition is initiated after the embryo has formed a few hundred cells and enters a period of rapid growth; once growth is under way, the embryo draws heavily upon the endosperm (Smart and O'Brien, 1983). However, this is not to rule out the existence of any interaction between the early division-phase embryos and the endosperm. It was noted earlier that the primary endosperm nucleus gets a head start over the zygote in initiating mitotic activity; one might be tempted to link this to a dependency of the embryo on the endosperm beginning with the very first division of the zygote. If early division-phase embryos utilize the nutrient substances of the endosperm, what structural modifications might be involved in this function? Most of the attention has been centered on the cell wall of the suspensor, for it appears to be the most likely site of communication with the endosperm metabolites. Electron microscopic investigations of the development of the endosperm in tandem with that of the embryo have shown that the outer wall of the suspensor produces projections of wall material that abut into the endosperm. This is true of *Capsella bursa-pastoris* (Schulz and Jensen, 1969, 1974), pea (Marinos, 1970a), *Diplotaxis erucoides* (Simoncioli, 1974), *Phaseolus coccineus* (Yeung and Clutter, 1979) and *Vigna sinensis* (Hu, Zhu and Zee, 1983), although the number of outgrowths present varies considerably in the different species, as does the complement of organelles associated with them. It is relevant to point out that a plasma membrane lines the wall invaginations and thus greatly increases their surface area. This confers transfer cell functions on the wall outgrowths involving active solute transport. Since the direction of flow of solutes cannot be determined from electron micrographs, we have to consider the possibility that the wall outgrowths might facilitate either the absorption of nutrients from the endosperm into the suspensor or the transport of metabolites from the degenerating suspensor into the endosperm.

Other observations have suggested that the endosperm may attract metabolites from the ovular tissues for possible use by the growing embryo. We can delve into the evidence for this point by considering the ultrastructural image of the embryo sac wall and the nearby cells of the ovular tissues. As seen in pea (Marinos, 1970a), sunflower (Newcomb and Steeves, 1971), *Stellaria media* (Newcomb and Fowke, 1973), cotton (Schulz and Jensen, 1977), *Haemanthus katherinae* (Newcomb, 1978), *Medicago sativa* (alfalfa) (Sangduen, Kreitner and Sorensen, 1983a) and *Vigna sinensis* (Hu et al., 1983), the inner wall of the embryo sac at the micropylar or chalazal pole is specialized as transfer cells and produces wall invaginations into the endosperm. In barley (Norstog, 1974) and *Vigna* (Hu et al., 1983) wall invaginations similar to those of transfer cells are present on the outer walls of endosperm cells. The growth of the embryo sac after fertilization also has profound effects on the ovular tissues at its micropylar or chalazal ends. To a large degree the response of these cells to the expanding volume of the embryo sac appears to be disintegration by enzyme activity, lysis or pressure resulting in the release of free cytoplasm or nuclei, as demonstrated in *Capsella* (Schulz and Jensen, 1969, 1974), *Eranthis hiemalis* (Pacini, Cresti and Sarfatti, 1972), barley (Norstog, 1974), *Diplotaxis erucoides* (Pacini, Simoncioli and Cresti, 1975) and pea (Hardham, 1976). The general interpretation from these observations is that the wall outgrowths on the embryo sac facilitate the passage of cytoplasmic nutrients from the lysed ovular cells into the endosperm. Attractive as this idea may be, it should be pointed out that the flow of nutrients through plant cell walls is a dynamic process, but we are forced to base our inferences from electron microscopic images that portray the scene in a static manner.

It is difficult to find structural evidence for the supply of nutrients from the vegetative parts of the plant to the growing embryo. From physiological experiments it is clear that seeds that accumulate large quantities of storage reserves in the embryo or in the endosperm act as powerful sinks for metabolites from other parts of the plant. Considering the amount of storage reserves synthesized by developing seeds, the vascular system of the ovule might be expected to play a role in translocating the precursors to the embryo or the endosperm. The best insight into the structural basis for the passage of nutrients from the vegetative parts of the plant into seeds is based on an anatomical study of developing pea seeds by Hardham (1976). This work showed that by the time storage protein synthesis is initiated in the cotyledons, most of the endosperm is consumed and that this coincides with a marked increase in the amount of vascular tissues in the ovule and the funiculus as well as in the number of phloem transfer cells in the funiculus.

Chemical nature of the endosperm

To conclude this section, we shall briefly consider the chemical nature of the endosperm metabolites that support the growth of embryos in their natural setting. From the viewpoint of nutrition of the embryo, analysis of the endosperm by standard biochemical methods has been, for all practical purposes, limited to that of *Cocos nucifera* (coconut). The liquid endosperm of coconut, popularly known as coconut milk, is the morphological equivalent of a nuclear endosperm. It is a highly nutritious mixture of vitamins, sugars and growth hormones such as auxins, gibberellins and cytokinins. As will become evident in Chapter 4, coconut milk nonspecifically promotes growth of cultured embryos and is effective in inducing undifferentiated callus growth in cultured phloem explants of carrot. Two other sources of endosperm that are potent in inducing callus growth in cultured explants are those of maize and *Aesculus woerlitzensis* (horse chestnut). Maize endosperm at the milky stage contains several amino acids and growth hormones including the cytokinin, zeatin. The growth-promoting activity of horse chestnut endosperm is probably due to IAA, adenyl compounds, *myo*-inositol and certain glycosides (Raghavan, 1976b). Thus far, no single compound or groups of compounds have been shown to duplicate the effects of these endosperm sources in their entirety in standard bioassays. However, it seems reasonable to assume that the growth-promoting effects of the endosperm on embryos are probably mediated by hormones either acting alone or in combination. To substantiate the involvement of growth hormones of endospermic origin in the growth of embryos *in vivo*, it is necessary to relate changes in the concentration of hormones in this tissue to specific phases of growth of the embryo. Evidence of this kind is now available for several plants and has been particularly well characterized in pea (Eeuwens and Schwabe, 1975).

Endosperm is the major storage tissue of cereal grains and seeds of a number of plants such as those of castor bean. In cereal grains, starch and proteins constitute the main forms of endosperm reserves, whereas in castor bean endosperm, lipids predominate. Accumulation of storage reserves typically begins with the cellularization of the endosperm and often dominates the development of the grain or seed thereafter. Indeed, during this period the biochemical machinery of the developing seed is primarily committed to the synthesis, transport and deposition of storage materials in the endosperm.

The extensive use of cereal grains as dietary components has stimulated a great deal of research on the molecular biology of storage protein accumulation in the endosperms of maize, wheat and barley. The storage proteins of maize are collectively known as zein, whose major components, as revealed by SDS-acrylamide gel electrophoresis are 22-kilodalton (kd) and 19-kd proteins. Normal genotypes of maize accumulate these proteins in the endosperm between 10 to 40 days after pollination (Jones, Larkins and Tsai, 1977b). Some information on the developmental regulation of the synthesis of zein proteins has come from analyses of *in vitro* translation products directed by mRNA of normal and mutant (*opaque*-2) endosperms. While mRNA isolated from the normal genotype directs the synthesis of major zein components *in vitro,* no detectable amounts of 22-kd protein are synthesized in the presence of mRNA from the mutant (Jones et al., 1977a). This is apparently mediated in part by a modification of the transcription pattern, as the lack of 22-kd zein in the mutant is reflected in a low level of zein mRNA (Pedersen et al., 1980). Thus the mutation in *opaque*-2 exerts its effect at the level of zein gene transcription.

It will take us further afield from the theme of this book to discuss the considerable work that has been done on the processing and maturation of storage proteins of cereals, organization of genes that code for the proteins and regulation of their synthesis; the interested reader will find an up-to-date summary of this work in the review by Brown, Ersland and Hall (1982). Whereas endosperms of several plants, including those of cereals, have been successfully grown as continuously proliferating tissues (Srivastava, 1982), it is not established whether the cultured tissues synthesize storage reserves as efficiently as the intact ones.

Formation of accessory embryos

The normal process of sexual reproduction in angiosperms is occasionally substituted or supplemented by an asexual process to give rise to embryos with diploid or haploid number of chromosomes. Depending on the cells or tissues from which asexual embryos are born, the process is described under various names such as polyembryony, apomixis, apogamy, diploid or haploid parthenogenesis. These terms and their variants are defined by Battaglia (1963) and Maheshwari and Sachar (1963) with appropriate examples and are not considered here. This section describes a few cases of spontaneously formed (adventive) embryos and experimentally induced asexual embryos from cells of the embryo sac and of the ovule, setting the stage for a discussion of the formation of embryo-like structures in tissue cultures in Chapters 5 and 6.

Adventive embryogenesis

Spontaneous formation of proembryo-like structures by cleavage and proliferation of the zygote or early division-phase embryos is fairly widespread in members of the Orchidaceae and in species of *Erythronium*, *Tulipa* and *Primula* (Maheshwari and Sachar, 1963). In *Eulophia epidendraea*, additional embryos are formed by the branching of the filamentous proembryo, by the production of bud-like outgrowths from the proembryo or by a selective activation of cells of an irregularly dividing zygote. Irrespective of the different pathways of their origin, adventive embryos exhibit a certain measure of uniformity in appearance (Swamy, 1943). A common origin of adventive embryos is by division of cells of the suspensor; examples are afforded by species of *Exocarpus*, *Lobelia*, *Garrya* and *Isotoma* (Maheshwari and Sachar, 1963; Bhatnagar and Johri, 1972). A case of adventive embryogenesis of the endosperm authenticated by histological and cytological data is reported in *Brachiaria setigera* (Muniyamma, 1977).

Adventive embryos are also formed by the division of synergids and antipodals with or without fusion with a male nucleus. Whereas the numerous reported cases of embryogenesis by synergids and antipodals discussed by Maheshwari and Sachar (1963) are spontaneous in origin, it was recently shown that when unpollinated ovaries of barley and rice are cultured in a synthetic medium, the synergids or antipodals begin to divide and form haploid embryos (Huang, Yang and Zhou, 1982; Tian and Yang, 1983).

Unlike adventive embryos derived from the antipodals and synergids, embryos originating from ovular tissues like the nucellus and integument are always diploid. The most interesting cases of adventive embryogenesis are those arising from the nucellus. Nucellar polyembryony is an important method of asexual reproduction in many cultivated species of *Citrus* and other members of Rutaceae and *Mangifera* (Anacardiaceae); the phenomenon has also been reported in members of Gramineae (Shanthamma and Narayan, 1977), Calycanthaceae, Capparaceae, Euphorbiaceae, Meliaceae, Malphigiaceae, Cucurbitaceae, Myrtaceae (Rangaswamy, Sethi and Shrotria, 1980) and Buxaceae (Naumova and Willemse, 1982). The nucellar embryos in *Citrus* are customarily said to originate following pollination and fertilization. However, we should hesitate to generalize the role of pollination and fertilization in nucellar embryogenesis in *Citrus*, as nucellar embryos have been observed in seeds without an embryo sac or without evidence of pollen tube entry; nucellar tissues from unpollinated ovules have also been shown to regenerate embryos in culture. According to Esen and Soost (1977), pollination and fertilization are not essential for initiation of nucellar embryos in ovules of certain cultivars of *Citrus sinensis* and *C. reticulata*, but their subsequent development is clearly dependent upon double fertilization and endosperm development.

Nucellar embryogenesis in *Citrus sinensis* and *C. reticulata* shows remarkable likeness to zygotic embryogenesis which also occurs simultaneously in the same ovule. The presumptive embryogenic cell of the nucellus

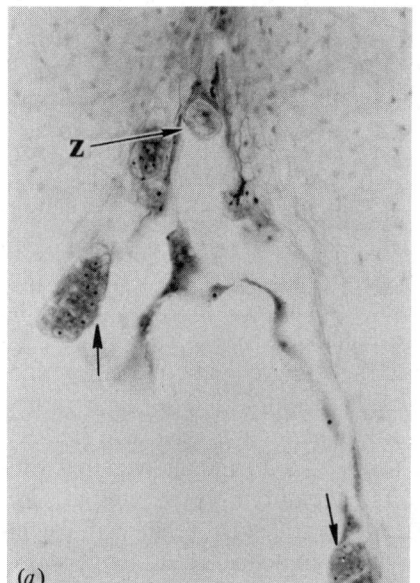 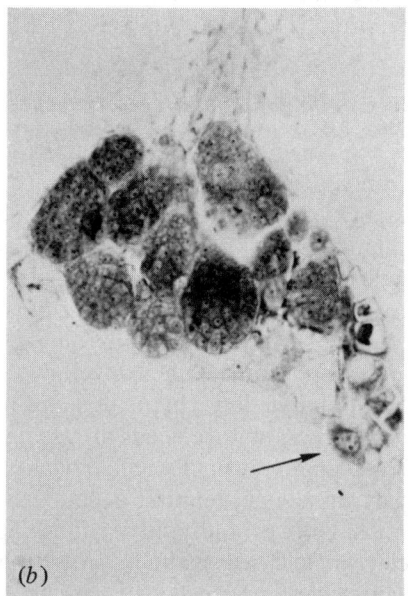

FIGURE 2.7. Nucellar embryogenesis in *Citrus reticulata*. (*a*) Micropylar end of a seed showing the undivided zygote (Z) and nucellar embryos (*arrows*). (*b*) Globular nucellar embryos crowding the micropylar end of the embryo sac. Arrow points to a one-celled progenitor of the nucellar embryo. (From Esen and Soost, 1977; photographs supplied by Dr. R. K. Soost.)

is easily distinguished from other nucellar cells by its large size and dense cytoplasm. Later, this cell isolates itself from the neighboring nucellar cells by severing plasmodesmatal connections and forming a new wall within the original primary wall. As a globular embryo is formed from the embryo mother cell, it goes through typical stages of zygotic embryogenesis and pushes its way into the central cell (Fig. 2.7). In a mature seed it is not unusual to find several embryos competing with one another and with the zygotic embryo to complete development in a chaotic environment, drawing upon the limited resources of the seed (Esen and Soost, 1977; Wilms, van Went, Cresti and Ciampolini, 1983). Ontogenesis of nucellar embryos in other species of *Citrus* and cultivated varieties of *Mangifera indica* (mango) follows the same trend as in *C. sinensis* and *C. reticulata* (Maheshwari and Rangaswamy, 1958). In *Opuntia dillenii*, following fertilization, the egg is depleted of its contents and begins to disintegrate. Although in most ovules the same fate overtakes the polar nuclei, in occasional ovules in which the primary endosperm nucleus begins to divide, nucellar cells are activated to produce embryos (Maheshwari and Chopra, 1955). Formation of adventive embryos from the cells of the integument is rare but has been reported (Devi and Pullaiah, 1976).

In vivo *induction of accessory embryos*

When considering the conditions favoring *in vivo* formation of accessory embryos in the ovule, it must be noted that in some cases this latent capacity may be revealed only under the influence of certain extracellular stimuli. One noteworthy example in which accessory embryos have been induced *in vivo* by specific treatments is *Eranthis hiemalis* (Ranunculaceae). Embryogenesis in *Eranthis* is uneventful and the mature seed, at the time of shedding, contains only a heart-shaped embryo subtended by a long suspensor. Further development of the embryo leading to differentiation of cotyledons, hypocotyl and radicle is accomplished during germination. When mature seeds are treated with naphthaleneacetic acid (NAA), 2,4-D or 2,4,5-trichlorophenoxyacetic acid (2,4,5-T) and allowed to germinate, in a small number of seeds the original embryo regenerates an additional embryo and twin seedlings appear. Treatment of seeds with colchicine, maleic hydrazide, isopropyl-phenylcarbamate, acid buffer and X-rays causes varying degrees of damage to embryos, but the suspensor or small groups of cells or single cells of the damaged embryo proliferate to form viable embryos (Haccius, 1963; Haccius and Reichert, 1964). The immediate cause for the formation of accessory embryos by these various treatments is injury to the apical meristem, which releases uninjured cells from apical dominance to express their latent potentialities.

Concluding comments

From the array of developmental characteristics of the embryo and endosperm touched on in this chapter, a few generalities stand out. One is that embryogenesis itself, being a period of intense gene activity, follows the same basic principles in both monocots and dicots. Differences in embryo ontogenesis observed mainly relate to the early division sequences that occur before histogenesis and organogenesis are initiated. We are still far from understanding the real functional meaning of the different pathways by which the basic body plan of the embryo is established. The fact that the same pattern of embryo ontogenesis occurs in members of widely different families suggests that the different ontogenetic patterns have some evolutionary significance.

We have a fairly good picture of endosperm development in a wide variety of angiosperms. However, this information does not seem to shed light on the mechanism by which the endosperm renders nutritive assistance to the growing embryo. The processes at work in the nutrition of embryos by the endosperm must be understood within the framework of the mechanism of absorption of metabolites by living cells.

The general import of the production of accessory embryos seems to be in the realization that the principles involved in the transformation of the specialized egg into the embryo have their counterparts in the other cells of the embryo sac as well as in the ovular cells. The groundwork for much of what we know about the development of accessory embryos in plants was laid decades ago. During the intervening years, spectacular success has been achieved by the use of tissue culture methods in inducing the formation of embryo-like structures from virtually any part of the plant. Together with the newer experimental approaches, the production of accessory embryos can be expected to contribute to our understanding of the expression of embryogenic competence of cells.

3
Cellular and biochemical aspects of embryogenesis

Much of the work reviewed in Chapter 2 concerned the variety of developmental patterns displayed by angiosperm embryos and the degrees of differentiation they attain. Against this background of information we must now turn our attention to the question how cells of the embryo attain the different states of genetic, cellular and biochemical commitment so that the patterns of differentiation conform to the adult body plan characteristic of embryos of each species. Consideration must also be given to the question of how cells of the embryo acquire specialized physiological properties so that they can contribute to total organismic function, as well as how these functions are dissipated as the embryo ages.

This chapter concentrates on the basic cellular changes involved in the development of the embryo from its single-celled beginning. A large number of studies have been carried out in recent years on the structural and biochemical characteristics of developing embryos and their relationship to morphological development and function. One aspect of these studies has been concerned with the ultrastructural changes that portray the transformation of the zygote into a full-term embryo with a view toward identifying the changes that may be associated with functional differentiation. A second focus has been on the macromolecular synthetic activity of embryos during progressive embryogenesis. An analysis of these investigations on the organogenetic part of the embryo and the suspensor forms the subject matter of the first two sections of this chapter. Next we shall attempt to define the parameters that govern the accumulation of storage proteins in the embryos and consider the mechanisms by which their synthesis is controlled. It is hoped that together these approaches will contribute to an understanding of the recondite problems of developmental expression during the progression of the zygote into an embryo. The final section briefly discusses the basis for the loss of viability or functional capacity of embryos.

The egg, zygote and early division-phase embryos

After fertilization, the egg is transformed into a zygote, which embarks upon one of the most critical periods in its development, as it is partitioned into cells that ultimately make up the body of the embryo and adult organism. While the zygote possesses structural and functional properties of a typical cell, it is also highly differentiated in anticipation of its developmental tasks. In order to understand the structure of the zygote in all its complexity, this section begins with a look at the unfertilized angiosperm egg. Our minimal image of the egg has been molded by events that dominate this cell prior to or at the time of fertilization. This makes it possible to interpret the zygote with the hindsight gained from the knowledge of prefertilization events.

The egg

To begin with, the angiosperm egg is a small cell, but it goes through a growth phase in order to attain a large mature size. It is during the growth period that the structural blueprint of the egg is laid down to the point that, once fertilized, it can proceed to construct the embryo. Our knowledge of the structure of the egg has been greatly enhanced since the initial studies of Jensen (1964) and van der Pluijm (1964) on cotton and *Torenia fournieri*, respectively. The ultrastructural appearance of eggs of maize (Diboll and Larson, 1966), *Crepis tectorum* (Godineau, 1966), *Linum usitatissimum* (Vazart, 1969), *L. catharticum* (D'Alascio-Deschamps, 1973), *Capsella bursa-pastoris* (Schulz and Jensen, 1968a), *Myosurus minimus* (Woodcock and Bell, 1968), *Epidendrum scutella* (Cocucci and Jensen, 1969a), *Petunia hybrida* (van Went, 1970), *Quercus gambelii* (Mogensen, 1972), sunflower (Newcomb, 1973a), *Plumbago zeylanica* (Cass and Karas, 1974; Russell, 1982), *Oenothera lamarckiana* (Jalouzot, 1975), *Stipa elmeri* (Maze and Lin, 1975), *Nicotiana tabacum* (tobacco) (Mogensen and Suthar, 1979), *Agave parryi* (Tilton and Mogensen, 1979), *Spinacia oleracea* (spinach) (Wilms, 1981) and wheat (You and Jensen, 1985) has revealed that the arrangement of organelles is somewhat variable from one species to another. In the most detailed of these studies on cotton (Jensen, 1964, 1965) and *Capsella* (Schulz and Jensen, 1968a), the egg appears as a strongly polarized cell with a large vacuole toward the micropylar end and an aggregation of cytoplasmic elements toward the chalazal end. The total amount of cytoplasm present in the egg is meagre and is spread in a thin layer surrounding the vacuole except near the chalazally located nucleus. Plastids, mitochondria and dictyosomes are randomly and parsimoniously distributed in the cytoplasm. Strands of endoplasmic reticulum (ER) are relatively abundant in the egg of cotton, where they seem to partially enclose other organelles. Occasional strands of ER also appear unique in containing an internal network of tubes probably formed by invagination of the inner membrane of ER. By contrast, the egg of *Capsella* has very little ER, which occurs in the form of short, randomly oriented strands. Eggs of both cotton and *Capsella* also contain liberal supplies of ribosomes that exist predominantly as monosomes. From the point of view of functional significance, the ultrastructural simplicity of the mature egg, in particular the comparative poverty of its cytoplasmic organization, tends to suggest that it is a quiescent cell whose metabolism is at a low ebb.

While the cytoplasmic blueprint of the egg is organized, a major change also occurs outside the cytoplasm. This change is principally concerned with the cell wall and, unlike other general features of the angiosperm egg uncovered by light microscopists, this is a discovery of the electron microscope era. A very suggestive feature of the egg of a number of plants, including *Torenia* (van der Pluijm, 1964), cotton (Jensen, 1965), *Crepis* (Godineau, 1966), maize (Diboll and Larson, 1966), *Linum* (Vazart, 1969), *Petunia* (van Went, 1970), *Quercus* (Mogensen, 1972), tobacco (Mogensen and Suthar, 1979) and spinach (Wilms, 1981), is that a continuous wall is present only around the micropylar half of the cell, the chalazal portion being covered by the plasma membrane alone. An intermediate situation is found in *Capsella* (Schulz and Jensen, 1968a), *Epidendrum* (Cocucci and Jensen, 1969a), *Plumbago* (Cass and Karas, 1974), *Agave* (Tilton and Mogensen, 1979), *Ornithogalum* (Tilton, 1981a) and *Glycine max* (Folsom and Peterson, 1984), in which isolated deposits of wall material dot the chalazal part of the egg. Viewed within the context of fertilization, the naked or partially naked chalazal part of the egg is of considerable significance in that it facilitates entry of the sperm as well absorption of food materials from the central cell.

The zygote

Once fertilization has been accomplished, alterations in the arrangement of organelles in the egg are straightforward, yet many properties of the egg change during the first few hours after fertilization. Even though many of the postfertilization changes occur in the egg cytoplasm, physical changes occurring outside the cytoplasm also play a role in the evolution of the zygote. For example, in cotton (Jensen, 1968), *Hibiscus costatus, H. furcellatus* (Ashley, 1972) and *Lagerstroemia speciosa* (Raghavan and Philip, 1982) there is a dramatic decrease in the size of the egg after fertilization to one-half its size before fertilization; in tobacco the fertilized egg experiences a shrinkage as well, but it is less striking (Mogensen and Suthar, 1979).

No generalizations can be made about the major ultrastructural upheavals that occur in the egg after fertilization. In cotton, cytoplasmic organelles like plastids, ER, mitochondria and ribosomes begin to flow about the egg within hours after fertilization and take up new positions at the chalazal end, where they form an investment around the nucleus. This shift accentuates the polarity already present in the unfertilized egg. Observations of the egg at different times after fertilization have also left no doubt that the increase in the total amount of ER as well as in the tube-containing ER, the increase in size and number of starch grains in the plastids, the elaboration of additional cristae in the mitochondria and the generation of a new population of ribosomes and polysome formation from existing ribosomes are part of the basic response of the egg to fertilization. Special attention has also been focused on the chalazal part of the egg, which as we have seen lacks a cell wall in the unfertilized state. Laying down and consolidation of a new wall at the chalazal end is an important milestone in the evolution of the zygote; increased activity of dictyosome vesicles coincident with cell wall regeneration seems to concur with this observation (Jensen, 1968). In several species of *Rhododendron* and *Ledum groenlandicum* (Ericaceae) a callose wall is laid down around the zygote during the first two days after fertilization (Williams, Knox, Kaul and Rouse, 1984). The fact that the wall essentially insulates the newly formed zygote from the influence of neighboring cells of a different genotype probably has some significance in the subsequent induction of the sporophyte development phase.

Some plants have sufficient density of ribosomes in the egg that their aggregation into polysomes is the most obvious change that occurs after fertilization. Examples in which this occurs are maize (Diboll, 1968), *Capsella* (Schulz and Jensen, 1968a), *Epidendrum* (Cocucci and Jensen, 1969b), *Linum usitatissimum* (Deschamps, 1969) and *L. catharticum* (D'Alascio-Deschamps, 1981). In *Quercus*, fertilization of the egg is associated with an increase in the number of lipid bodies and with a change in ER from rough to smooth; the basis for these changes may be one of adaptive adjustment by synthesis of storage materials and membrane systems for the division of the zygote (Mogensen, 1972). Little change is noted, however, in the redistribution of organelles in the egg and zygote of tobacco (Mogensen and Suthar, 1979).

It is clear from the above description that several concomitant changes coordinated by many assembly lines must be taking place during the formation of the zygote, but they differ among different plants. One consequence of these changes is that the zygote acquires high metabolic activity and essentially becomes a repository of developmental information that can play a role in morphogenesis. To a large extent this is reflected in the subsequent division of the zygote to form the embryo.

Becoming multicellular

Except in the Asterad and Chenopodiad types of embryo development, after the first division of the zygote, the small terminal cell at the chalazal end produces the organogenetic part of the embryo, and the large cell at the micropylar end gives rise to the suspensor or becomes vesicular in nature (see Chapter 2). Thus, in most angiosperms, polarity displayed by the zygote profoundly affects the fate of the daughter cells formed, as if an unequal partitioning of the cytoplasmic materials of the cell elicits different modes of nuclear expression. Here we are concerned with the organogenetic part of the embryo; a detailed consideration of the suspensor is taken up in the next section.

Ultrastructural and histochemical studies have shown that the marked degree of asymmetry displayed by the two-celled embryo is complemented by subtle variations in the distribution of organelles and in the concentration of macromolecules. For instance, in cotton (Jensen, 1964), *Capsella* (Schulz and Jensen, 1968a) and *Quercus* (Singh and Mogensen, 1975) the terminal cell of the two-celled embryo is similar to the chalazal part of the zygote in that it has a dense cytoplasm enriched with organelles, whereas the basal cell is relatively impoverished of organelles. In the two-celled cotton embryo, the terminal cell is also characterized by the presence of a large number of plastids and large and variably shaped mitochondria, whereas the basal cell has a preponderance of vacuoles of various sizes and shapes (Jensen, 1964). In cotton (Jensen, 1964) and *Vanda* (Alvarez and Sagawa, 1965) a functional differentiation between the terminal and basal cells is evident on the basis of a higher concentration of RNA in the former. By contrast, in *Stellaria media* (Pritchard, 1964), *Capsella* (Schulz and Jensen, 1968a) and *Limnophyton obtusifolium* (Shah and Pandey, 1978), RNA concentration is higher in the basal cell than in the terminal cell. Apparently, an increased accumulation of macromolecules does not infallibly predict in all cases the division of a cell in the embryogenic pathway.

Subsequent divisions of both cells affect the structural and cellular characteristics of the embryo in various ways. A proclivity to acquire ribosomes is inherent in the terminal cell of the three-celled embryo of *Capsella;* this cell also stains more intensely for proteins and nucleic acids than do the suspensor cell and the basal cell. On the other hand, ER and dictyosomes are more abundant in the latter cells than in the terminal cell (Schulz and Jensen, 1968a). As the terminal cell forms the globular embryo and the suspensor cell forms the filamentous suspensor, ribosomal density is distinctly greater in the cells of the embryo than in the suspensor. The presence of starch and lipid bodies in the terminal cell of the three-celled embryo and their disappearance during the division of this cell have suggested that the storage bodies probably function as a source of energy for mitosis (Schulz and Jensen, 1968b). In *Capsella,* up to the globular stage of the embryo, the hypophysis with its highly vacuolate cytoplasm, low ribosome profile and generous amount of ER, starch and lipids is the only cell that displays basic ultrastructural differences from the cells of the rest of the embryo. The hypophysis is apparently determined in its future course of development, as ribosomal density and concentrations of nucleic acids and proteins in the cells of the root cap and protoderm derived from the hypophysis appear intermediate between those of embryo cells and suspensor cells. The earliest stage of embryogenesis in which cytological differences between cells are noted in anticipation of tissue and organ differentiation is the heart-shaped embryo. Here we are led to a picture in which cells of the procambium and ground meristem appear more highly vacuolate than those of the protoderm (Schulz and Jensen, 1968b).

Ultrastructural evidence for cell specialization during embryogenesis in other plants is, on the whole, less complete than in *Capsella*. Early embryogenesis in barley involves an increase in the population of ribosomes and mitochondria in all cells. As the embryo proceeds through two or three division cycles, there is a progressive decrease in the distribution of vacuoles toward the cells of the suspensor (Norstog, 1972a). In sunflower, cells of the embryo have more ribosomes than do those of the suspensor, giving these cells a more electron-dense appearance (Newcomb, 1973b). According to Singh and Mogensen (1975), cells of the embryo and suspensor of *Quercus* appear to amplify the ultrastructural features of the terminal and basal cells, respectively, of the two-celled embryo. As the embryo develops to the globular stage, suggestive of increased metabolic activity, the organelle population undergoes a complete reorganization. Increases in plastid number and complexity, in mitochondrial number, in dictyosomal activity, in ribosomal aggregation and in the amount of ER are a conspicuous part of the processes that peak in the late globular embryo. By contrast, the number and distribution of these organelles in the suspensor cells remain relatively unchanged.

From the heart-shaped to the mature stages, no generally characteristic ultrastructural changes augur the onset of tissue and organ differentiation. However, histochemical studies indicate that RNA begins to accumulate in higher concentrations in the cells of the presumptive cotyledons than in the cells of the rest of the embryo (Rondet, 1962; Pritchard, 1964; Schulz and Jensen, 1968a; Vallade, 1970; Norreel, 1972; Syamasundar and Panchaksharappa, 1976; Shah and Pandey, 1978). Cotton embryos show a striking correlation between tissue and organ differentiation and the activity of succinic dehydrogenase, a key enzyme of aerobic respiration. Generally, only cells in those parts of the embryo that are actively growing or differentiating display high enzymatic activity; at different stages of embryogenesis, these include the suspensor, cotyledons, hypocotyl, shoot apex, subapical region and the developing radicle (Forman and Jensen, 1965). Alcohol dehydrogenase is involved in anaerobic metabolism, and multiple forms of this enzyme are generally present in cells. Differentiation of the axis and cotyledons of embryos of *Phaseolus vulgaris* is associated with the appearance of specific isozymes of alcohol dehydrogenase in these organs (Boyle and Yeung, 1983). Since the embryo grows in an environment in which the occurrence of anaerobic metabolism can hardly be doubted, it is not unreasonable to invoke different functional roles for organ-specific isozymes of alcohol dehydrogenase to stimulate embryo growth. The enzymology of functional differentiation is much less clear cut in embryos of other plants studied (Bhalla, Singh and Malik, 1979, 1980b,c; Singh, Bhalla and Malik, 1980a).

Overall, in terms of structural and cellular changes, the picture of the embryo becoming multicellular does not yet begin to reflect the complexity it attains in a developing seed. This is because the changes described thus far appear to have emerged in a complete vacuum of biochemical and molecular data. The following discussion, limited to the problems of nucleic acid synthesis, is intended to partially fill this void.

Biochemical trends during embryogenesis

During early embryogenesis, the developmental program of the embryo is geared to the production of new cells. DNA synthesis must keep pace with the production of cells so that each new cell will have a fixed amount of DNA. Comparative studies using Feulgen microspectrophotometry and autoradiography of ^{3}H-thymidine incorporation have shown that the estimated DNA content of the nucleus (known as the C value) of the egg is within or close to the 1C range expected of a haploid cell. The fertilized egg most often displays only a 2C DNA complement, whereas in two- to four-celled embryos DNA values range from 2C to 4C (Woodard, 1956; Bennett and Smith, 1976; Kowyama, 1983). An abnormally high DNA value (16C) has been reported in the zygote of *Hordeum distichum* cv. Hannchen, whereas in two- and three-celled embryos DNA values range from 4C to 8C (Mericle and Mericle, 1970). Considerable variation also exists in the DNA content of the zygote nucleus (3C to ~6C) and of nuclei of four-celled embryos (~5C) of *Petunia hybrida* (Vallade, Cornu, Essad and Alabouvette, 1978). These reports indicate that some reservations are in order to the traditional view of DNA constancy in the egg and sperm and in the product of their fusion.

Changes in DNA content of embryos during the period of their rapid growth have received but little experimental attention. The only work in which DNA contents of rapidly dividing embryos have been related to cell number is that of Fisher and Jensen (1972), who found that in cotton the amount of DNA per cell remains fairly constant from the early heart-shaped to cotyledonary stage embryos. During this period, the cell number per embryo increases from 3,400 to ~100,000. In cereals such as *Hordeum distichum, Triticum vulgaris* and rice, DNA content per embryo or per cell does not appear to change during the middle and later periods of embryogenesis even as dehydration preparatory to seed maturation sets in (Duffus and Rosie, 1975; Zhu, Shen and Tang, 1980a,b). Of course, these observations do not tell us whether DNA values of the embryo cells are maintained at the 2C level typical of diploid cells.

The study of storage protein accumulation during embryogenesis has done much to stimulate interest in the dynamics of DNA changes during later periods of growth of embryos, when major increases in fresh and dry weights occur. Embryos of various legumes and in particular, the cotyledons have been favored materials for this work. After an initial period of growth by cell divisions, subsequent growth of the cotyledons is marked by cell expansion and concomitant accumulation of enormous quantities of reserve proteins, together with carbohydrates or oils. Scharpé and van Parijs (1973) found that cells of developing cotyledons of *Pisum sativum* continue to duplicate DNA after the cell number has stabilized and that cells of full-grown cotyledons have DNA levels averaging between 32C and 64C. According to D. L. Smith (1973), two features distinguish the changes in DNA contents of cells of developing cotyledons of *P. arvense*. First, numerical increases in DNA values of the nuclei ranging from 8C to 32C occur during the first few days of growth by cell expansion. Second, during the later period of growth of cotyledons, DNA remains in a stable form in some cells as they attain values as high as 64C (Fig. 3.1). The result of these changes is that the DNA content per cell of the cotyledon continues to increase during its growth by cell expansion. A continued increase in the nuclear DNA content of cells also occurs during the cell-expansion phase of cotyledon development of *Vicia faba* (Wheeler and Boulter, 1967; Millerd and Whitfeld, 1973), cotton (Walbot and Dure, 1976) and soybean (Dhillon and Miksche, 1983). Some investigators (Millerd and Whitfeld, 1973; Walbot and Dure, 1976) have established that continued synthesis of DNA in the absence of chromosomal duplication accounts for the mechanism underlying modal increases in DNA values of developing cotyledons. This phenomenon, which is an abbreviated version of the mitotic cycle, is designated endoreduplication, or endoreplication. Evidence based on buoyant density and reassociation kinetics of DNA has eliminated the possibility of a selective amplification of certain gene sequences during endoreduplication. Results in accord with this finding have come from a comparison of the annealing behavior of cloned genomic fragments for soybean embryo storage proteins with an excess of DNA isolated from soybean embryo and soybean leaf. The presence of identical smooth curves resulting from hybridization reactions of the two DNA samples is taken as proof of the absence of selective gene amplification during continued DNA synthesis in the cotyledon cells (Goldberg, Hoschek, Ditta and Breidenbach, 1981a). If the cloned gene was being amplified selectively, the an-

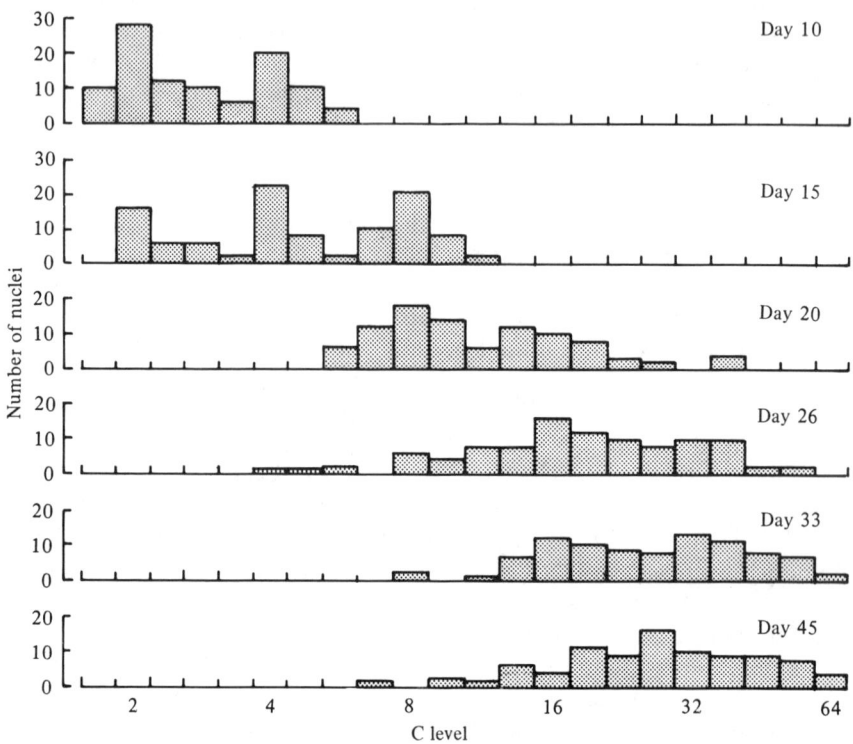

FIGURE 3.1. Frequency histograms showing DNA levels of nuclei at different stages of development of cotyledons of *Pisum arvense*. (From D. L. Smith, 1973.)

nealing curve with excess of embryo DNA would have been different from that of leaf DNA and a lot more complex with distinct transitions at different *Cot* (the product of DNA concentration at the beginning of reassociation experiment and time) values as illustrated by Figure 3.2. That the storage protein genes are not selectively amplified has also been confirmed by the similarity of results from blot hybridization of genes for the major soybean storage protein, glycinin, and a 15-kd protein region with gel blots of endonuclease-digested DNA from embryos and leaves (Fischer and Goldberg, 1982).

Another fascinating biochemical aspect of embryogenesis deals with RNA and protein metabolism. RNA continues to accumulate in developing embryos so that by the completion of embryogenesis its amount would have increased many times the initial value. Embryos of *Vicia faba* (Wheeler and Boulter, 1967; Manteuffel, Müntz, Püchel and Scholz, 1976), *Phaseolus vulgaris* (Walbot, 1971), *Hordeum distichum* (Duffus and

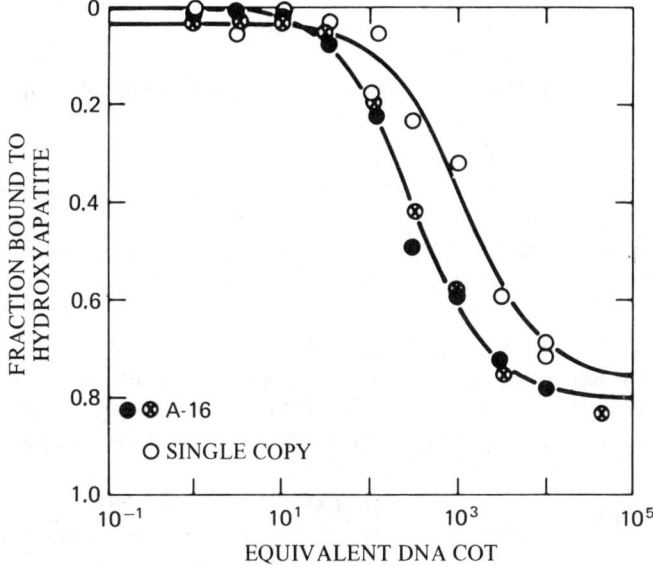

FIGURE 3.2. Hybridization of cloned ^{3}H-DNA probe A-16, isolated from soybean embryos with soybean leaf (●) and embryo (○) DNA. The kinetics of hybridization of single-copy ^{32}P-DNA with embryo DNA is also shown. (From Goldberg et al., 1981a.)

Rosie, 1975), rice (Zhu et al., 1980a) and *Triticum vulgaris* (Zhu et al., 1980b), among others, illustrate a rising level of RNA through most periods of growth of embryos. A series of studies by Walbot and associates have shown that RNA and protein metabolism of embryos of different ages and of specific parts of the same embryo are profoundly interrelated. When the organogenetic part of the embryo and the suspensor of *P. coccineus* are analyzed separately, RNA and protein contents and the rates of synthesis of these macromolecules per cell are high in the early heart-shaped embryos and decline thereafter. The decrease is probably attributable to the inability of the nucleolus to synthesize new RNA to keep up with the rapid tempo of mitosis in the embryo. Generally, on a per-cell basis, at comparable stages of development, the biosynthetic capacity of the cells of the embryo is much less than that of the nondividing cells of the suspensor (Walbot, Brady, Clutter and Sussex, 1972; Sussex, Clutter, Walbot and Brady, 1973). In *P. vulgaris,* on a per-embryo basis, the synthesis and accumulation of RNA increase in parallel with the increase in cell number. While RNA content of the embryo remains constant during its maturation phase, the rate of RNA synthesis declines to zero. Since protein accumulation in the embryo continues after the cessation of RNA synthesis, the possibility exists that these proteins are coded on RNA synthesized at an earlier period in the ontogeny of the embryo and stored for later use (Walbot, 1971).

Observations on RNA metabolism of cotyledons of developing embryos (Galitz and Howell, 1965; Beevers and Poulson, 1972; Poulson and Beevers, 1973; Scharpé and van Parijs, 1973; D.L. Smith, 1973; Walbot, 1973; Millerd and Whitfeld, 1973; Millerd and Spencer, 1974; Davies and Brewster, 1975) have generally supported the results from whole embryos. Indeed, some variations in the timing of peak accumulation of RNA have been noted in the cotyledons of various species, but for reasons of space they are not discussed here. The changing pattern of RNA synthesis in developing cotyledons prompts the question: What are the conditions that determine transcription and translation processes during cotyledon development? Poulson and Beevers (1973) found that although RNA content of pea cotyledons remains fairly constant during the period of seed maturation, the capacity of cotyledons for RNA synthesis declines considerably during this period. This observation correlates well with the fact that ribosomal preparations made from cotyledons of mature embryos show a decreasing abundunce of polysomes (Beevers and Poulson, 1972). The apparent decrease in polysomal content of cotyledons of mature embryos poses a fundamental problem. Is it caused by a reduction in the total ribosome content, or is it an expression of intrinsic change in the property of ribosomes? Cotyledons of developing pea embryos show three periods of protein synthetic activity: a reduced rate of protein synthesis in very young embryos, an accelerated rate in rapidly growing embryos and a declining rate in mature and dehydrated embryos. Although the rate of protein synthesis in a cell-free system extracted from cotyledons of mature and dehydrated embryos is much lower than that from cotyledons at the stage of rapid growth, addition of an artificial messenger more than restores the protein synthetic activity of the former. On the basis of these data, the reduced protein synthetic activity of cotyledons of mature embryos can be attributed to the nonavailability of sufficient mRNAs.

A genetic approach was used by Davies and Brewster (1975) to analyze RNA synthetic activity in developing pea cotyledons. These investigators found that when plants having widely different ribosomal RNA (rRNA) contents in the cotyledons are crossed, cotyledons of F_1 hybrid embryos have rRNA content identical to that of the maternal parent. Two models have been proposed to account for this result. The first supposes that some aspect of gene activity such as transcription or processing of RNA is under the control of the maternal genome. The second postulates that a selective action of maternal alleles occurs in the hybrid. Further work is necessary to choose between these two alternatives.

Expression of genes that code for proteins is initiated by the transcription of DNA into RNA by a reaction involving the participation of the enzyme RNA polymerase. Since the nuclear DNA content of cotyledonary cells increases during the period in which major changes in RNA accumulation occur, a logical question is whether increased RNA synthesis results from an increased transcriptional activity of the nucleus. Scharpé and van Parijs (1973) as well as Broekaert and van Parijs (1978) suggested that the presence of multiple gene copies per cell might lead to an enhanced rate of RNA and protein synthesis in the cotyledons. However, Millerd and Spencer (1974) showed that cotyledons of a particular variety of pea do not synthesize additional RNA polymerase in proportion to the increasing amount of nuclear DNA available as a template, despite the fact that the template activity of chromatin isolated from cotyledons of different ages is unaffected. Although this observation basically eliminates a gene dosage effect on RNA synthesis in the cotyledons, work by Cullis (1976) showed that an increase in RNA polymerase activity of the nucleus occurs during the period of endoreduplication in the cotyledons of certain genotypes of pea. This might suggest that some of the extra DNA synthesized is used as template for RNA synthesis. Cullis (1978) later showed that the extent of utilization of extra copies of DNA is associated with the rate of growth of cells of the cotyledons. For example, use of DNA above 2C level as template for RNA synthesis is achieved when plants are raised at 15°C, allowing for slow growth of cotyledons. In plants grown at 30°C, at which temperature cotyledons grow rapidly, RNA polymerase activity remains at the level characteristic of 2C DNA. These apparent differences in the template activity of chromatin from two sets of cotyledons are attributable either to differences in the number of cistrons active in transcription under both conditions or to differences in the specific activity of RNA polymerase.

In summary, the available evidence appears to indicate no firm ground for the idea that, except under special conditions, transcription in the cells of the cotyledon is related to the availability of extra copies of DNA. The endoreduplicated nucleus has a severalfold increase in the information content of the cell, but we have no knowledge of how this is used. This finding opens up the question of the role of endoreduplication of the nuclear genome in the genetically assigned function of the cotyledons.

Before we leave the subject of nucleic acids, it is pertinent to emphasize that the role of DNA and RNA in embryogenesis goes beyond that of providing the information to be translated as well as the machinery to implement translation. For example, actively growing cells not only synthesize RNA but also degrade some of the RNA synthesized as part of their metabolism. These events lead to changes in the nucleotide pool and in the concentrations of the ribose sugar, which serves a variety of functions in the metabolism of the cell. Little is known about these functions in developing embryos, but there is clearly an urgent need for studies on

58 Cellular and biochemical embryogenesis

nucleic acid metabolism in developing embryos that do not involve information processing and transfer.

The suspensor

As an object of our studies of functional differentiation during embryogenesis, the suspensor has two important advantages: (1) it is an ephemeral organ that does not form a part of the mature embryo, and (2) morphological and structural adaptations for specific functions find their highest expression in the suspensor. Chapter 2 described the modifications of the suspensor in representative species; the present section attempts to summarize the information accrued from the ultrastructural and biochemical cytology of the suspensor and their relevance to the function of this organ. With regard to both the alignment of the organelles and the density of their distribution, suspensors of different species thus far investigated as well as the different cells of the same suspensor are astonishingly diverse. Other aspects of the suspensor such as the number of cells present and their life-span also vary in different species. Current knowledge on the ultrastructure of the suspensor is therefore best presented as case histories.

Ultrastructure of the suspensor

Capsella will serve as the first case. For the purposes of this discussion, three stages have been identified in the life of the suspensor of *Capsella:* (1) the octant embryo when the suspensor consists of six cells; (2) the globular embryo when the suspensor attains its maximum cell number and appears as a filament of ten cells interposed between the basal cell and the embryo, and (3) the heart-shaped embryo when the suspensor attains its maximum length and genetically permissible life-span. Throughout its development, the suspensor is functionally connected with both the embryo and the basal cell by numerous ER-containing plasmodesmata. Although the suspensor cells harbor the usual array of organelles, ribosomes appear to be especially concerned with cellular interactions in this organ. Fitting with this interpretation is the observation that, at the octant stage of the embryo, the suspensor cells have a respectable number of ribosomes that show an increasing gradient toward the chalazal end. At the globular stage of the embryo, ribosomes gradually dissipate in the cytoplasm of the suspensor cells and completely disappear from cells at the heart-shaped stage of the embryo. Symptomatic of the inevitable failure of the cellular machinery of the suspensor is the onset of cytoplasmic degeneration at the heart-shaped stage and the eventual collapse of cells. As alluded to in Chapter 2, the invaginations of the outer wall of the suspensor cells of *Capsella* have been a focal point of ultrastructural investigation. They develop from the outer lateral walls of certain suspensor cells of the globular embryo and increase in number and complexity at the heart-shaped stage. The presence of wall projections is a feature that the basal cell shares with the suspensor cells. The inner wall of the micropylar and lateral parts of the basal cell are also thrown up into an elaborate network of invaginations abutting into the cytoplasm. In fact, these invaginations first appear in the wall at the micropylar end of the zygote and increase in number, peaking in the basal cell at the heart-shaped embryo stage (Schulz and Jensen, 1968a, 1969).

As will become evident, a paradigm for defining the structure and nuclear cytology of the suspensor in angiosperm embryos is *Phaseolus*. In *P. coccineus*, the suspensor is polarized from the early proembryo stage onwards with respect to mitotic activity. Cell divisions in the basal cells of the suspensor mostly cease by the proembryo stage, but they continue in the chalazal region. The suspensor attains its maximum number of about 200 cells at the heart-shaped stage of the embryo; at the cotyledonary stage it reaches its maximum size (Fig. 3.3). The considerable amount of substance invested in the suspensor, however, does not enhance its longevity (Yeung and Clutter, 1978).

Cell specialization in the suspensor of *Phaseolus coccineus* begins with the appearance of wall invaginations as early as the mid-proembryo stage. Soon afterwards, the cells become committed to a developmental pathway characterized by a disproportionate increase in the density of ribosomes, plastids, mitochondria, dictyosomes and smooth ER, which continues until the cotyledon stage. In considering the ultrastructure of the wall of the suspensor during development, it is significant that an increase in the number of invaginations parallels the increase in organelle number and that there is also a raft of organelles, particularly ER, dictyosomes and mitochondria in close proximity to the wall ingrowths. These observations further strengthen a role for the wall ingrowths in the exchange of nutrients (Yeung and Clutter, 1979).

Stellaria media is of interest because the suspensor in this case is formed from the terminal cell of the two-celled embryo (Caryophyllad type), but the ultrastructural profile presented by the basal cell and cells of the suspensor is no less complex. For example, the basal cell displays massive wall projections at the micropylar end along with a preponderance of mitochondria near them. An important step in the conditioning of the basal cell and cells of the suspensor to their function is that certain ultrastructural components like microbodies, plastids and extensive whorls of tubular ER which are absent from the cells of the embryo are common to the basal cell and cells of the suspensor. The plastids of the suspensor have an unusual morphology with numerous tubules and electron translucent inclusions. There is also a decreasing gradation in the complexity of plastids and in the

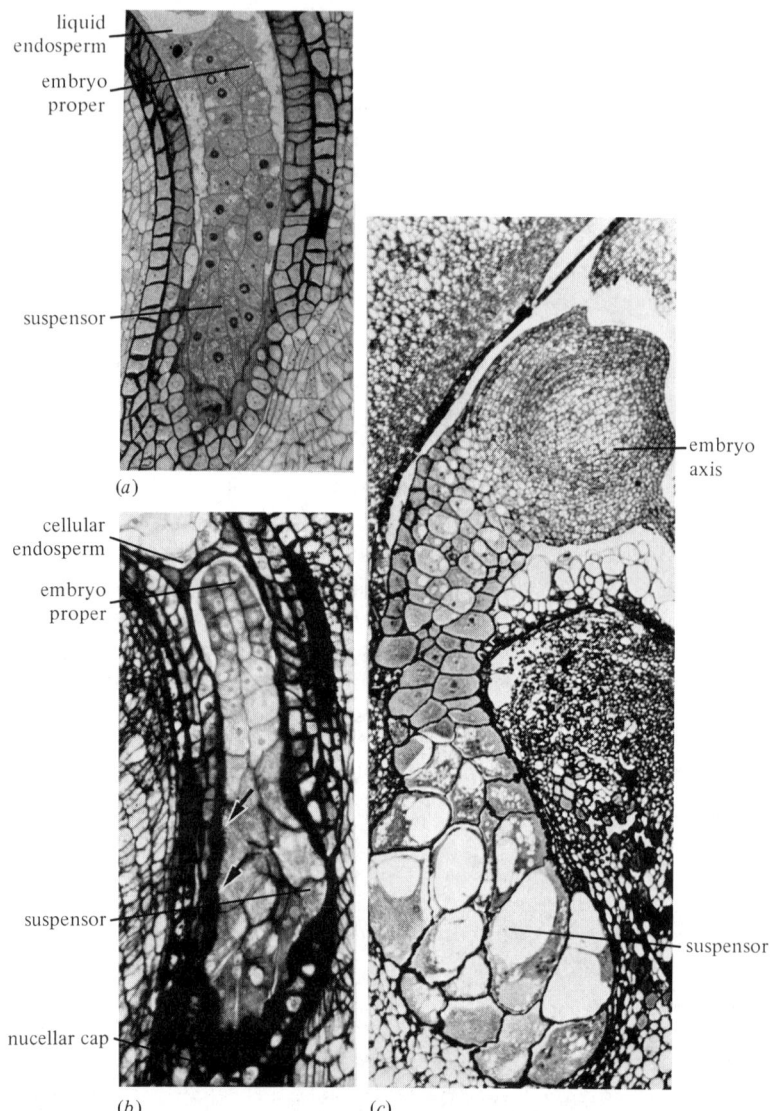

FIGURE 3.3. Development of the embryo and suspensor in *Phaseolus coccineus*. (a) An early proembryo showing very little structural differentiation between the embryo and suspensor. (b) A late proembryo showing wall ingrowths (*arrows*). (c) The heart-shaped embryo with suspensor. (From Yeung and Clutter, 1978; photographs supplied by Dr. E. C. Yeung.)

number of microbodies from the basal cell to the chalazal suspensor cells (Newcomb and Fowke, 1974). Plastids found in the suspensor of *Pisum sativum* (Marinos, 1970b), *Phaseolus vulgaris, P. coccineus* (Schnepf and Nagl, 1970; Yeung and Clutter, 1979), *Ipomoea purpurea* (Ponzi and Pizzolongo, 1972, 1973), *Tropaeolum majus* (Nagl and Kühner, 1976), *Medicago sativa* and *M. scutellata* (Sangduen et al., 1983a) give every appearance of having undergone some degree of specialization unique to each species.

Wall labyrinths of the type described above appear to be a widespread phenomenon and have been documented in suspensor cells of *Phaseolus vulgaris* (Schnepf and Nagl, 1970), *Pisum sativum* (Marinos, 1970a), *Diplotaxis erucoides* (Simoncioli, 1974), *Tropaeolum majus* (Nagl, 1976b), *Brassica napus* (rapeseed) (Tykarska, 1979), *Vigna sinensis* (Hu et al., 1983), *Medicago sativa, M. scutellata* (Sangduen et al., 1983a) and *Alyssum maritimum* (Prabhakar and Vijayaraghavan, 1983). Based on the transfer cell morphology of the suspensor cells, there is now considerable and varied evidence to support the solute absorption theory of suspensor function, first articulated by Schulz and Jensen (1969). This theory envisages that metabolites from the endosperm or from the surrounding diploid cells of the ovule are delivered to the embryo through the suspensor, the wall invaginations facilitating the transfer by increasing the surface area of absorption. The plasmodesmata criss-crossing the embryo–suspensor complex may serve to maintain an open communication for the flow of solutes between the suspensor and the embryo. That the path of nutrients to the developing heart-shaped embryos of *Phaseolus coccineus* and *P. vulgaris* is through the suspensor has become evident from the work of Yeung (1980). This work demonstrated that if ^{14}C-sucrose is administered through pods or to isolated embryos, much of the radioactivity appears in the suspensor and in the suspensor end of the embryo. Although new knowledge of the wall architecture of the suspensor of angiosperm embryos has pushed the frontiers of our understanding of the function of this organ to a higher level, until more is known about the gradient properties of the suspensor and its surrounding cells, the mechanism of nutrient incorporation into the embryo by way of the suspensor will remain unclear.

Nuclear cytology of the suspensor

In view of the many aspects of the nuclear cytology of the suspensor, as well as the number of observations to be collated, each is considered piece by piece to provide a balanced appreciation of the importance of each cytological observation in the overall picture.

The evolution of DNA content of a cell is remarkable for its conservatism. Yet, it is now recognized that pockets of cells in multicellular organisms exhibit variations in DNA content which are important in gene regulation and differentiation. One variation that frequently occurs in eukaryotic cells and that has been noted in suspensor cells of angiosperm embryos is endoreduplication of the genome. A number of surveys (Hasitschka-Jenschke, 1962; Corsi, Renzoni and Viegi, 1973; Nagl, 1974, 1978; Freed and Grant, 1976; Viegi, Pagni, Corsi and Renzoni, 1976) have found suspensor cells with endoreplicated genomes in *Alisma lanceolata, A. plantago-aquatica, Brassica nigra, Cucubalus baccifer, Dianthus chinensis, Echinodorus tenellus, Eruca sativa, Gagea lutea, Geranium phaeum, Lotus carmeli, L. pedunculatus, L. purshianus, Lupinus* sp., *Melandrium rubrum*, several species of *Phaseolus, Potomogeton densus, Sophora flavescens, Trapa natans, Tropaeolum majus* and *Tunica saxifraga*. It is worthy of note that endoreduplication is not ubiquitous in suspensor cells of angiosperms and thus does not appear to be an indispensible phase of suspensor differentiation.

The level of endoreduplication observed varies widely in suspensor cells of various plants. On a general level, however, we can state that because the suspensor is not a storage organ of the embryo, it contains a lot more DNA in the genome of its cells than is necessary to code for all the proteins it needs. On the basis of nuclear volume or quantitative Feulgen assay, the highest values for nuclear DNA content so far reported for suspensor cells are 8,192C for *Phaseolus coccineus* (Brady, 1973) and 4,096C for *P. vulgaris, P. hysterinus* and *P. multiflorus* (Nagl, 1974); the lowest values are 16C for *Sophora flavescens* and 32C for *Geranium phaeum* and *Tunica saxifraga* (Nagl, 1978). In *P. coccineus* (Brady, 1973; Nagl, 1974) and *Eruca sativa* (Corsi et al., 1973) the outcome of endoreduplication in the different cells of the suspensor is somewhat different, but its consequences in terms of gene regulation are probably the same. For example, for both species, a moment's comparison between the values for nuclear volume or DNA content of the cells of the suspensor away from the embryo and close to the embryo illustrates a range of variation. In *P. coccineus*, one can recognize a progressive increase in the level of endoreduplication beginning with low degree in the cells at the junction between the suspensor and embryo proper, medium degree in the cells in the neck region of the suspensor and very high degree of endoreduplication in the largest cells of the basal region of the suspensor (Table 3.1). A similar trend is seen in the nine- to ten-celled suspensor of *E. sativa* embryos of different stages of development, with DNA values ranging from approximately 10C to 75C; here, the cell next to the basal cell, rather than the basal cell, has the highest DNA content.

A study of the chromosomal constitution of endoreduplicated suspensor cells of *Potomogeton densus* and *Alisma lanceolatum* by Hasitschka-Jenschke (1959) revealed that these cells have a striking chromatin structure. Their nuclei are shown to contain bundles of interphase chromosomes known as polytene chromosomes possessing increased diameter and

Table 3.1. *DNA content of suspensor cell nuclei of* Phaseolus coccineus, *based on data from nuclear volume and quantitative Feulgen cytophotometry*

Location of cells	DNA contents, C values	No. of nuclei	Volume (μm^3)	No. of nuclei	DNA content, Feulgen cytophotometry units ± SD[b]
Embryo	2	14	277	13	11.3 ± 1.27
Suspensor, close to embryo	4	17	529	20	22.4 ± 1.17
	8	14	1,412	12	47.4 ± 2.55
	16	11	2,230	16	87.7 ± 8.40
	32	16	4,761	32	169.7 ± 14.7
	64	19	12,488	9	348.0 ± 32.2
Suspensor, median part	128	21	26,261	–[a]	–[a]
	256	22	52,857	6	1,410 ± 90
	512	17	116,227	20	2,640 ± 340
Suspensor, basal cells	1,024	18	240,250	18	5,810 ± 510
	2,048	22	479,453	28	10,500 ± 745
	4,096	9	826,550	23	21,900 ± 1,860
	8,192	–[a]	–[a]	7	47,100 ± 4,003

[a]No value given.
[b]Standard deviation.
Source: Modified from Nagl (1974).

greater length than is usually seen. A later investigation by Nagl (1962) demonstrated polyteny in the suspensor cells of *Phaseolus coccineus*. Subsequently, giant polytene chromosomes have been observed in suspensor cells of other species of *Phaseolus* (Nagl, 1974), *Gagea lutea* (Hasitschka-Jenschke, 1962), *Alisma plantago-aquatica* (Bohdanowicz, 1973), *Eruca sativa* (Corsi et al., 1973), *Tropaeolum majus* (Nagl, 1976a), some species of *Lotus* (Freed and Grant, 1976) and *Brassica nigra* (Viegi et al., 1976). In *P. coccineus* and *P. vulgaris,* the most complex of the polytene chromosomes are found in the large cells in the basal region of the suspensor. The constant appearance of these chromosomes in the basal suspensor cells and the ease with which they can be identified in cytological preparations have made these plants favorite objects in which to study changes associated with gene expression in the suspensor. Cytological analysis of chromosomal patterns has led to the identification of 11 pairs of polytene chromosomes each in the basal suspensor cells of *P. coccineus* (Nagl, 1967) and *P. vulgaris* (Nagl, 1969a). Standard Feulgen staining and Giemsa C-banding techniques have shown that each giant chromosome is subdivided into heterochromatic and euchromatic regions and that many parallel chromatid strands run longitudinally along the chromosomal axis (Nagl, 1978).

The precise number of bands in the polytene chromosomes is not clear, and in some chromosomes the bands are also not distinct. That the polytene chromosomes may be so altered by the environmental conditions of growth of the donor plants as to impair their ability to produce bands was suggested by the results of an experiment in which a distinct banding pattern was evident in chromosomal preparations of *Phaseolus vulgaris* suspensor cells made from plants raised at a lower temperature than usual (Nagl, 1969a). The banded state is reversible, as demonstrated by returning plants to a warm temperature. Accelerated banding is also seen in polytene chromosomes prepared from plants grown in short days or in long days as opposed to plants grown in a 12-hour light – 12-hour dark regimen (Nagl, 1973b). Another treatment that causes banding is culture of suspensors in a medium enriched with 0.5 mM to 0.25 M $CaCl_2$ (Nagl, 1970b). It is difficult to imagine how these various treatments can trigger development of banding patterns in the polytene chromosomes. By autoradiography of incorporation of ^{3}H-uridine, a precursor of RNA synthesis, Nagl (1969b) showed that whereas polytene chromosomes in their normally extended state have a high affinity for the label, these same chromosomes in their low-temperature-induced condensed and banded state are inactive in incorporating the isotope. Moreover, treatment of suspensors with actinomycin D, an inhibitor of DNA-dependent synthesis of RNA (i.e., mRNA), results in condensation and imperfect banding of polytene chromosomes as well as inhibition of RNA synthesis in them; these observations suggest that morphological changes in polytene chromosomes are related to transcriptional activity.

The most interesting polytene chromosomes of the suspensor cells of *Phaseolus* are the two pairs (I and V) that possess nucleolus-organizing centers which are recognized by their ability to organize a nucleolus. Nucleolar organizers represent sites at which the chromosome has separated into individual fibers or groups of fibers that traverse the nucleolus, creating a visible puff or bulge in the process and terminating distally at the surface of the nucleolus as a satellite (Nagl, 1965). Like the banding pattern on the polytene chromosomes, the nucleolus-associated regions of the chromosomes also display a temperature-dependent variability in their structure. Thus, the dispersed condition of the chromatid fibers, leading to an incipient bulge in the nucleolus, can be perpetuated at optimum temperatures, but high or low temperatures are inimical to such change. Under the latter conditions, a marked condensation of the chromatin occurs as the DNA strands seem to fold back into the nucleolus-organizing center to form a single thread (Nagl, 1970a).

A further correspondence between the banding of the chromosomes and intranucleolar chromatin threads is demonstrated by the observation that the latter actively incorporate ^{3}H-uridine into RNA and that actinomycin D, which induces condensation of the chromatin fibers in the nucleolus organizing center, inhibits RNA synthesis (Nagl, 1969b). It is now well established that in higher eukaryotes genes for rRNA are clustered at the nucleolus-organizing regions. By molecular hybridization of radioactive rRNA to its complementary DNA *in situ*, it has been shown that the nucleolus organizing regions of polytene chromosomes of *P. vulgaris* (Brady and Clutter, 1972) and *P. coccineus* (Brady and Clutter, 1972; Avanzi, Durante, Cionini and D'Amato, 1972) contain ribosomal cistrons. Other cytological features of genes coding for rRNA have also been described, although by the use of less precise methods. For example, some bands in the nucleolus organizing center of chromosome pairs I and V of *P. coccineus* that contain DNA sequences complementary to ribosomal genes behave differently at successive stages of embryogenesis by undergoing disproportionate and nonsynchronous DNA synthesis (Forino, Tagliasacchi and Avanzi, 1979). According to Cionini and Avanzi (1972), the nucleolar organizers of these chromosomes appear first to be labeled after a brief exposure to ^{3}H-actinomycin D; since actinomycin D binds to guanine, the ribosomal cistrons of the suspensor polytene chromosomes might be expected to be rich in guanine-cytosine. Another point of interest is that loci corresponding to rRNA cistrons are not associated exclusively with the nucleolus organizing region, as they have also been identified in the nonnucleolar regions of chromosome pair I and in chromosome pair II, which lacks a nucleolus organizing center (Avanzi et al., 1972; Durante et al., 1977).

Another aspect of the polytene chromosomes of suspensor cells of *Phaseolus* that deserves mention is the behavior of the nucleolus and extrusion of micronucleoli. The nucleolus is recognized by its dense granular matrix and bright vacuoles; at extremes of temperatures, it suffers both a decrease in size and loss of vacuoles. Reflecting its high metabolic activity under normal conditions of growth of the plant, the nucleolus also extrudes many micronucleoli. Additional micronucleoli are liberated from the euchromatic and heterochromatic regions as well as from the loops of the polytene chromosomes, to be described later (Nagl, 1970a, 1973a).

It is pertinent to mention here that polytene chromosomes have been very well characterized both structurally and functionally in the salivary glands of several dipteran larvae. One of the most important alterations of insect polytene chromosomes is the lateral expansion of some of the bands into bulges known as puffs. These outgrowths generally emanate as small enlargements of the bands, accompanied by decondensation and unwinding of the chromatid strands. Occasionally puffs take a dynamic form and are visualized as extensive swellings accompanied by extrusion of the chromatin material from the axis in the form of loops. In the polytene chromosomes of suspensor cells, the most conspicuous examples of puffing are seen in *Phaseolus coccineus* and *P. vulgaris*. Although puffs occur randomly in varying states of regression in the different heterochromatic and euchromatic regions of the chromosomes, corresponding regions of pair II (or pair VI) of polytene chromosomes of *P. vulgaris* exhibit a precise regularity in their occurrence. In these chromosomes, one puffed region is limited by heterochromatin on both sides, while the other puff merges into a euchromatic region at one end (Nagl, 1969c). Puffs represent sites at which DNA synthesis and transcription occur. Recent work by Tagliasacchi, Forino, Frediani and Avanzi (1983) and Tagliasacchi et al. (1984) on the distribution of puffs and pattern of DNA and RNA synthesis on the puffs on chromosome pairs VI and VII of *P. coccineus* has extended cytological information in two important directions. First is the demonstration that some DNA-synthesizing puffs appear only at specific stages of embryogenesis. This finding has opened up the possibility of a stage-specific variation in the replication of DNA fractions in the same polytene chromosome of the suspensor cell. A second point of general interest is that by using the criterion of autoradiographic localization of ^{3}H-thymidine and ^{3}H-uridine it has become clear that separate groups of DNA- and RNA-synthesizing puffs are not present; rather, all puffs are capable of synthesizing both molecules. It is likely that puffs that synthesize DNA initially follow it up by transcription.

The extensive pattern of spontaneous loop formation displayed by insect polytene chromosomes does not have a counterpart among chromosomes of suspensor cells. However, polytene chromosomes of *Phaseolus coccineus* suspensor cells form facsimiles of loops after momentary exposure of plants to an elevated temperature (Nagl, 1970a) or within several hours after changing the photoperiodic conditions of plant growth (Nagl, 1973b). That radioactive uridine is incorporated into loops vividly demonstrates the commitment of these outgrowths to transcriptional activity (Nagl, 1974).

Because of the disproportionate accumulation of DNA in the nuclei of suspensor cells, it is scarcely surprising that DNA replication patterns in these cells have been subject to intensive study. In particular, the question of whether polytene chromosomes undergo extra DNA synthesis (DNA amplification) has been bedeviled by controversy, beginning with the observations of Avanzi, Cionini and D'Amato (1970) on ^{3}H-thymidine incorporation in suspensor cells of *Phaseolus coccineus*. Although the number of suspensors used in this study was not enormous, in every instance most cells incorporated the label exclusively in the heterochromatic regions of polytene chromosomes, while in a small number of cells, the label appeared all over the nucleus. DNA : histone ratios of the different regions of the chromosomes showed that the extra DNA synthesized is complexed with histones in varying proportions suggesting possible differences in transcriptional activity in the same chromosome (Diez and Cionini, 1971). Overall, these results have been interpreted to indicate that certain genes in the heterochromatic regions undergo DNA amplification in addition to the scheduled DNA synthesis attributable to endoreduplication. This step must involve considerable rearrangement in a small part of the genome, which is replicated more often than the remainder. As shown in a later study (Cremonini and Cionini, 1977), DNA amplification is prevalent in the highly polytene chromosomes of cells in the micropylar region of the suspensor, which are in advanced stages of development, while a geometrical increase in DNA amount associated with endoreduplication is restricted to cells in the early stages of suspensor development.

The concept of DNA amplification in the suspensor chromosomes was questioned by Brady (1973), who found that DNA amounts of polytene chromosomes of *Phaseolus coccineus* suspensor cells increase in discrete geometric series to yield nuclei containing up to 8,192C DNA. This finding suggests that each polytene chromosome consists of many identical copies of the genome, all of which replicate at successive cell cycles. Brady and Clutter (1974) interpreted ^{3}H-thymidine incorporation into heterochromatin as indicative of late DNA synthesis rather than of DNA amplification. Their defense of this assertion included the observation that DNA synthesis in the polytene chromosomes of *P. coccineus* suspensor cells is concomitant with DNA despiralization and completes within 4 to 6 hours.

Results in broad agreement with those of Avanzi et al. (1970) were obtained by Lima-de-Faria et al. (1975), who followed density shifts in DNA isolated from *P. coccineus* suspensor cells in cesium chloride gradients and obtained evidence for both amplification and underreplication of DNA. The presence in the suspensor cells of a class of DNA with buoyant density that has no parallel with DNA from other parts of *P. coccineus* suggests DNA amplification. On the other hand, a lower saturation value in hybridization experiments with total suspensor DNA and rRNA appears to indicate underreplication, especially of rRNA genes. The overall impression left by these analyses is that at the level of resolution offered by density shifts a persuasive case can be made for the occurrence of differential DNA replication in the suspensor polytene chromosomes of *P. coccineus*. An explanation for the discrepancy in the results from the two laboratories will probably require further understanding of the DNA replication pattern in this system at the cytological and molecular levels.

The mechanics of DNA replication pattern has also been studied to some extent in the suspensor cells of *Tropaeolum majus*. The presence of a few polytene chromosomes with very little heterochromatin along with others with large amounts of heterochromatin in the nuclei of certain suspensor cells attests to the possibility of heterochromatin underreplication in the former and the full extent of polytenization in the latter. Data from Feulgen DNA measurements and *in situ* hybridization experiments are also consistent with the notion of DNA underreplication (Nagl, Peschke and van Gyseghem, 1976; Deumling and Nagl, 1978).

In summary, the general conclusion derived from our current understanding of the nuclear cytology of suspensor cells is that certain features of the polytene chromosomes are suggestive of intense transcriptional activity. However, before the potential of the polytene chromosomes as an assay system for gene action in the suspensor cells can be realized, additional work is needed to determine what proportion of RNA is transcribed as informational RNA and whether all polytene chromosomes transcribe the same mRNA or different mRNAs are associated with different chromosomes in the same nucleus. Thus, we will gain further understanding of the mechanism by which gene action is regulated in the suspensor and the ways by which their products contribute to suspensor function.

Biochemical indices of suspensor activity

Several cytochemical, biochemical and physiological changes have been described as part of the differentiation program of the suspensor. Embryos of *Brassica campestris* have figured in some investigations on the cytochemical localization of peroxidase, succinic dehydrogenase, malate dehydrogenase, aldolase, glucose 6-phosphate dehydrogenase, glutamate dehydrogenase, cytochrome oxidase and phosphorylase. Although activities of the various enzymes in the basal cell and suspensor cells change according to independent patterns, it can be said in a general way that enzymes are in short supply in suspensors of early division-phase embryos and are maximally active at the globular to heart-shaped stages of embryos (Malik, Vermani and Bhatia, 1976; Malik, Singh and Thapar, 1976). The timing of increased enzyme activity in the suspensor is interesting in light of the fact that this organ ceases to grow in the globular to heart-shaped embryos; an explanation may lie in the need for high metabolic activity in the suspensor for the absorption and translocation of nutrient substances to the growing embryo.

Possibly the most distinctive cytochemical changes occur in the suspensor during its programmed cell death. This is initiated at the micropylar end of this organ after it is fully formed. The presence of specialized plastids in the suspensor cells of *Stellaria media* (Pritchard and Bergstresser, 1969), *Phaseolus coccineus* and *P. vulgaris* (Nagl, 1976c; Gärtner and Nagl, 1980) has provoked ideas on the role of acid phosphatase activity in the plastids during autolysis of the suspensor. In *P. coccineus* and *P. vulga-*

ris, enzymatic activity associated with the plastids of the senescing suspensor progresses from the cells at the micropylar end toward the chalazal end. Activities of phosphatases and other hydrolytic enzymes in the suspensor of *Tropaeolum* are also of interest in this context. Attention was drawn in Chapter 2 to the presence of two suspensor haustoria in *T. majus*, which penetrate the chalazal and placental regions of the carpel, respectively. Studies by Nagl (1976c, 1977) and Gärtner and Nagl (1980) have shown that self-digestion of the suspensor is initiated by degenerative changes in the mitochondria and that the earliest activity of acid phosphatase is found in the chalazal suspensor haustorium. Other enzymes, such as alkaline phosphatase and acetylesterase, are also active in the suspensor proper and in the chalazal haustorium, becoming maximally effective at the early cotyledonary stage, when suspensor autolysis is well under way (Singh, Bhalla and Malik, 1980b). These hydrolytic enzymes may be active in the lysis of cell organelles of the suspensor. The lysed materials of the organelles can be looked upon as a possible energy source for use by the maturing embryo; an increase in the pool of free amino acids reported in the degenerating suspensor of *Tropaeolum* provides circumstantial evidence of the nature of the substances transmitted to the embryo (Singh et al., 1980b).

Carbohydrates, fats and proteins have been localized by cytochemical means in the suspensor cells of so many plants that it is not worthwhile to list them. We shall limit ourselves to the comment that these inclusions have invariably been considered as storage forms of metabolites absorbed by the suspensor. However, as dark CO_2 fixation by suspensor cells of *Tropaeolum majus* has been demonstrated (Bhalla et al., 1980a), we have to accept that some of the carbohydrates present in the cells may have been synthesized through a nonphotosynthetic route.

Studies on RNA and protein metabolism have provided evidence of high transcriptional and translational activity of the suspensor cells. Walbot et al. (1972) and Sussex et al. (1973) explored the changes in RNA and protein content and in the rates of synthesis of these molecules in the suspensor of *Phaseolus coccineus* surgically removed from the embryo. Their results showed that expressed on a per-cell basis, RNA content and the rate of RNA synthesis are low during the early development of the suspensor, increasing to a maximum at the late heart-shaped or early cotyledon stage of embryos and then declining. Protein content and the rate of protein synthesis are also low in the suspensor cells of early-stage embryos but increase to substantially high levels in those of late-stage embryos. The patterns of RNA and protein metabolism of the suspensor are very different from those displayed by the embryo proper; moreover, the suspensor also contains more RNA and proteins per cell and synthesizes these molecules more efficiently than do embryo cells of the same age. Apparently, the situation in *P. coccineus* is not atypical, as cells of the suspensor and its haustorium in *Tro-*

paeolum majus also display higher transcriptional and translational activity than do embryo cells (Bhalla et al., 1981).

It should be clear that major increases in the synthesis of RNA and proteins occur in the suspensor cells after they have quit dividing. Having seen that polyteny is frequent in suspensor cells of certain angiosperms, we are bound to ask whether the increased RNA synthetic activity is caused by the increased number of gene copies present in the polytene cells. Only one bit of evidence limited to *Phaseolus coccineus* is available to probe this question. The basic problem in estimating the template activity is to determine the average number of gene copies present in the nucleus. In the method used by Clutter, Brady, Walbot and Sussex (1974), values for average number of diploid gene copies were obtained by dividing the average total DNA content of a suspensor cell by the average diploid amount of DNA. The agreement between figures based on spectrophotometric and biochemical determinations of DNA contents proves that comparisons of template activity measured in terms of RNA per hour per unit of DNA between diploid and polytene chromosomes are valid. The point that has emerged from this study is that the template activity of the suspensor cells at the late heart-shaped or early cotyledon stage of embryos is higher than is expected of the number of diploid gene copies present. This observation suggests that polyteny might account for increased RNA synthesis of the suspensor cells, although it remains to be resolved whether this occurs by differential template activity or by selective gene amplification.

The only other relevant information regarding the biochemical activity of the suspensor concerns the production of growth hormones. From the functional point of view, certain ultrastructural features of suspensor cells have been considered suggestive of hormone synthesis, but only recently has direct evidence been forthcoming to support the view that the suspensor produces GA, cytokinins and auxins. Let us first concentrate on GA activity, evidence for which is derived from the work of Alpi, Tognoni and D'Amato (1975). This study showed that in the heart-shaped embryos of *Phaseolus coccineus,* the total amount of GA-like substances present in the suspensor is several times greater than in the embryo proper. An abrupt change occurs in the cotyledon-stage embryos when there is a decrease in GA content of the suspensor and a concomitant increase in the content of this hormone in the embryo proper. In suspensors of *Cytisus laburnum* and *Tropaeolum majus,* GA-like substances are present even in higher concentrations than in *P. coccineus* suspensor (Picciarelli, Alpi, Pistelli and Scalet, 1984). Changes in the cytokinin status of the suspensor also bear certain similarities to changes in GA activity (Lorenzi et al., 1978). In fact, one could consider cytokinin activity as a stringent marker of embryogenesis in *P. coccineus* as there is an abundance of biologically active cytokinins in the suspensor during early embryogenesis and a low activity in the embryo proper; a complete reversal of the status of the active cytokinins in the suspensor and embryo occurs in the cotyledon-stage embryos. Changes in the auxin content of the suspensor have drawn little attention, although it appears that the suspensor contains higher amounts of auxin than does the embryo proper (Alpi et al., 1975; Przybyllok and Nagl, 1977). Finally, support for the view that the suspensor contains high levels of growth hormones has come from an unexpected quarter. Spontaneous growth of a callus from the suspensor of *P. coccineus* cultured in a medium lacking growth hormones has been interpreted as possibly caused by the high level of endogenous hormones in its cells (Bennici, Cionini and D'Amato, 1976).

What is the function of hormones in the suspensor? One prevalent notion is that in the early stages of its development, the embryo does not bear the burden of synthesizing the hormones necessary for its growth, but that this function is taken up by the suspensor, which apparently transmits the hormones to the embryo. As a caveat for this interpretation, it should be mentioned that we do not know whether early-stage embryos have a developmental program dependent on hormones received from the suspensor. Nonetheless, in support of this view, it has been shown that heart-shaped embryos of *Phaseolus coccineus* deprived of their suspensor achieve very little growth in culture unless the medium is supplemented with GA or cytokinins (Cionini, Bennici, Alpi and D'Amato, 1976; Bennici and Cionini, 1979; Yeung and Sussex, 1979). The basic question of whether the suspensor has the capacity to synthesize growth hormones has been examined in *P. coccineus*, which showed that a cell-free system obtained from the suspensor synthesizes precursors of GA as well as GA_1, GA_5 and GA_8 from radioactively labeled substrates (Ceccarelli, Lorenzi and Alpi, 1979, 1981a,b). In another line of work, high levels of IAA detected in the suspensor of *Tropaeolum majus* (Przybyllok and Nagl, 1977) have been correlated with changes in peroxidase activity of the suspensor and embryo cells, suggesting a possible transfer of IAA from the former to the latter (Singh et al., 1979). This correlation is somewhat typical of organs that synthesize or utilize IAA for growth. If we may generalize from a recent study (Brady and Walthall, 1985), it is suggested that the role of hormones such as GA in the suspensor is in enhancing or maintaining the protein level in the embryo.

In conclusion, it cannot now be naively assumed that the primary function of the suspensor is to orient the embryo in close proximity to the source of nutrients in the embryo sac or in the ovule. Other roles have been proposed for the suspensor, but they have barely begun to be explored. The evidence presented in this section can be used to defend at least three views relating to suspensor function. One is that wall invaginations absorb food materials from the surrounding tissues and transmit them to the growing embryo. A second view based on the polytene nature of chromosomes of suspensor cells and their high template activity is that cells of the suspensor synthesize some gene products essential for embryo growth. Finally, there is evidence for the accumulation of hormones in the suspensor and the hypothesis that they are translocated to the growing embryo. The information we have currently available in support of these and other views on the functions of the suspensor fails to elucidate the mechanism by which these functions are accomplished.

Storage protein synthesis

The termination of the early proliferative phase of growth marks a key point in angiosperm embryogenesis; the tempo of mitosis has reached its peak, and the embryo is poised to enter a new developmental pathway. This pathway involves a long period of growth by cell expansion when synthesis and accumulation of an acervate complex of storage proteins occur, leading to increases in fresh and dry weights of the embryo. The proteins are stored within the cell in discrete protein bodies and remain virtually unchanged in character and composition during the period of quiescence or dormancy of embryos. The other attribute of storage proteins coincides with the subsequent requirement of the seed in the sense that as it begins to germinate, these proteins are hydrolyzed and serve as the source of carbon and nitrogen skeletons to the growing embryo. Storage proteins are also relatively insoluble in physiological salt solutions so as to be precipitable *in situ* as they are synthesized. The synthesis of storage proteins represents a dramatic perturbation of the metabolism of the cells of the embryo programmed for the production of housekeeping proteins to one concerned with the production of specialized proteins. No other cells in the embryo can match the commitment of cells of the cotyledons to the synthesis and accumulation of gene products.

Many aspects of synthesis of storage proteins during embryogenesis have been investigated using embryos of various agronomically important legumes such as peas and beans. In these plants, proteins, together with reserve carbohydrates or oils, are synthesized mainly in the cotyledons, which form the bulk of the embryo. As alluded to earlier, in legumes, storage protein synthesis is a stellar event that occurs coincident with the cessation of cell division and the onset of cell expansion in the cotyledons. In many legumes examined, reserve protein bodies are initially found in the central vacuoles of the parenchymatous cells of the cotyledons. Where the proteins are synthesized and how they cross the tonoplast are not entirely clear. Recent evidence seems to suggest that ER is the primary site of protein body synthesis and that the proteins synthesized are deposited as discrete masses on the tonoplast. Release of the protein mass is accomplished by the invagination of the tonoplast and pinching of the tonoplast with the enclosed proteins (Bollini and Chrispeels, 1979; Yoo and Chrispeels, 1980).

Developmental changes in storage protein synthesis

For many years, the only approach to the study of changes in storage protein content of embryos was by biochemical analysis. More recently, various sensitive immunological techniques have been used to detect small amounts of specific proteins as well as their subunits in embryos at relatively early stages of development. As a result, a reasonably clear picture is beginning to emerge of the pattern of storage protein synthesis and accumulation during embryogenesis in several plants.

Cotyledons of *Vicia faba* and *Pisum sativum* accumulate two major, high-molecular-weight proteins, legumin and vicilin, with the synthesis of vicilin generally preceding that of legumin (Graham and Gunning, 1970; Millerd and Spencer, 1974). In embryos of *V. faba* analyzed at different stages of development, no legumin is detected in cotyledons less than 5 mm long. Below 7 mm length, cotyledons contain less than 1 μg of legumin, whereas cotyledons more than 10 mm long show a severalfold increase in legumin content (Millerd, Simon and Stern, 1971). Fractionation of legumin and vicilin into component polypeptides has shown that they consist of three and four subunits, respectively. The subunit composition of vicilin is found to change throughout embryogenesis, whereas that of legumin remains constant. This suggests that the synthesis of individual subunits is probably under separate genetic control. Despite the fact that vicilin accumulation in *V. faba* cotyledons begins before that of legumin, starting in a modest way, legumin synthesis overtakes vicilin synthesis, thereby accounting for the ability of cotyledons to store nearly four times more legumin than vicilin (Wright and Boulter, 1972). This is also true of the synthesis of storage protein subunits during embryogenesis in pea. As seen from the SDS-acrylamide gel electrophoretic profile of proteins of cotyledons of developing pea embryos (Fig. 3.4), vicilin subunits are present in appreciable amounts at earlier stages than others, whereas legumin subunits and convicilin (a minor storage protein fraction) are prominent later in embryogenesis (Gatehouse et al., 1982). More recent experiments involving hybridization of transcription products synthesized by nuclei isolated from pea cotyledons of different ages with cDNA encoding for vicilin and legumin have essentially confirmed this finding (Evans, Gatehouse, Croy and Boulter, 1984). In attempts to monitor storage protein synthesis under controlled conditions in culture, it has been found that cotyledons excised from young embryos of pea, containing no trace of vicilin, begin to synthesize this protein in culture (Millerd, Spencer, Dudman and Stiller, 1975). While legumin synthesis is not initiated in excised and cultured cotyledons of the same age (Millerd et al., 1975), cotyledons of cultured young embryos begin to synthesize this protein and accumulate it in higher quantities than cotyledons of embryos of similar fresh weight *in vivo*

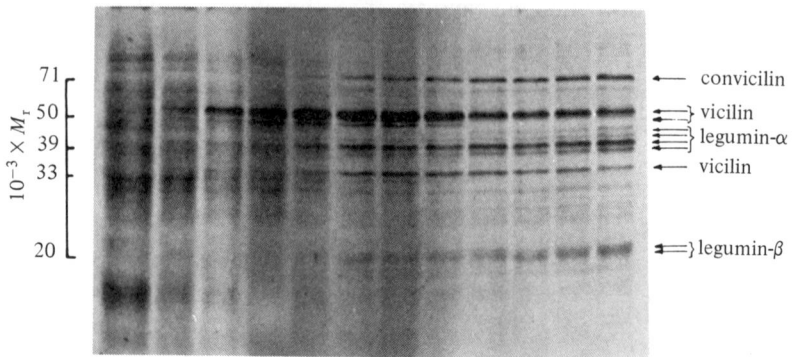

FIGURE 3.4. SDS-polyacrylamide electrophoretic separation of subunits of legumin and vicilin from cotyledons of pea embryos of different ages, as indicated by days after flowering. Samples analyzed at 33 days are mature seed embryos. M_r, relative molecular mass. (From Gatehouse et al., 1982; reprinted by permission from *Biochem. J.* 208: 119–127. © 1982, The Biochemical Society, London.)

(Domoney, Davies and Casey, 1980). These results are consistent with the notion that switching on of genetic information for storage protein synthesis in cotyledons is independent of any signal provided by the parent plant.

Striking differences occur in the composition of storage proteins of cotyledons of soybean during embryogenesis. The major storage proteins of soybean are a 7S glycoprotein, conglycinin, and a nonglycosylated 11S protein, glycinin; approximately 1 to 2 percent of the seed protein mass also includes another glycoprotein, lectin. Conglycinin has three subunits, α, α' and β, whereas glycinin has at least four acidic and four basic subunits. Apparently the subunits of these proteins do not appear in the cotyledons at the same stage of development. Whereas the acidic and basic subunits of glycinin and α and α' subunits of conglycinin begin to accumulate coincident with the cessation of cell division in the cotyledons, conglycincin β and certain acidic subunits of glycinin (A-4) do not appear until the cotyledons are mature (Gayler and Sykes, 1981; Meinke et al., 1981; Sengupta, Deluca, Bailey and Verma, 1981). It has also been shown that, in contrast to the cotyledons, the embryo axis of soybean contains very little of A-4 subunit of glycinin as well as of conglycinin subunits β and α' but has an additional subunit of the latter protein. As the function of storage proteins in the embryo axis is not clearly understood, the significance of these differences in the subunit composition of storage proteins in the cotyledons and embryo axis appears enigmatic.

Storage proteins of two other legumes, French bean and *Vigna unguicu-*

lata (cowpea), also warrant brief descriptions in this summary. A globulin designated G1 (also known as phaseolin), along with a smaller amount of another globulin, G2, form the predominant components of storage proteins of French bean cotyledons. The difference between G1 and G2 seems to be related to their requirement for high (G1) or low (G2) salt for solubility. Three subunits have each been identified for G1 and G2 globulins; the polypeptides specified for G1 have molecular weights of 43, 47 and 53 kd, and those for G2 have molecular weights of 43, 32 and 30 kd. Electrophoretic characterization has shown that G1 subunits maintain their ratio in cotyledons of different ages, as if their synthesis is closely coordinated. By contrast, the 30-kd subunit of G2 does not appear until the cotyledons are mature (Sun, Mutschler, Bliss and Hall, 1978). The protein fraction of cowpea is composed of a 7S globulin and a legumin-like 11S globulin. From a developmental point of view, a small increase noted in the number of serologically and electrophoretically detectable components of these proteins in the cotyledons of mature embryos is probably significant (Carasco, Croy, Derbyshire and Boulter, 1978).

Unlike the process occurring in legumes, in cotton major increases in storage protein content occur during the cell-division phase of growth of cotyledons, which spans a period of fresh weight increase from 50 mg to 90 mg in the embryo. The abundant storage proteins of cotton are a glycosylated 52-kd α-globulin and a nonglycosylated 48-kd β-globulin. These proteins are selectively carved out of precursor sets of about 60- to 69-kd proteins, which indulge in complex processing to yield them (Dure and Chlan, 1981; Dure and Galau, 1981). Storage proteins of *Brassica napus* are constituted of a legumin-like 12S neutral globulin designated cruciferin and a cluster of basic proteins (1.7S) designated napin, both of which are distributed in the embryo axis as well as in the cotyledons. Whereas both proteins begin to accumulate at about the same rate in the embryo during most of its cell-expansion phase, subsequently a major shift occurs in the species of proteins synthesized, resulting in a higher ratio of cruciferin to napin in the mature embryo (Crouch and Sussex, 1981; Crouch, Tenbarge, Simon and Ferl, 1983; Finkelstein and Crouch, 1984). Although storage reserves of cereals are found mainly in the endosperm, immunological studies have demonstrated the presence of embryonal antigens in the embryos of maize; when the embryo axis and scutellum are analyzed separately, a differential antigen spectrum can be seen in each organ (Khavkin, Misharin, Ivanov and Danovich, 1977). The antigenicity of the protein has been attributed to a globulin that exists in two fractions. Each fraction consists of subunits, some of which appear earlier than others during embryogenesis (Cross and Adams, 1983). Presence of storage proteins in embryos of cereals could possibly reflect the existence of some mechanism to compensate for any deficiency in the endosperm reserves.

Expression of storage protein genes

From the point of view of gene expression, the need for synthesis of specific mRNAs to account for the change in the developmental program of the cotyledons from histodifferentiation to one concerned with the production of specialized proteins is obvious. Over the past few years, a large-scale effort has been mounted to examine the molecular aspects of storage protein gene expression in cotyledons of various plants, and much has been learned primarily in French bean, soybean and cotton. Toward an understanding of the regulation of storage protein synthesis, there are a variety of questions to be considered. When does mRNA synthesis, which codes for proteins, begin in the embryo? If mRNA is synthesized at an early stage of embryogenesis, what is the fate of this message during the maturation phase of the embryo? How is the expression of genes for protein synthesis coordinated during embryogenesis? Is gene action for storage protein synthesis controlled at the transcriptional level or at the translational level? Experimental work designed to gain answers to these questions has largely been conducted by molecular biologists, who have employed molecular probes as well as synthetic and recombinant DNA probes to find direct evidence for gene activity.

The concept of differential gene expression for storage protein synthesis is supported by the characterization of mRNA from cotyledons of different stages of development or during the phase of maximum protein accumulation. When the template activity of the polysomal fraction or of purified poly(A)RNA fraction from mature French bean cotyledons is assayed in a cell-free system, segments of G1 protein are found in the translation products. Immunoprecipitation with monovalent antibody for G1, combined with polypeptide mapping, has led to the characterization of an mRNA fraction coding for 47-kd and 43-kd polypeptide subunits of G1 protein (Sun, Buchbinder and Hall, 1975; Hall et al., 1978). In a similar way, comigration of *in vitro* translation products with purified protein subunits and polypeptide mapping have verified that mRNA isolated from soybean cotyledons codes for subunits α and α' of conglycinin (Beachy, Thompson and Madison, 1978; Beachy, Barton, Thompson and Madison, 1980). A difficulty in interpreting these results is that mRNA from cotyledons of neither plant matches the efficiency of protein synthesis *in vivo* in the total number of subunits of each protein synthesized *in vitro*. Nonetheless, the conclusion is clear: In both French bean and soybean, mRNAs coding for certain storage proteins of the cotyledons appear to be present in the cells when they are maximally active in protein accumulation.

In another experimental approach, cDNA probes made from mRNA for storage proteins of cotton cotyledons have been used to monitor changes in the population of specific storage protein mRNAs during embryogenesis by competitive hybridization with mRNA. The extent of hybridization between cDNA and embryonic mRNA populations is a measure of their complementarity to the original storage protein mRNA. The possibility that cotyledons of cotton embryos (50 mg fresh weight) undergoing active cell divisions are most enriched for storage protein mRNA has been raised by Dure and Galau (1981); this is inferred from the observation that when embryos are pulsed with radioactive amino acids, 70-kd (a glycosylated intermediate of 52-kd), 52-kd and 48-kd protein sets are the most heavily labeled in fluorographs. Kinetics of reassociation of cDNA made from cotyledons of 50-mg embryos to mRNAs isolated from cotyledons of older (100 mg fresh weight) embryos and of mature seed embryos has shown that mRNA sequences for abundant storage proteins decrease and suffer almost total obliteration as embryos progress from the cell division phase into phases of cell expansion and maturation. Whether it is relevant that this appears to be solely attributable to the cessation of transcription of these genes is hard to say. An obvious conclusion from these results is that mRNA for storage proteins of cotton cotyledons is active only during a transient phase of embryogenesis. In a similar study using cDNA made from soybean embryos at the stage of maximum storage protein accumulation (mid-maturation stage), Goldberg et al. (1981b) showed that the appearance of mRNA sequences for storage proteins in high frequency closely corresponds to the period of their maximum accumulation and that the messages decay in late embryogeny. These superabundunt mRNAs, which number 7 to 10, are absent from leaf polysomes, indicating that they are specific for embryos and, because their reassociation with total leaf DNA proceeds faster than that observed with single-copy leaf DNA, it appears that these mRNAs are encoded by low-frequency repetitive DNA sequences.

In the first investigation, and in many ways the most perceptive work of its kind, to study changes in mRNA structure during embryogenesis, Goldberg et al. (1981a) followed the expression, in soybean embryos of different ages, of four genes for certain subunits of glycinin and conglycinin of soybean cotyledons. These genes are initially selected from a cDNA library of mRNA sequences prepared from mid-maturation-stage embryos by cloning into plasmid pBR322. They are screened by the colony hybridization procedure against cDNA prepared from mid-maturation-stage embryos, followed by restriction endonuclease mapping. Hybridization experiments between cloned messages and poly(A)RNA that compare the changes in the prevalence of cloned probes at different stages of embryogenesis have shown that mRNAs for storage proteins do not appear until the stage of cotyledon initiation and that even at this stage they are present in very low levels. The period from the cotyledon stage (5 to 10 mg fresh weight) to the mid-maturation stage (150 to 200 mg fresh weight) of em-

Cellular and biochemical embryogenesis

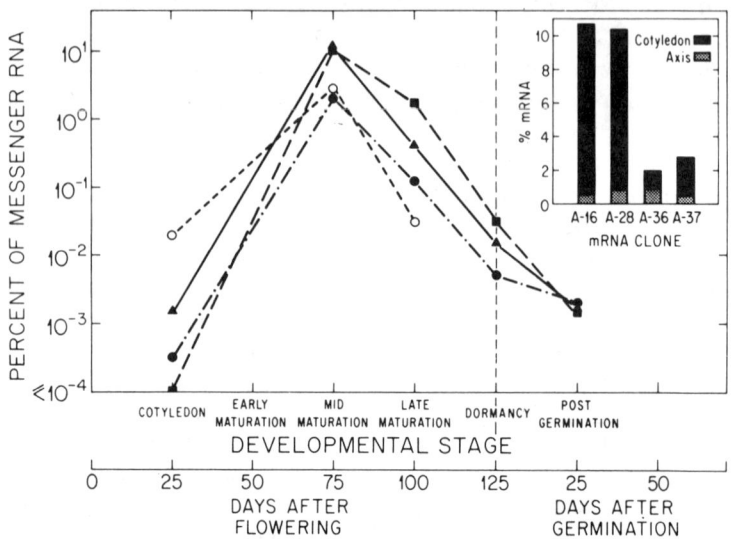

FIGURE 3.5. Changes in the expression of four genes (A-16 -▲-, A-28 -■-, A-36 -●- and A-37 -○- for storage proteins of soybean cotyledons during embryogenesis and germination. Inset shows the concentration of cloned messages in mid-maturation stage cotyledon and axis mRNA populations. (From Goldberg et al., 1981a.)

bryogenesis is marked by a 100- to 100,000-fold increase in the prevalence of cloned messages; this is followed by the disappearance of the messages as dehydration sets in and the embryos become mature (Fig. 3.5). A more recent study has shown that accumulation and decay of lectin genes during embryogenesis also follow a course analogous to that of genes for glycinin and conglycinin (Goldberg, Hoschek and Vodkin, 1983). Are genes for the synthesis of storage proteins in soybean embryos controlled at the transcriptional level or at the translational level? As the cloned transcripts for glycinin and conglycinin do not hybridize with leaf poly(A)RNA, it is clear that, with respect to the presence or absence of storage protein genes, the difference between embryos and leaves is transcriptional in nature. In a somewhat similar investigation, using cDNA clones complementary for purified glycinin and conglycinin mRNAs, Meinke et al. (1981) showed that mRNA for conglycinin β appears in soybean embryos several days before actual accumulation of this protein begins. This implies that translational restraints may control the expression of individual subunits of a major storage protein at specific stages of embryogenesis.

In comparable investigations on pea, changes in legumin and vicilin mRNA species in cotyledons of different ages have been found to correlate well with the rate of synthesis of these polypeptides (Gatehouse et al., 1982). Moreover, the levels of specific transcripts for legumin and vicilin detected in the cotyledons appear to increase or decrease with changes in mRNA levels, indicating transcriptional control of gene expression (Evans et al., 1984)

These ideas regarding the expression of storage protein genes in legumes have been supported by more extensive studies on cotton by *in vitro* translation of mRNA and hybridization of mRNA with cDNA clones for specific storage proteins (Galau, Chlan and Dure, 1983; Dure et al., 1983). A key result of these investigations has been the identification of one subfamily of genes coding for 48-kd proteins and two subfamilies of genes coding for 52-kd proteins. Equally important is the finding that all mRNA subsets for 48-kd protein and one subset for 52-kd protein constitute about 15 percent of the mRNA mass during the period of maximum protein accumulation, whereas the mRNA subset for the other 52-kd protein accounts for only 5 percent of the mRNA during this period. When the levels of mRNA are measured in embryos of a wide range of developmental stages, the picture observed is one of coordinate expression characterized by the presence of gene families in very small embryos, rapid increase in their concentration during the period of accelerated embryo growth and precipitous decline during late embryogenesis.

In spite of much recent information about the control of gene expression for storage protein synthesis in developing embryos, little is known about the trigger for gene activation. It is the regulation of this switch that initiates gene action as part of the process of development. A role for the plant hormone abscisic acid (ABA) in the modulation of activity of storage protein genes has been suggested from experiments on the precocious germination of embryos, to be described in the next chapter. In precociously germinating embryos of *Brassica napus,* synthesis of cruciferin is found to continue at a slow rate; however, under conditions that suppress precocious germination such as high osmoticum or presence of ABA in the medium, cultured embryos accumulate this protein to the same extent as embryos *in situ* (Crouch and Sussex, 1981). Synthesis of napin is also slowed down in precociously germinating embryos and reinstated by ABA. When cloned cDNAs from embryos supplied with exogenous ABA are selected by colony hybridization using cDNA prepared from embryos grown with or without ABA, the differential screening results in the identification of clones containing napin gene sequences. Hybridization of the genomic clones with embryo RNA showed that the level of napin mRNA is much higher in embryos cultured in the presence of ABA than in its absence. This observation would agree with a model in which regulation of napin synthesis in embryos by ABA is controlled at the level of mRNA accumulation (Crouch et al., 1983). The action of ABA in inhibiting precocious germination in cultured embryos and embryo axes of *Phaseolus vulgaris* is also paralleled by the synthesis of glycoprotein G1 (Sussex and Dale, 1979; Long, Dale and Sussex, 1981). However, in pea embryos cultured in the presence of a high osmoticum in the medium sufficient to forestall precocious germination, addition of ABA does not enhance accumulation of storage proteins (Davies and Bedford, 1982). Similarly, although ABA prevents precocious germination of cotton embryos, this is not accompanied by maintenance of the levels of 52-kd and 48-kd proteins in the cotyledons (Dure and Galau, 1981). These reports have reduced the clarity of the picture of a role for ABA in the expression of storage protein genes in developing embryos.

Loss of viability of embryos

Thus far we have considered embryogenesis as a positive process that proceeds from the egg to the mature embryo and then comes to a standstill. Although in many species the mature embryo enclosed in the seed is relatively long lasting under suitable storage conditions, it is not in a completely static and unchanging state – it is altering slightly and beginning to lose its viability. Within the context of the biochemical changes taking place during the positive phase of embryogenesis, it is interesting to look at the causes that lead to loss of embryo viability.

Ultrastructural, cytological and biochemical changes occurring in embryos of seeds undergoing deterioration in storage have been followed in an attempt to determine the causes for the loss of embryo viability. As seen in the electron microscope, several disruptive changes are associated with the loss of viability of embryos. These changes include damage to the membrane-lined organelles, dispersion of ribosomes, disappearance of dictyosomes, disintegration of chloroplasts and rupture of lipid bodies and their coalescence into large droplets (Berjak and Villiers, 1972; van Staden, Gilliland and Brown, 1975; Vishnyakova et al., 1976). A point of view that has emerged from a study of the nuclear cytology of cereals is that grain senescence is caused by the accumulation of toxic or mutagenic compounds in the embryo, resulting in chromosomal aberrations and inhibition of growth. This possibility is supported by the finding that the incidence of spontaneous chromosomal damage in isolated embryos of *Triticum durum* aged for 3 years is not significantly different from that detected in embryos of aged intact caryopses. Toxic compounds generated in the endosperm during storage also cause cytological damage independently in the embryo, as shown by the occurrence of a higher frequency of chromosomal aberrations when a young embryo is transplanted onto an endosperm of an aged seed than into the endosperm of a young seed (Floris and Anguillesi, 1974). The reason that chromosomal damage in the embryo should hinder germination is uncertain, but it carries the implication that the repair system is inactive in the cells of the aged embryo and, by the time seeds are allowed to germinate, the damage may be too extensive and beyond repair. A measure of the accumulation of toxic substances in the endosperm is given by the response of a young embryos transplanted onto the endosperm of an aged caryopsis. In this experiment, which involved germination of young embryo encased in the endosperm of caryopses allowed to age for periods up to 5 years, it was found that the growth made by the transplants progressively diminishes with increasing age of the nurse endosperm (Floris, 1970). One of the strongest indications that this is attributable to the synthesis of specific substances in the aged endosperm is the observation that a fraction separated from the methanol extract of aged *T. durum* endosperm causes inhibition of growth and chromosomal aberrations in the cells of normal *T. durum* embryos (Floris, Giovannozzi-Sermanni and Meletti, 1972). The identity of the inhibitory substance in *T. durum* caryopses remains to be established. Inhibitory substances resembling phenolic compounds, absent in viable rice grains, have been identified in their nonviable counterparts (Chatterjee et al., 1976).

Early biochemical studies showed that as seeds age, mitochondrial function is extinguished in the embryos. Consequently, it was speculated that the loss of an ATP-generating system in the embryo is a major factor contributing to the loss of viability. Without discounting the importance of an energy coupling system for growth processes in the embryo, more recent studies, however, indicate that the story is more complex.

An important question is the extent to which the protein-synthesizing machinery of the nonviable embryo remains functional. A major impairment in the protein-synthesizing ability of cells is to be expected as nonviable embryos hardly exhibit any growth during imbibition preparatory to germination. But in which component of the protein-synthesizing machinery does the lesion occur and, from this information, is it possible to make meaningful predictions about the factors responsible for the failure of protein synthesis in nonviable embryos? The possibility that ribosomal and postribosomal supernatant fractions isolated from nonviable embryos are defective has now been supported by experiments in which the ability of mixtures of ribosomal and supernatant fractions from viable and nonviable embryos to synthesize proteins has been determined (Roberts, Payne and Osborne, 1973; Bray and Chow, 1976a,b; Dell'Aquila, de Leo, Caldiroli and Zocchi, 1978). Other experiments (Roberts et al., 1973; Brocklehurst and Fraser, 1980) have focused on the structural and biochemical defects of ribosomal and supernatant fractions of nonviable embryos. Fractionation of total RNA extracted from nonviable rye embryos indicates that the major RNA components of the ribosome, 18S and 25S RNA, are increasingly susceptible to breakdown to low-molecular-weight fragments with decreasing embryo viability. Such disintegration of rRNA is seen as symptomatic of interference by endogenous nucleases. On the other hand, with one major exception, there is no loss in the functional capacity of transfer RNA (tRNA) systems, aminoacyl-tRNA synthetases and transfer factors of the supernatant fraction of nonviable embryos. The exception is in the activity of transfer enzyme (transferase I), which determines the guanosine triphosphate (GTP)-dependent enzymatic binding of aminoacyl-tRNA to the ribosome. The inactivation of this enzyme is also paralleled by the impairment of protein synthesis, suggesting that this may be a lesion in the loss of viability of rye embryos (Roberts and Osborne, 1973).

Available evidence also supports the idea that diminution in the capacity for protein synthesis of nonviable embryos is associated with a decline in the synthesis of new RNA. According to Bray and Dasgupta (1976), there is a preponderance of low-molecular-weight RNA synthesis as well as a much-reduced synthesis of 4S, 5S and rRNA during the first hour of imbibition of nonviable *Pisum arvense* embryos. An analogous situation exists during imbibition of nonviable rye embryos (Sen and Osborne, 1977). The question of the role of any overall damage to the genetic material in the loss of embryo viability, already inferred from cytological observations, has been somewhat slow in being resolved, but it appears that cleavage of DNA into low-molecular-weight fragments occurs during loss of viability of rye embryos in storage (Cheah and Osborne, 1978). These observations offer a basis for the hypothesis that loss of integrity of the genetic material may cause reduced template activity or faulty transcription, leading to the senescence of embryo cells.

Concluding comments

The changes that occur during progressive embryogenesis in angiosperms are clearly important, with many interesting ramifications. Which of the two poles of the unfertilized egg is polarized does not appear to be important, but cytoplasmic organization of the egg after fertilization ensures that the chalazal part containing much of the cytoplasm maintains the overall responsibility of forming the embryo. Until recently, plant embryologists have been preoccupied with the organogenetic part of the embryo, almost to the exclusion of the suspensor. This situation has been remedied, and new information on the structure and biochemical cytology of the suspensor has helped reflect informatively on its possible functions.

Two key cellular events that occur during embryogenesis appear to be endoreduplication and accumulation of gene products – storage proteins – in the cells of the cotyledons. Although the role of endoreplicated DNA in the accumulation of storage proteins has not been established, our understanding of the synthesis of storage proteins at the molecular level has increased considerably in recent years. Other studies, especially in soybeans and French beans (which are beyond the scope of this chapter), have focused on the fine structure of the genes coding for storage proteins. It is not unreasonable to expect that within the next few years genes for storage proteins from most of our economically important crop plants will be sequenced to enhance our understanding of gene expression for their synthesis.

In a sense, loss of viability should be looked upon as part of the developmental program of the embryo that prevents it from attaining immortality. A growing body of evidence suggests that there exists within the genome a program that initiates degradative changes in the embryo with time. An understanding of the basis of these changes will enhance our efforts to forestall the appalling waste that occurs during storage of seeds.

4
Experimental embryogenesis

It is clear from the material presented in Chapters 2 and 3 that until recently the bulk of the effort in the study of embryos has been directed toward describing the developmental, structural and cellular aspects of transformation of the zygote into a full-fledged embryo in an undisturbed normal environment. Within the framework of embryogenic development, it is important to recognize that organs and structures of the embryo that we identify grossly or in histological preparations are the end products of dynamic processes and complex intercellular interactions.

This chapter describes the results of experiments designed to explain why an embryo develops the way it does in its privileged location in the embryo sac. These experiments are based mainly on the perturbation of the normal development of the embryo and, according to the currently popular jargon derived from similar experiments performed on animal embryos, this approach, aimed to discover the mechanism underlying normal embryogenic development, is known as experimental embryogenesis. Broad in scope, it includes not only manipulation of the orderly development of the embryo, but also of somatic cells and pollen grains in culture that results in the formation of embryo-like structures and analysis of embryogenesis at all levels of complexity ranging from molecular to organismal. As developed in this chapter, the focus of experimental embryogenesis is on the manipulation of normal embryogenic development, involving isolation and culture of embryos of various stages of development, embryo morphogenesis in cultured ovules and analysis of mutants impaired in embryogenesis. Other areas of experimental embryogenesis are treated in separate chapters.

Seed embryo culture

The cultivation of embryos excised from ovules and seeds under aseptic conditions in a medium of known chemical composition constitutes the basic methodology of embryo culture. By growing embryos outside the environment of the ovule it is possible to identify the special nutritional requirements essential for continued growth, differentiation and morphogenesis of embryos, which it is difficult to determine while they are enclosed in the ovule or in the seed. Ideally, one would hope to follow progressive embryogenesis *in vitro,* starting with a zygote, but this objective has not been attained. Most of the work has been done on the culture of relatively mature and differentiated embryos excised from quiescent seeds. The logic behind this drive was the hope that a knowledge of the nutrient requirements for growth of excised seed embryos under controlled conditions would provide clues regarding the nature of the storage materials utilized during seed germination. The culture of embryos excised from dormant seeds, as contrasted with those of quiescent seeds, is another area of interest particularly in determining the factors involved in overcoming seed dormancy. A third topic for discussion concerns the culture of proembryos with a view toward determining the changing characteristics of the physiological environment of the embryo sac during the early stages of embryogenesis. It is necessary to emphasize here that the field of embryo culture has a long history; scores of papers dealing with the nutritional requirements of embryos excised from seeds of various plants have been published. Rather than attempt to deal encyclopedically with these studies, the approach taken here is to focus on certain key areas of research designed to elucidate the requirements for growth and differentiation of embryos in culture. Comprehensive treatments of embryo culture are to be found in two recent reviews (Raghavan, 1980; Raghavan and Srivastava, 1982).

The work of Hannig (1904) on the culture of embryos excised from seeds of *Raphanus* and *Cochlearia* is generally accepted as marking the birth of plant embryo culture. Basing his formulation on data from mineral nutrition of plants, Hannig included in his medium macronutrient salts and cane sugar and obtained transplantable seedlings from cultured embryos. In retrospect, reasons for the use of mature embryos excised from seeds rather than immature embryos excised from ovules for the first successful embryo cultures are not difficult to envisage. With a few exceptions, the embryo enclosed in the seed is a fully developed bipolar axis bearing one or more cotyledons. At opposite ends of the axis are found the plumule consisting of the shoot apical meristem and one or two embryonic leaves, and the radicle complete with a root apical meristem covered with a root cap. Therefore, at time of culture, the embryo consists of differentiated primordial organs the cells of which are poised to undergo division and elongation, leading to the progressive development of the shoot and root axes of the seedling. The relatively large size of seed embryos and the ease with which they can be isolated without recourse to micromanipulative techniques have also contributed in a great measure to their use in the pioneering studies.

Culture of seed embryos of various plants into miniature seedlings is now common place. Judging from the inherent overall simplicity of their growth requirements, the main import of these studies is that seed embryos are independent and autotrophic and are able to grow when supplied with a limited diet containing a few inorganic salts and sucrose. The critical stage at which the embryo attains autotrophy is clearly a fundamental property of the species in question. A recent work (Batygina and Vasilyeva, 1981) is revealing in the different stages at which autotrophy becomes apparent in embryos of certain plants; they range from the stage at which the plumule is well developed, as in *Nelumbo nucifera* (lotus) and *Vicia faba,* the stage of scutellum differentiation in *Triticum aestivum,* heart-shaped stage embryo in *Paeonia anomala* and allied species and the stage of globular proembryo in *Dactylorhiza maculata.* In lotus, in which detailed observations have been made, the precise configuration of changes that herald autotrophy includes an altered carbohydrate metabolism in the embryo, digestion of the endosperm and degeneration of the inner integument in the micropylar region of the ovule.

Nutritional requirements

Here we shall consider the importance of some components of the medium such as carbohydrate and nitrogen sources on the growth of cultured embryos. Although supplementation of the growth medium with a carbon energy source is not essential to induce growth of seed embryos, it is the general experience that growth and survival of embryos are markedly enhanced by its presence. Several attempts have been made to evaluate the effectiveness of different carbohydrates for the growth of embryos. Sucrose is unsurpassed, among others, for embryos of *Datura stramonium* (van Overbeek, Siu and Haagen-Smit, 1944), *Capsella bursa-pastoris* (Rijven, 1952), *Hordeum distichum* (Cameron-Mills and Duffus, 1977) and maize (Burghardtová and Tupý, 1980). Even with the ideal sugar in the medium, growth of embryos is not straightforward and, as shown in *H. vulgare,* the response of isolated embryos to sucrose may be markedly influenced by the genotype of the donor plant (Dunwell, 1981b). In cultured embryos of *H. distichum,* the stability of the pool of accumulated sucrose, as well as substrate specificity and sensitivity of sucrose uptake to metabolic inhibitors have suggested that active transport might play a pivotal role in the assimilation of this sugar by embryos *in vivo* also (Cameron-Mills and Duffus, 1979).

This brief mention of the effect of sugar on the growth of embryos cannot help but raise the question as to whether the compound has any morphogenetic effects on cultured embryos. Superficially at least, exogenous addition of an appropriate sugar to the medium appears to promote growth of the root of excised embryos of *Phaseolus vulgaris* × *P. acutifolius* hybrid (Honma, 1955) and *Citrus* (Ozsan and Cameron, 1963), and of both root and shoot of embryos of *Triticum sativum* (Augsten, 1956) and *Elaeis guineensis* (oil palm) (Buffard-Morel, 1968). The failure of a concentration of sugar effective in enhancing root growth to promote shoot growth is attributable to the requirement of a higher sugar optimum by the root, as demonstrated in embryos of *Datura stramonium* (Rietsema, Satina and Blakeslee, 1953), coconut (de Guzman, del Rosario and Eusebio, 1971) and pea (Davis, 1983). The favorable effect of sugar on root growth of coconut embryos is also dependent on the composition of the mineral salt medium, a high nitrogen-containing medium such as Murashige-Skoog medium being particularly effective in this respect (del Rosario and de Guzman, 1978). When we consider the fact that root and shoot primordia are already laid down in the embryo at the time of culture, it is obvious that the role of sugar in embryo growth has to be viewed in a limited context.

Seed embryos are able to thrive moderately well in a medium containing nitrates or ammonium salts as the sole source of nitrogen. However, because amino acids are essential for protein synthesis, considerable research has been undertaken on the relative effectiveness of various amino acids on the growth of embryos. A significant outcome of these studies is the demonstration that the amide glutamine is a relatively efficient source of nitrogen for growth in culture of embryos of several plants (Paris, Rietsema, Satina and Blakeslee, 1953; Rijven, 1956, 1960; Matsubara, 1964). As shown by Rijven (1956), the growth-promoting effects of glutamine are especially striking in comparison with those of asparagine in short-term experiments on cultured embryos. The specificity for glutamine displayed by embryos of *Capsella bursa-pastoris* in this work is fairly well defined, as shown by the observation that glutamic acid or its decarboxylation product, γ-aminobutyric acid does not substitute for glutamine in promoting growth of embryos (Rijven, 1955).

One puzzling effect of amino acids is the existence of mutual antagonism and synergism between different compounds on the growth of embryos. Sanders and Burkholder (1948) called attention to these phenomena with the finding that a mixture of 20 amino acids put together in the proportion present in casein hydrolyzate is as effective as that compound in promoting growth of preheart-shaped and early heart-shaped embryos of *Datura innoxia* and *D. stramonium*. This occurs despite the fact that the mixture consists of readily utilized, partially utilized and toxic compounds. Moreover, the favorable effect of the complete amalgam is not reproduced by incomplete mixtures of amino acids or by single amino acids. Mutual synergism and antagonism between amino acids have also been described in the growth of seed embryos of *Heracleum sphondylium* and of cereals such as *Avena sativa* (oat), barley, wheat, *Sorghum vulgare,* rye, maize and rice (Wright and Srb, 1950; Stokes, 1953; Harris, 1956; Miflin, 1969; Green and Phillips, 1974; Green and Donovan, 1980; Bright, Wood and Miflin, 1978).

Perhaps the most detailed demonstration of interaction between amino acids has been provided in cultured embryos of wheat, barley (Bright et al., 1978) and maize (Green and Donovan, 1980). Growth of these cereal embryos is inhibited by the simultaneous addition of lysine and threonine to the medium; the inhibition is alleviated by methionine, homoserine or homocysteine. From these data it has been proposed that the concerted action of lysine and threonine results from feedback inhibition of enzymes of the aspartate pathway, thus starving the embryo of methionine or its metabolic intermediates. These observations are of exceptional interest for possible selection of feedback-sensitive mutant embryos of cereals.

In embryo culture studies, some attention has also been given to the morphogenetic effects of amino acids. Embryos of *Heracleum sphondylium* attain their nearly normal form in a medium containing arginine, while shoot and cotyledon development is retarded in the presence of glutamic acid. In a medium containing glycine, seedlings appear somewhat stunted and distorted with normal cotyledons (Stokes, 1953). In maize embryos, tryptophan, tyrosine, threonine and homoserine are found to inhibit root growth (Green and Donovan, 1980). Although more or less similar morphogenetic changes may be seen in embryos fed with different amino acids, it follows that there is no necessary correlation between these changes and the metabolism of different amino acids in the embryo.

Hormonal effects

To the embryologist plant hormones such as auxins, gibberellins, cytokinins, ethylene and abscisic acid are of special interest, quite apart from their widely recognized role in the modulation of growth and development of the whole plant. Some of these hormones are undoubtedly present in the endosperm and have occasionally been invoked in the regulation of embryo growth in the ovule. As the endosperm is used up by the developing embryo, restrictions on the behavior of cells of the meristematic regions tighten, resulting in cessation of growth of the embryo. Analysis of the effects of hormones on the growth of seed embryos is thus concerned with situations in which exogenous hormones intervene to control the behavior of cells of the shoot and root meristems, which have essentially become quiescent or dormant.

The effects of auxins on growth and morphogenesis of embryos have been studied extensively, and, despite some unexplained discrepancies, it can be said in a general way that low concentrations of auxins promote growth of embryos and that high concentrations inhibit their growth. In seed embryos with well-differentiated embryonic organs (e.g., *Phaseolus vulgaris*), the plumule has a higher auxin optimum than the radicle with respect to promotion and inhibition of growth; the region of the embryo axis between the plumule and radicle has an intermediate optimum (Furuya and Soma, 1957). A secondary effect of high concentrations of auxin is the production of callus. In recent studies, the factors involved in the formation of calluses from embryos of cereals and grasses under the influence of 2,4-D have received some attention. In maize (Green and Phillips, 1975), oat (Cummings, Green and Stuthman, 1976), wheat (Maddock, Lancaster, Risiott and Franklin, 1983) and *Poa pratensis* (Kentucky bluegrass) (McDonnell and Conger, 1984) genetic differences account for the ability of cultured embryos of more or less the same age to produce differentiating cultures; in maize the response of embryos of different ages to 2,4-D is also dependent on the particular genotype.

In barley (Bayliss and Dunn, 1979) varietal differences, as well as growing conditions of the plant from which embryos are excised, seem to determine the course of auxin-induced regeneration of callus. The source of origin of auxin-induced callus also appears to differ in embryos of different species as well as in embryos of different ages of the same species. In mature oat embryos, the callus originates from the radicle, the scutellum undergoing necrosis, while in very young embryos the entire explant is transformed into a callus without any necrosis of tissues (Cummings et al., 1976). In maize, irrespective of embryo age at excision, calluses are most successfully induced from the scutellum (Green and Phillips, 1975). This organ also appears to be the most proliferative in immature embryos of barley (Dale and Deambrogio, 1979), whereas in more mature embryos growth occurs from the mesocotyl (Granatek and Cockerline, 1978; Bayliss and Dunn, 1979). The ease of regeneration of particular organs of cultured embryos into a callus does not mean that other parts of the embryo are incapable of regeneration; apparently in embryos of different plants, certain cells that have increased potential for uncontrolled growth possess a selective advantage and tend to overgrow cells of other embryonic regions. Although 2,4-D has traditionally been used to induce callus growth in cultured embryos, high concentrations of other auxins such as 3,6-dichloro-2-methoxybenzoic acid (dicamba), 2-(2-methyl-4-chlorophenoxy)propionic acid (2-MCPP), NAA, 4-amino-3,5,6-trichloropicolinic acid (picloram), γ-(2,4-dichlorophenoxy)butyric acid (2,4-DB), 2,4,5-T and 2,4,5-trichlorophenoxypropionic acid (2,4,5-TP) are apparently superior to 2,4-D for callus growth in wheat embryos (Zhou and Lee, 1984). As we shall see in Chapter 5, calluses regenerated from embryos cultured in the presence of auxin have high morphogenetic potencies and differentiate into embryo-like structures.

The effects of GA on the growth of cultured seed embryos fit the expectations from experiments with seedling plants and are reminiscent of symptoms of precocious germination. One of the dogmas of GA action on plants is that roots are relatively insensitive to this hormone; however, in cultured embryos of *Capsella bursa-pastoris*, GA promotes elongation of the root meristem and production of lateral roots (Veen, 1963). If one considers the effects of GA on embryos of *Capsella* of different ages, it becomes apparent that older embryos produce roots with abundant laterals whereas younger embryos produce roots with fewer laterals (Raghavan and Torrey, 1964). Differences in GA sensitivity are also noted in embryos of cotton (Dure and Jensen, 1957). The only noticeable effect of the hormone on immature embryos (37 mg fresh weight) is in enhancing cotyledon maturation and cell elongation; on the other hand, GA accelerates cell division, cell elongation and axis growth in older embryos (64 mg fresh weight). The underlying basis for age-dependent sensitivity of cultured embryos to GA has not been determined. A reasonable guess is that the hormone can induce growth only in embryos in which meristems are sufficiently well developed.

Only scattered observations on the morphogenetic effects on embryos of cytokinins have been reported. An interesting modification of embryo growth related to the possible role of cytokinins is to be found in the effects of the purine, adenine, which causes a spectacular growth of callus from the scutellum of cultured embryos of *Pennisetum typhoideum* (Narayanaswami, 1959). Although the shoot and root axes of the embryo do not respond to adenine, extirpation of the shoot meristem eliminates the effects of adenine on the scutellum. In intact embryos, adenine-induced callus growth is enhanced by the addition of IAA to the medium. From these observations it appears that, rather than acting alone, adenine interacts with some substance diffusing from the shoot apex, possibly an auxin, in causing scutellar outgrowth on cultured embryos. Following the somewhat inconclusive data of Veen (1963) on kinetin-induced inhibition of root growth in cultured embryos of *Capsella bursa-pastoris*, Raghavan and Torrey (1964) showed that the degree of inhibition of root growth depends on the age of embryos and conditions of culture. In dark cultures, relatively low concentrations of kinetin, which inhibit root elongation in heart-shaped embryos, are only marginally inhibitory in slightly older embryos and without effect in torpedo-shaped and mature embryos. In light, kinetin inhibits root elongation in embryos of all ages but promotes precocious leaf expansion and callus growth. In cultured barley embryos kinetin has been shown to induce the formation of embryo-like structures from the region of the epiblast (Norstog, 1970).

The era of evaluating the nutrient and hormonal requirements of cultured embryos of quiescent seeds is now past, and it is worth summing up the overall results. The major conclusion is that the seed embryo is a remarkable evolutionary structure that can grow into a normal seedling in the absence of crucial exogenous nutrients or hormones. Although addition of hormones undoubtedly modifies growth of cultured embryos, no evidence yet exists that demonstrates or excludes the involvement of a specific hormone in the morphogenesis of embryos of a wide range of plants. In an operational sense, there is little difference between the effect of hormones on embryonic organs and corresponding organs of a seedling plant, so it seems reasonable to expect that mechanisms of hormone action in cultured embryos may include those that are apparent in seedling plants.

Embryo culture and seed dormancy

In contrast to quiescent seeds that germinate when they are provided with adequate moisture, sufficient oxygen for normal aerobic metabolism and a temperature within physiological limits, dormant seeds fail to germinate under these conditions, or do so with great difficulty. Viewed in physiological terms, dormancy may be imposed in seeds by a requirement for special temperature or light regimens or by mechanical causes such as the presence of a hard seed coat, immaturity of the embryo or the failure of water and gases to penetrate the seed coats. Inability of excised embryos to germinate under conditions favorable for germination generally indicates that dormancy is not attributable to failure of entry of water or gases or to mechanical resistance of seed coats but is the result of an inherent physiological state of the embryo. In many plants, this type of dormancy may be overcome by storage of seeds under special temperature conditions, a treatment known as afterripening or by provision for a specific wavelength of light.

This brings us to the main concern of embryo culture in seed dormancy and an important question in experimental embryogenesis in general. What is the role of growth hormones in imposing developmental arrest of the embryo and in terminating it? There is impressive evidence to show that GA and ABA intervene to regulate dormancy, but the mechanism of their action remains enigmatic. For example, addition of GA to the medium has been shown to promote germination of embryos excised from nonafterripened or partially afterripened seeds of *Euonymus europaeus* (Beranger-Novat and Monin, 1971), *Corylus avellana* (hazel) (Jarvis and Wilson, 1977), *Pyrus malus* (apple) (Côme and Durand, 1971) and *Sorbus aucuparia* (Bianco and Bulard, 1977). The inference from this work is that the synthesis of endogenous gibberellins in the embryo is an essential part of dormancy-breaking mechanism in seeds that require afterripening. This is also borne out by the discovery of significant increases in GA content of seeds and embryos during afterripening (Isaia and Bulard, 1978; Côme and Thévenot, 1982). Experiments on the growth of cultured embryos of dormant and nondormant strains of *Avena fatua* (wild oat) have provided a slightly different picture of the role of endogenous GA in dormancy regulation. According to Andrews and Simpson (1969), embryos excised from dormant seeds fail to germinate in a liquid medium in which embryos of nondormant strains germinate. Since addition of GA to the medium induces germination of dormant embryos, their inability to germinate under normal conditions has been interpreted as due to the action of an inhibitor that blocks the availability of GA. The finding that dormant embryos permitted to percolate on an agar medium germinate even in the absence of GA is in agreement with this interpretation.

Information relating to the site and mode of action of GA in overcoming embryo dormancy has stemmed from the work of Jarvis and Wilson (1977) on hazel. The observation that GA stimulates growth of isolated embryos when applied directly on the embryo axis rather than on the cotyledons would seem to indicate that the hormone has a direct effect on the former in breaking dormancy. This is also consistent with the data that labeled GA is transported less readily from the cotyledons to the embryo axis than in the reverse direction. Later work (Jarvis, Wilson and Fowler, 1978) showed that embryo axes cultured in a medium containing sucrose and inorganic ions germinate readily in the absence of GA. The mode of action of GA has accordingly been construed to provide carbohydrates for axis growth perhaps by influencing sucrose metabolism. A delayed action of GA in breaking dormancy is assumed to be localized in the cotyledons, where reserved lipids are possibly converted into sucrose.

Although various attempts have been made to determine the extent of involvement of cytokinins in overcoming embryo dormancy, the results have been equivocal. On the other hand, there is a strong case for the involvement of ABA in some aspect of embryo dormancy, based mainly on the effects of this hormone on the growth of isolated embryos and changes in the levels of endogenous ABA in the embryo during afterripening of seeds. In embryos of afterripened seeds of apple (Rudnicki, Kamiński and Pieniażek, 1971; Durand, Thévenot and Côme, 1973), the dormancy syndrome is reinstated by treating them with ABA and, not surprisingly, the effects of ABA are reversed by GA or to some extent by cytokinins. The basis for growth inhibition of embryos by ABA has been substantiated by the finding that during afterripening of seeds there is a progressive decrease in free ABA content of embryos (Barthe and Bulard, 1978). The role of ABA in dormancy regulation in hazel is more difficult to explain. As shown by Jarvis and Wilson (1978), nondormant embryo axes are somewhat refractory to the inhibitory effects of ABA; a dose of this hormone, which prevents the growth of dormant embryo axes, is ineffective on nondormant embryo axes. Moreover, during dry storage of seeds, which induces embryo dormancy, there is a transient increase in ABA concentration in the embryo axes followed by a subsequent decline so that level of ABA in the fully dormant embryo is less than that of nondormant embryo. The significance of the initial rise of ABA level in the embryo axis is conjectural, although it has led to the general conclusion that once the dormant state has been achieved by a high level of ABA, the latter may not be necessary for maintenance of dormancy (Jarvis and Shannon, 1981).

Within the context of ABA action in maintaining domancy of the embryo, it seems likely that the inhibitor is localized in a specific part of the embryo. For this reason, attempts have been made to culture dormant embryos from which certain parts are surgically removed. In studies conducted along these lines, it has been shown that dormant embryos of *Fraxinus excelsior* (Bulard and Monin, 1963), apple (Thévenot and Côme, 1973) and hazel (Jarvis, 1979) cultured without cotyledons germinate normally.

The information presented above is a cross section of the work using excised embryos to study the regulatory processes and mechanisms during development, maintenance and release of dormancy in seeds. The specific contributions of experiments involving hormone applications to seeds in order to regulate dormancy and germination have not been included in this discussion, but these studies, as with isolated embryos, have merely indicated that growth-promoting and growth-inhibiting hormones play a vital role in dormancy regulation. These investigations have led to schemes proposed to explain embryo development, seed dormancy and germination in terms of changing levels of specific hormones. In one scheme, which has been the subject of some discussion (Villiers and Wareing, 1965; Wareing and Saunders, 1971) it is assumed that a growth-promoting hormone such as auxin is apparently utilized by the embryo during the early period of its development, whereas accumulation of an inhibitor such as ABA is the basis for induction and maintenance of dormancy of the embryo in the mature seed. Release from dormancy is thought to involve the disappearance of a large part of the inhibitor and the appearance of other growth-promoting hormones, such as GA or cytokinins, which potentiate embryo growth.

Close scrutiny of the scheme has revealed its inadequacies to account convincingly for the developmental strategies of embryos of a wide variety of angiosperms (Wareing, 1982). First, compared with the limited number of plants in which a correlation between the levels of endogenous growth hormones and state of development of the embryo has been established, there is a considerably large number of plants in which such a correlation has not been sought or is nonexistent. Second, analysis of endogenous hormone levels in many cases has been complicated by the use of seeds instead of embryos, as this has created uncertainty over what proportion of the total hormone is synthesized in the embryo and what proportion is transported from other parts of the seed. Third, we simply do not know whether young embryos have evolved sophisticated mechanisms for the synthesis of growth hormones. In summary, we can say that there is much ground to be covered combining embryo culture techniques with molecular

94   *Experimental embryogenesis*

and biochemical approaches to unravel the role of hormones in the growth, developmental arrest and germination of embryos.

We shall conclude this discussion with a reference to dormancy due to the presence of underdeveloped embryos exhibited by seeds of certain angiosperms. The classic examples of seeds with rudimentary embryos at the time of shedding are the orchids, and the considerable work done on the germination of orchid seeds has been recently reviewed (Arditti, 1979). However, it is doubtful whether orchid seeds with their diminutive embryos exhibit any true dormancy, as the seeds begin to germinate easily in association with a mycorrhizal fungus or in a sugar-enriched medium in the laboratory. More than 40 genera of angiosperms are known in which seeds exhibit a form of dormancy induced by the presence of rudimentary embryos at maturity (Martin, 1946). The only example of this kind in which dormancy mechanisms have been well studied by embryo culture technique is *Ilex;* the key experimental results and conclusions from this work may be summarized as follows. Hu (1975) found that heart-shaped embryos excised from seeds of several *Ilex* species resume embryogenic growth immediately after culture in a mineral salt medium containing sucrose, as compared with a period of nearly 1 year required for completion of embryogenesis in the seed under normal germination conditions. On this basis it has been suggested that inhibitors present in the endosperm are responsible for the arrested growth of the embryo in the seed. In accordance with this suggestion, it was found that an embryo cultured besides its own endosperm achieves little growth. A remote possibility that rudimentary embryos are unable to utilize nonreducing sugars was discounted by the demonstration that these sugars substitute for sucrose in inducing embryo growth in culture (Hu, Rogalski and Ward, 1979). Another interesting observation, shown in Figure 4.1, is that growth of embryos in culture is inhibited by light (Hu, 1976); from this, the main conclusion pertinent to our discussion is that in addition to inhibitors of endospermic origin, light-sensitive factors present in the embryo also mediate in its dormancy.

Embryo culture, comparative morphology and morphogenesis

The value of evidence from embryo culture investigations to the interpretation of the morphology of embryonal parts has not been fully appreciated. The chief problem is the difficulty of reconciling conventional morphology with the apparently atypical behavior of tissues and organs under experimental conditions. As we have seen in Chapter 2, graminean embryos are characterized by the presence of such special structures as coleoptile, scutellum, epiblast and coleorhiza, which set them apart from other monocots. Neither the coleoptile nor the scutellum is very well understood morphologically, and opinion is divided as to whether they are homologous to

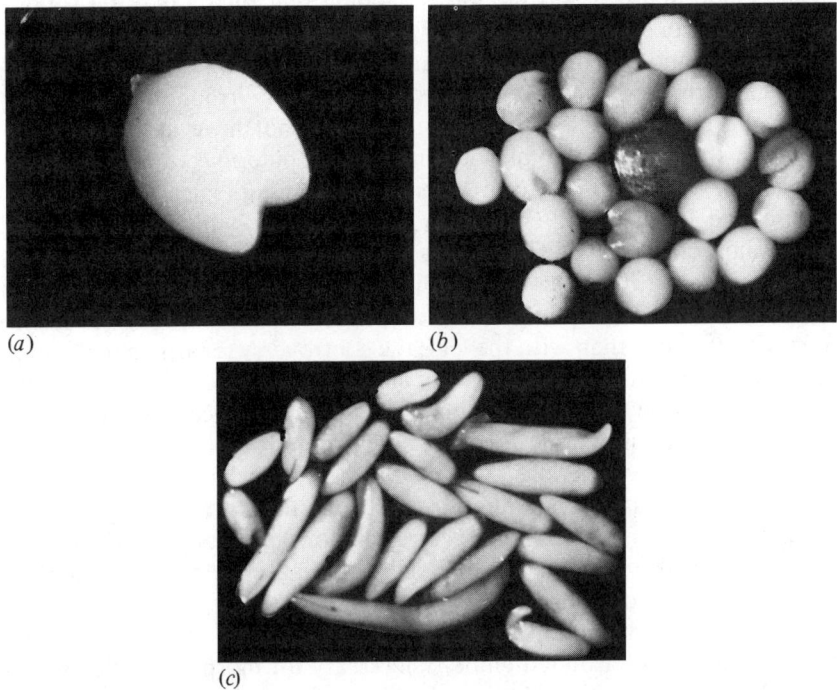

FIGURE 4.1. Growth of embryos of *Ilex opaca* in culture. (*a*) Excised embryo before culture. (From Hu, 1976.) (*b*) Embryos cultured under a 16-hour photoperiod for 11 days. (*c*) Embryos cultured in complete darkness for 11 days. (Photographs supplied by Dr. C. Y. Hu.)

foliar organs. An important study in this context is that conducted by Norstog (1969), who found that when barley embryos are grown in a medium containing kinetin, greening of the coleoptile and scutellum is accompanied by the appearance of typical foliar hairs on their abaxial epidermal surface. The fact that these controversial structures of the embryo can express leaf characteristics under experimental conditions is considered strong evidence of their foliar nature. The epiblast and coleorhiza have also been variously interpreted by morphologists. The former is often considered equivalent to a rudimentary second cotyledon, whereas the latter is identified as part of the suspenser or of the hypocotyl, or, more realistically, as the suppressed primary root (Esau, 1965). Some observations on the behavior in culture of embryos of normal wheat grains and wheat grains subjected to massive doses of γ-irradiation (γ-plantlets) are of validity in the interpretation of the morphology of the epiblast and coleorhiza. With the knowledge that leaf hair originates as an unequal division of the epidermal cell and the understanding that such a division is not a prerequisite for root hair formation, Foard and Haber (1962) found that the epiblast and coleorhiza of cultured wheat embryos occasionally produce hairs that resemble root hairs. Moreover, γ-irradiation, which reduces the number of leaf hairs formed, does not have any detrimental effect on the developmental potential of the epidermal cells of the root, epiblast or coleorhiza to form hairs. At the physiological level, the response of coleorhiza and epiblast to IAA, GA and 2-chloroethyltrimethyl ammonium chloride (CCC) appears to be strikingly different from those displayed by the first leaf. Although these results support the contention that the epiblast and coleorhiza are closely related and have more in common with the root than with the leaf, the controversy regarding the homology of these organs continues (Negbi and Koller, 1962).

Successful growth of isolated embryos in culture is conditioned by the presence of the embryonic root and shoot primordia, buttressed in some cases by massive cotyledons. While culture of the whole embryo permits the expression of latent potentialities of its organs, it is relevant within the present context to consider the reverse relationship, that is, whether morphogenesis of the embryo can proceed in the absence of certain organs. Among the questions that address the relationship of the different parts of the embryo to its morphogenesis in culture are the following: Is complete organization of the isolated embryo necessary for morphogenetic expression? To what extent does removal of parts of the embryo influence the development of the remaining embryo? How may such morphogenetic patterns be interpreted? Partial answers to some of these questions have been obtained from an analysis of the growth in culture of surgically operated embryos.

A principal interest in these investigations is the role of the scutellum and cotyledons in the growth of cultured embryos of grasses and dicots, respectively. The experiments that involved culture of embryos from which the scutellum or cotyledons were surgically removed led to the view that continued growth of the embryo axis depends on the presence of these embryonic organs. The role of the scutellum established in some early work on the culture of embryos of *Zea mays* (Andronescu, 1919) has been reinforced in a study of more recent vintage (Tilton, 1981b). This latter study has shown that the growth of the hypocotyl is influenced by the scutellum and that removal of the hypocotyl impedes growth of the rest of the embryo. These effects of the scutellum and hypocotyl are thought to be caused by the production of a growth stimulus by these embryonal parts, but the nature of the stimulus needs to be established. Among dicots, some interesting observations on the morphogenesis of decotylated embryos of *Cassytha filiformis* have been made by Rangaswamy and Rangan (1971). The unusual feature of the embryo of *Cassytha* is the presence of two massive cotyledons that enclose the embryo axis and that can be fully or partially severed without injury to the latter. By culturing mutilated embryos following various decotylating patterns, these investigators showed that the amount of cotyledonary tissue extirpated from the embryo determines the extent of growth attained by the plumule. Generally, the larger the portion of the cotyledon removed, the more inhibited was the growth of the plumule. The role of cotyledons in the growth of French bean embryos was formalized as a quantitative concept in a study in which the increase in fresh weight of the embryo axis during culture is found proportional to the amount of cotyledonary tissue left on it. Two possible explanations for the effects of cotyledons are that they either supply critical metabolites for the growth of embryo axis or provide surface area for the uptake of nutrients. The former explanation is favored for growth of French bean embryo, since the embryo axis displays enhanced growth when planted near a cotyledon in the same medium or when the medium is fortified by a homogenized brei of the cotyledons (Monnier, 1982). However, a series of experiments on the growth of partially afterripened apple embryos oriented in various positions in the medium appear to support a role for the cotyledons in transmitting nutrients to the elongating embryo axis. Here, growth of the embryo axis is appreciably accelerated when it is planted with cotyledons in contact with the medium; by contrast, orientation of the embryo in various other positions such as with only one end in contact with the medium slows down growth (Thévenot and Côme, 1971).

These studies indicate that as in the seedling plant, correlative influences from other parts determine the final form of the embryo in culture. Work completed to date has done little more than suggest that these correlations are mediated by chemical stimuli. But, in surgical interruption of the embryo system, it would be unwise to look upon the extirpated part as the source of the stimulus without taking into account the role of wound hormones which obviously originate in the part undergoing morphogenesis.

Proembryo culture

Proembryos bathed in the liquid endosperm in the embryo sac are different from seed embryos in that nutrient substances of the endosperm can freely diffuse into proembryos to sustain their growth. Although utilization of metabolites of endospermic origin by the growing proembryo has not been satisfactorily proved, the close association between the proembryo and the liquid endosperm probably justifies the view that proembryos lack biosynthetic capacities and are therefore heterotrophic in nature.

The small size of proembryos and their heterotrophic dependence on the endosperm present an interesting contrast to the large, autotrophic seed embryos and suggest that nutritional requirements for *in vitro* growth of proembryos are bound to be more exacting than those of seed embryos. Central to this issue is the nature of additives to the medium necessary to support survival and continued growth of proembryos. Considering the interest in the culture of proembryos, it is not surprising that deviations occur at different levels in the requirements for growth in culture of proembryos of different plants. They manifest themselves as requirements for special nutrients of endospermic origin, modifications of physical conditions of culture, provision for hormones and other substances and the need for an attached suspensor.

Effect of coconut milk and other plant extracts

The history of proembryo culture divides rather strikingly into two periods – the first up to 1940 and the second dating from 1940 to the present. Although the design of nutrient media for proembryo culture proceeded along empirical lines during the pre-1940 epoch, only limited success was obtained in attempts to culture proembryos of selected plants in media containing IAA, amino acids, yeast extract, casein hydrolyzate and other complex products. The implication of these results is that the optimum conditions for *in vitro* growth of proembryos are those that imitate closely the composition of the endosperm or the milieu of the embryo sac (reviewed by Raghavan, 1976b).

The strategic lead in the culture of proembryos in the post-1940 period came from the work of van Overbeek, Conklin and Blakeslee (1942) on the physiology of growth of *Datura stramonium* embryos. These workers found that whereas torpedo-shaped and heart-shaped embryos of *Datura* grow well in a mineral salt medium supplemented with vitamins, glycine, succinic acid and adenine, still smaller embryos (200 to 500 μm long) fail to grow or grow feebly before turning into undifferentiated callus. A dramatic increase in the growth of these embryos occurs when the medium containing organic addenda is enriched with nonautoclaved coconut milk (Fig. 4.2). A phenomenon observed in this work was the hormonal nature of the substance from coconut milk, which was designated an embryo factor. By repeated fractionation and elimination of toxic principles, the embryo factor was obtained in a sufficiently pure form to promote growth of embryos in a dilution as low as 1 : 19,000 on a dry weight basis as compared with 1 : 110 for the crude milk (van Overbeek, Siu and Haagen-Smit, 1944). van Overbeek and colleagues did not attempt to characterize the active principles in the embryo factor; as shown by subsequent analysis of coconut milk, *myo*-inositol, auxins and cytokinins probably constitute the bulk of the active principle. Since this pioneering study, periodic reports of successful growth of proembryos of other plants secured by the addition of

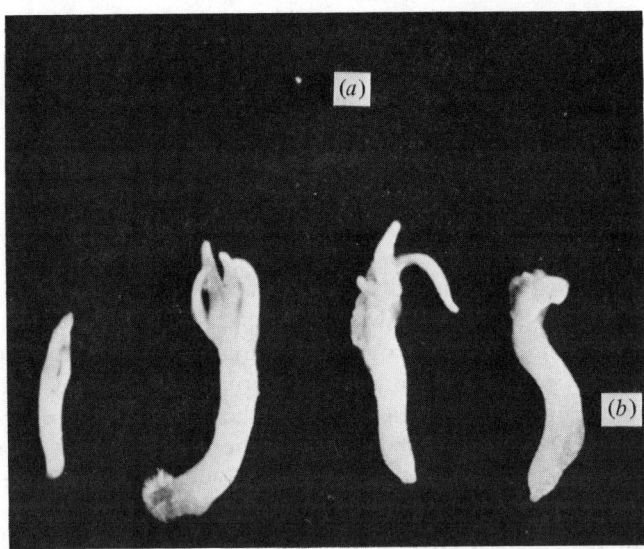

FIGURE 4.2. Effect of coconut milk on the growth of proembryos (140-μm diameter) of *Datura stramonium*. (*a*) Original size of the embryo. (*b*) Embryos of the same size cultured on an agar medium containing mineral salts, sucrose, vitamin mixture and nonautoclaved coconut milk. (From van Overbeek et al., 1942.)

water extracts of date, banana, wheat gluten hydrolyzate and tomato juice (Kent and Brink, 1947), diffusates from young seeds of *Lupinus luteus* and *Sechium edule* (Matsubara, 1962) and endosperm of *Cucumis sativus* (cucumber), *Cucurbita maxima* and *C. moschata* (pumpkin) (Nakajima, 1962) have appeared.

Synthetic media have been concocted by some workers to replace the undefined factors of coconut milk and other plant extracts with chemically defined substances. Norstog (1961) showed that undifferentiated barley embryos which generally fail to survive in White's medium can be successfully cultured by supplementing it with coconut milk. In later studies the requirement for coconut milk for growth and enhanced survival of barley embryos was met by manipulations of the mineral salt component and organic addenda of the medium (Norstog and Smith, 1963; Norstog, 1973). In another study, growth of proembryos of cucumber and pumpkin induced by their endosperm extracts was duplicated by a mixture of IAA, 1,3-diphenylurea and casein hydrolyzate (Nakajima, 1962). These results, which are restricted in the number of species studied, nevertheless represent important steps toward the formulation of completely synthetic media for the culture of proembryos.

Osmotic environment of the embryo sac and culture of proembryos

The inability of proembryos of certain plants to grow in media supplemented with substances of the embryo factor type has given rise to a consideration of the possible osmotic significance of the chemical environment of the embryo sac in embryo growth. It has been known from direct measurements of the osmotic concentration of the ovular sap that the liquid endosperm surrounding the proembryo has a very low (more negative) osmotic potential value that substantially decreases (becomes more negative) toward embryo maturity (Ryczkowski, 1960; Smith, 1973; Yeung and Brown, 1982). Such observations have suggested that proembryos of most if not all plants might be potential targets of growth regulation by the high osmotic pressure in the liquid endosperm rather than by the alleged embryo factors present in it. In regard to the *in vitro* culture of proembryos, this means that cells of proembryos being osmotically very active will take up water and expand in a medium of low osmolality; consequently growth processes may be hindered unless the nutrient medium is isotonic with the osmotic value of the cell sap. Paradoxically, the first clear demonstration of a regulatory role of low osmotic potential value of the nutrient medium in the growth of proembryos came from work on the growth of embryos of *Datura stramonium* (Rietsema, Satina and Blakeslee, 1953). Whereas mature embryos thrive in the complete absence of sucrose in the medium, earlier stages of embryos require progressively higher concentrations, until globular or pre-heart-shaped embryos require a medium containing 8 to 12 percent sucrose for continued embryogenic development. Interestingly enough, globular or pre-heart-shaped embryos respond exactly the same way to a medium containing 2 percent sucrose plus enough mannitol to be isotonic with 8 to 12 percent sucrose. This finding augments the possibility that high sucrose is functioning as an osmotic stabilizer, rather than as a carbon energy source. The same relationship has also been reported in the growth of proembryos of *Capsella bursa-pastoris* (Rijven, 1952; Veen, 1963), *Datura tatula* (Matsubara, 1964), flax (Preťová, 1974) and pea (Stafford and Davies, 1979). In physiological terms, the regulatory effect of high osmolality of the medium in the growth of cultured proembryos may be interpreted as a consequence of a more effective control of the flow of metabolites and inorganic ions into cells.

Hormonal control of growth of proembryos

An interesting example relevant to the role of hormonal substances is to be found in the growth *in vitro* of proembryos of *Capsella bursa-pastoris*. Although earlier workers (Rijven, 1952; Veen, 1963) cultured heart-shaped embryos in a liquid medium of high osmolality secured by the addition of 12 to 18 percent sucrose, Raghavan and Torrey (1963) were able to grow embryos of the same age in a relatively simple medium containing inorganic ions, vitamins and 2 percent sucrose solidified with agar. To grow still smaller globular embryos it was necessary to supplement this medium with a balanced mixture of low levels of IAA, kinetin and the related purine adenine. The addition of the hormone mixture eliminated the need for a medium of high osmolality, suggesting that the picture of osmotic regulation of growth of proembryos is somewhat misleading. Nonetheless, the need for a properly balanced mixture of hormones for the growth of the tiny embryos is in a large part replaced by supplementary osmoticum or by an increase in the concentration of major salts by a factor of 10. The role of high salt concentrations in the medium for growth of proembryos of *C. bursa-pastoris* has been brought into sharper focus by the work of Monnier (1976a). From a comparative analysis of different inorganic salt media, it was found that the high salt medium of Murashige-Skoog is superior although the small number of proembryos that survives in culture made the use of this medium a frustrating exercise. In further work, by determining the most effective level of each macronutrient and micronutrient salt of Murashige-Skoog medium for supporting embryo growth, a modification of this medium with increased levels of Ca and K and a decreased concentration of NH_4NO_3 that led to excellent growth and survival of proembryos was concocted. Enhanced growth and survival of proembryos were also obtained by supporting them on a bed of polyacrylamide instead of agar (Monnier, 1975) or by increasing the partial pressure of O_2 in the medium (Monnier, 1976b). If an extension of embryo culture work to include culture of still smaller early division phase embryos and ultimately the single-celled zygote is justified, this recognition of the role of both chemical and physical conditions of culture of proembryos offers a means of pursuing the study with a new series of experimental variables.

A role for the suspensor

In Chapter 3 we saw that cells of the suspensor possess structural modifications for the absorption of nutrients from the surrounding endosperm. A central question that has opened up in the analysis of growth requirements of proembryos of certain plants is whether there is a causal relationship between the presence of a suspensor and growth of the organogenetic part of the embryo. The simple approach to interfering with the influence of the suspensor is to sever its connection with the rest of the embryo and follow the growth of the latter. In experiments of this kind, it has been shown that continued growth of early heart-shaped embryos of *Eruca sativa* (Corsi, 1972) and *Phaseolus coccineus* (Nagl, 1974) demands the presence of an attached suspensor. Confirming the importance of the suspensor, Yeung and Sussex (1979) showed that in *P. coccineus,* even the suspensor detached and kept in close contact with the embryo in the culture medium is effective in promoting growth of heart-shaped embryos. In the case of *Capsella* it has been claimed that the suspensor enhances growth in culture of small embryos by acting as a cellular barrier against leakage of substances from the cells of the embryo and by preventing the entry of high concentrations of mineral substances into the embryo (Monnier, 1984).

Another line of evidence has shown that whereas intact precotyledonary stage embryos of *P. coccineus* grow in a mineral salt medium, chance of survival of embryos of the same age deprived of the suspensor is very remote, unless GA is present in the medium. GA, however, inhibits growth of suspensor-deprived postcotyledonary-stage embryos as compared with intact embryos of the same age grown in a hormone-free medium (Cionini et al., 1976; Yeung and Sussex, 1979). These responses of embryos to GA are correlated with a falling level of endogenous GA in the organogenetic part of precotyledonary-stage embryos and a rising level in the suspensor. In the postcotyledonary-stage embryos, as the suspensor undergoes senescence, it shows a sharp decline in the level of GA with a corresponding increase in level of the hormone in the embryo proper (Alpi et al., 1975). Further studies have demonstrated that whereas zeatin or zeatin riboside favors growth of suspensor-deprived precotyledonary-stage embryos of *P. coccineus,* they scarcely promote growth of postcotyledonary-stage embryos (Bennici and Cionini, 1979). Cytokinin autonomy in the latter is apparently attributable to the abundance of active cytokinins in the organogenetic part of the embryo, while in the precotyledonary-stage embryos active cytokinins are mainly found in the suspensor, and not in the embryo proper (Lorenzi et al., 1978). Overall, these results implicate the suspensor in the growth of proembryos of *P. coccineus,* its assigned function being the supplier of GA and cytokinins.

Growth of early division-phase embryos in cultured ovules

Apart from the need to formulate a suitable nutrient medium, one deterrent to the culture of early division-phase embryos and the zygote is the difficulty of isolating them from the ovule. From this point of view, the use of ovule culture has proved rewarding as an alternate method to study growth requirements of early division-phase embryos. A number of reports have convincingly demonstrated that ovules cultured at the two-celled proembryo stage or zygote stage complete normal development in culture resulting in viable seeds. The nutrient requirements for growth of ovules appear to be surprisingly simple and generally include a source of amino acids or hormones (Raghavan, 1980). Rarely, as shown in *Petunia hybrida,* the source of iron supplied in the medium is critical in initiating embryo development in cultured ovules (Niimi, 1974).

In cultured ovules of certain plants, growth of the extraembryonal tissues overtakes the growth of the enclosed embryo, so that perfectly respectable seeds with underdeveloped embryos result. This problem has been analyzed in cotton ovules with a view to formulate a medium for successful culture of ovules at the zygote stage. When 6-day postanthesis ovules enclosing 12-celled proembryos are cultured on White's medium supplemented with kinetin, the latter grow slowly and attain about one-third the size of *in situ* embryos (Joshi and Johri, 1972). In another study (Eid, de Langhe and Waterkeyn, 1973), somewhat similar results were obtained with 5-day postanthesis ovules bearing two- to ten-celled proembryos cultured in Murashige-Skoog medium. This work indicated, however, that the response of embryos enclosed in the ovules depends on the availability of moderate amounts of NO_3^- and NH_4^+ in the medium. On the basis of this observation, it was subsequently shown that ovules containing even the fertilized egg can be cultured with reasonable degree of success in a medium containing low concentrations of IAA, kinetin and GA along with 15 mM NH_4^+ ion, leading to the production of a large number of mature embryos (Fig. 4.3). Evaluation of the response of cultured ovules in different media showed that the poor development of the ovule in the basal medium results in space restrictions for growth of the embryo and the eventual failure of the latter to attain normal size. The addition of hormones enables the ovule to grow to its usual *in vivo* size, whereas NH_4^+ is critical for the continued division and growth of the zygote (Stewart and Hsu, 1977).

Various possible explanations for the successful growth of early division-phase embryos and zygotes enclosed in cultured ovules must be considered. First, it is possible that the extraembryonal tissues of the ovule supply some nutrients for the growth of the enclosed embryo or that nutrients absorbed by the ovule are easily translocated to the embryo. Second, injury at excision to the tiny embryo is avoided in the ovule culture method. Third, ovule culture offers no room for a possible osmotic shock that young embryos experience when they are removed from the embryo sac. However, the advantages of *in ovulo* embryo culture must be balanced against the fact that information derived from this approach has not defined the chemical or physical gradients that control the precise sequence of development of the zygote into an embryo.

Precocious germination

The developmental anomaly known as precocious germination of embryos merits brief consideration. Depending on the age of embryos at excision,

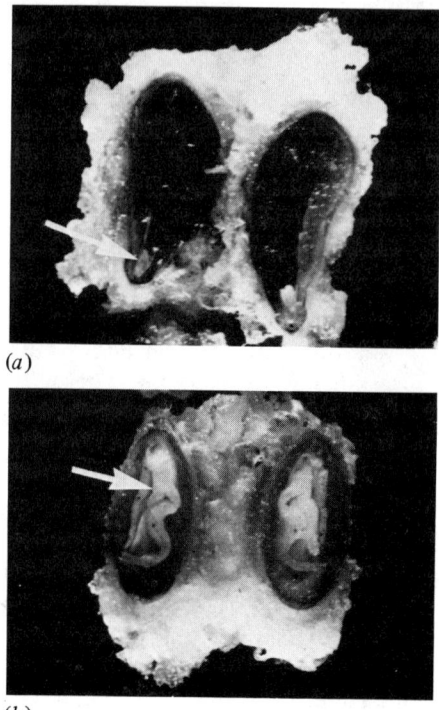

FIGURE 4.3. Growth of embryos in cultured ovules of cotton. Ovules are bisected to show embryos. (a) Growth in the basal medium supplemented with IAA, kinetin and GA. (b) Growth in the basal medium supplemented with IAA, kinetin, GA and NH_4^+. Arrows point to embryos. GA, gibberellic acid; IAA, indoleacetic acid. (From Stewart and Hsu, 1977; from negatives supplied by Dr. J. McD. Stewart.)

their growth in culture may take one of two paths, or even a combination of both. Mature embryos grown in culture generally exhibit a gradual increase in size and a proportionality of growth of the different organs to produce seedlings that look normal in every respect. This pattern of growth is attributable to the activity of meristems and primordia producing a population of small, nonvacuolate cells that undergo limited elongation and is sustained by food materials present in the embryo at the time of excision. In this respect, the growth of a mature embryo in culture is not unlike that seen in an embryo during germination of the seed. By contrast, cultured immature embryos inexplicably skip the later part of the embryogenic program and grow into rudimentary, weak seedlings by a process known as precocious germination. This apparent dichotomy in the growth patterns of mature and immature embryos in culture has repeatedly been noted since the early days of embryo culture investigations and has elicited frequent comment. In developmental terms, during precocious germination, the normal process of embryogenesis that would lead to a period of arrested growth of the embryo in the seed is replaced by an alternate developmental program that leads to germinative growth.

Most plants undergo a vulnerable period during embryogenesis when embryos germinate precociously; those cultured earlier to this sensitive period fail to grow, whereas older embryos germinate normally. As shown in *Phaseolus coccineus* and *P. vulgaris,* a combination of developmental and biochemical criteria offers the most meaningful way of expressing the germination potential of embryos. In these species, the critical period during which embryos develop the capacity to germinate precociously in culture is 16 to 20 days after anthesis; the developmental stage represented by this time scale corresponds to embryos poised to undergo major increases in fresh and dry weights and RNA content (Walbot, 1971; Walbot, Clutter and Sussex, 1972). A more precise means of detecting the germination instincts of cultured embryos, as opposed to embryogenic development, is to monitor the appearance or disappearance of specific proteins. For the work on embryos of *P. vulgaris,* the storage protein phaseolin was selected and its activity demonstrated by immunoelectrophoresis of radioactively labeled homogenates. In embryos and embryo axes younger than 20 days, the concentration of phaseolin is low but increases continuously during progressive embryogenesis and drops off during germination. Behavior in culture of excised embryo axes showed that while mature 29- to 30-day old axes germinate normally within 1 day after culture, 20- to 28-day old axes display a lag period before germination. However, embryo axes continue to synthesize phaseolin during the lag period, indicating that they are completing the embryogenic program in culture rather than undergoing slow germination (Long et al., 1981). These observations suggest that embryos may rely on the same biochemical marker to signal the onset of germination or continuation of embryogenic development.

Continuation of embryogenic development, in a morphological and biochemical sense, is also apparent during precocious germination of embryos of *Brassica napus*. At the morphological level, secondary cotyledons, rather than leaves, are formed at the shoot apex of germinating embryos whereas at the biochemical level the latter continue to synthesize and accumulate cruciferin as in normally developing embryos (Finkelstein and Crouch, 1984).

One approach employed to investigate the causes of precocious germination and its control is manipulation of the culture medium. This has its genesis in the view that immature embryos are already predisposed to evolve into full-term embryos while enclosed in the ovule, but upon excision and culture are prevented by restraints imposed by the nutrient medium. The first important utilization of the culture medium to control precocious germination was undertaken with barley embryos. Prolongation of embryonic growth in barley is characterized by enlargement of the scutellum and development of the leaf and root within their respective embryonic sheaths, whereas precocious germination involves coleoptile elongation, expansion of root, shoot geotropism and greening. Kent and Brink (1947) found that supplementation of the usual agar medium containing mineral salts and sucrose with casein hydrolyzate, tomato juice or other natural plant extracts forestalls precocious germination and prolongs embryogenic growth of immature barley embryos. The role of casein hydrolyzate in rescuing embryos from precocious germination suggested that one or more of the major components of this nutrient amalgam such as inorganic phosphate, sodium chloride or the amino acid complement might be involved in the process. In subsequent work, Zieber, Brink, Graf and Stahmann (1950) followed the growth of immature embryos in media to which these components were added singly or in combinations and showed that amino acids and sodium chloride, mixed in the proportions they are present in casein hydrolyzate, effectively duplicate the activity of the latter in suppressing germination. The design of this experiment permitted the conclusion that the inhibition of germination results from the high osmotic pressure generated in the medium by casein hydrolyzate. The apparent ability of immature embryos to develop normally in a medium supplemented with high concentrations of sucrose or mannitol in the absence of casein hydrolyzate also suggested that it is possible to prolong embryonic growth by increasing the osmolality of the medium. According to Norstog and Klein (1972), the tendency of immature embryos of barley to germinate precociously is diminished to a considerable degree by high intensities of light, moderately high temperatures and reduced O_2 tension. Addition of GA or kinetin, however, induces precocious germination in embryos cultured under these conditions, whereas ABA suppresses it and counteracts the effects of GA (Norstog, 1972b; Umbeck and Norstog, 1979). These interesting interactions suggest that endogenous hormones are crucial in controlling precocious germination of embryos during normal seed development.

Other investigations have provided more definitive evidence favoring a role for ABA in the control of precocious germination of embryos. Ihle and Dure (1970, 1972) examined the activity of certain proteases involved in the mobilization of stored reserves of cotyledons as an event connected with the germination of cotton embryos and showed that culture of immature embryos triggers precocious germination as well as development of enzyme activity. Because removal of the embryo from the confines of the ovule promotes precocious germination, it appeared that the ovule tissues might supply some regulatory molecules to the embryo that prevents its

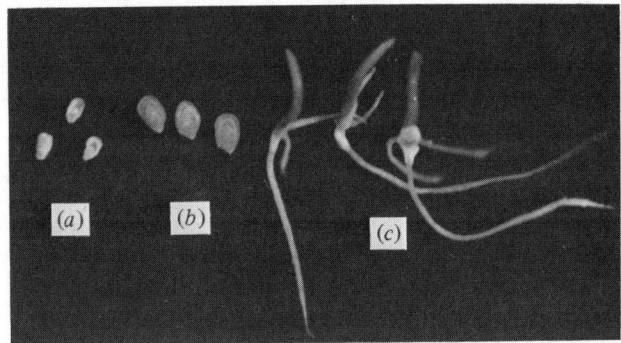

FIGURE 4.4. Inhibition of precocious germination of cultured embryos of *Triticum aestivum* by abscisic acid (ABA). (*a*) Embryos excised 18 days after anthesis, at the time of culture. (*b*) Embryos of the same age cultured in ABA for 5 days. (*c*) As in (*b*), but cultured in the absence of ABA. (From Triplett and Quatrano, 1982; photograph supplied by Dr. R. S. Quatrano.)

germination. This was found to be the case when it was discovered that addition of an aqueous extract of the ovule or ABA to the medium halts precocious germination of immature embryos and development of enzyme activity in them. It has been proposed that during embryogenesis in cotton, a sustained supply of ABA from the ovule tissues purposefully checks the germination instincts of the embryo until it becomes fully mature and dehydrated inside the seed. At this time the embryo is liberated from the chemical influence of the ovule tissue and germinates in a normal fashion when confronted with favorable conditions. *In vitro* culture of immature embryos of *Brassica napus* (Crouch and Sussex, 1981), *Phaseolus vulgaris* (Long et al., 1981), *Triticum aestivum* (Triplett and Quatrano, 1982) and *Glycine max* (Ackerson, 1984a) has also pointed to the necessity of incorporating ABA in the medium if precocious germination is to be avoided (Fig. 4.4).

At the molecular level, manifestations of the control of precocious germination by ABA are those that affect the synthesis of certain proteins in the embryos. In cotton, one set of proteins synthesized by the young embryo during the rapid phase of its growth is relatively stable and is detected in the mature seed embryo, while another set constituting an abundant class of proteins is found only in the mature embryo and disappears rapidly during germination. As shown in Figure 4.5, synthesis of the first set of proteins is induced prematurely in precociously germinating embryos. Immature embryos synthesize the second set of proteins prematurely in culture as well, but only in the presence of ABA in the medium (Dure, Greenway and Galau, 1981). Synthesis of the latter set of proteins has also

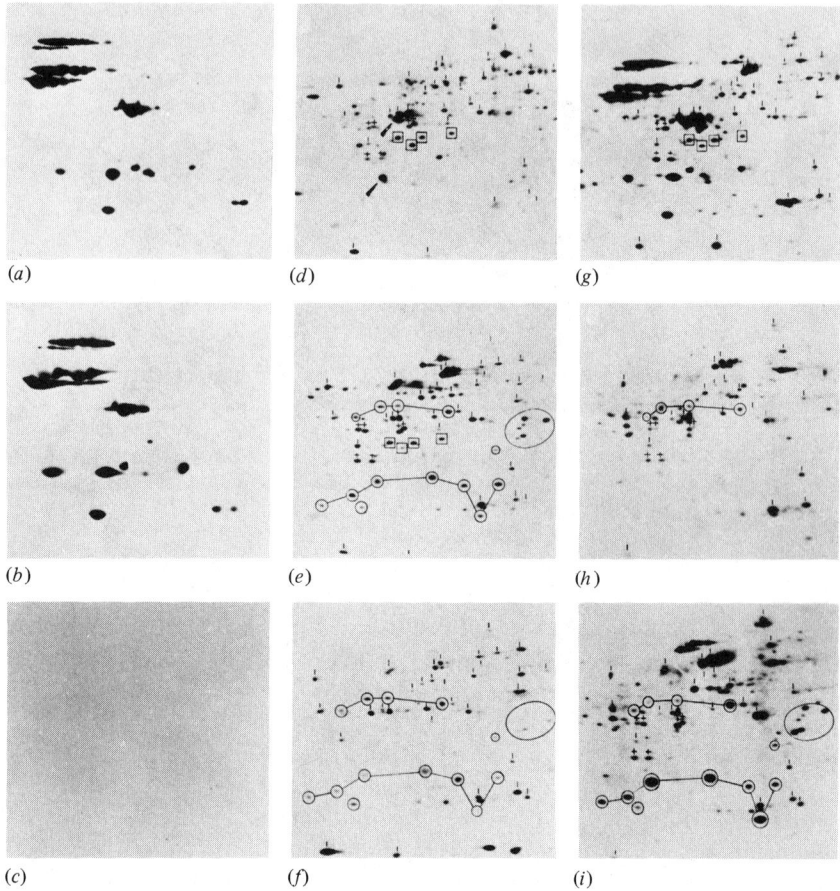

FIGURE 4.5. Two-dimensional electrophoretic patterns of proteins synthesized by cotton embryos during normal development and during precocious germination in the presence or absence of abscisic acid (ABA). (a–c) Fluorographs of pellet proteins from cotyledons of young, slightly older and fully mature embryos, respectively. (d–f) Fluorographs of soluble proteins from cotyledons of young, slightly older and fully mature embryos, respectively. In (d) the arrows indicate the two proteins largely confined to the pellet fraction, present as contaminants. Perpendicular tick marks (|) indicate the proteins that are synthesized by cotyledons at all stages of development. Proteins with crosses above them as well as those enclosed in boxes are found in cotyledons of young and slightly older embryos, but not in those of mature ones. Two sets of circled proteins are not synthesized by cotyledons of young embryos but are readily synthesized by cotyledons of slightly older and fully mature embryos. (g) Fluorograph of total proteins (i.e., a summation of pellet proteins and soluble proteins presented in a and d) synthesized by cotyledons of young embryos. (h) Fluorograph of total proteins synthesized by cotyledons of young embryos after 4 days of precocious germination. (i) Fluorograph of total proteins synthesized by cotyledons of young embryos incubated for 4 days in ABA. (Reprinted with permission from Dure et al., 1981. *Biochemistry* 20: 4162–4168. 1981. American Chemical Society; photographs supplied by Dr. L. S. Dure.)

been verified in a cell-free system by hybrid-arrested translation and hybrid-selected translation using cloned cDNA made from mature cotton cotyledon mRNA. Because immature embryos cultured in a medium containing ABA synthesize the same proteins as those synthesized in an *in vitro* system programmed by cloned probes, the latter come close to being a set of ABA-induced genes (Dure, Galau, Chlan and Pyle, 1983). The work of Choinski and Trelease (1978) has hinted that malate synthase activity develops in cotton embryos supplied with exogenous ABA, which prevents precocious germination.

The synthesis of storage proteins in precociously germinating embryos of *Phaseolus vulgaris* (Sussex and Dale, 1979; Long et al., 1981) and *Brassica napus* (Crouch and Sussex, 1981) is reinstated by the addition of ABA, as was noted in Chapter 3. Another example in which a marker has been identified for ABA action during precocious germination is wheat; here, ABA-induced inhibition of precocious germination of cultured immature embryos is associated with the accelerated synthesis of a family of proteins of which a lectin, wheat germ agglutinin, is a major component (Triplett and Quatrano, 1982). Some of these proteins are also synthesized by mature embryos; the identity of gene sequences coding for wheat germ agglutinin in mature embryos and in immature embryos supplied with ABA has been confirmed by hybridization between cDNA probes complementary to mRNA of both sets of embryos (Quatrano et al., 1983). A recent demonstration of the reversal of precocious germination and promotion of lectin biosynthetic activity in rice embryos by ABA bears an obvious parallel to the results from wheat embryos (Stinissen, Peumans and de Langhe, 1984). Overall these studies appear to indicate that precocious germination is somehow related to the inability of the early-stage embryo to synthesize certain proteins specific to the mature stage and that ABA prevents precocious germination by conferring this ability.

In an attempt to unravel the role of ABA in precocious germination, Karssen, Brinkhorst-van der Swan, Breekland and Koornneef (1983) have focused on the germination of embryos in an ABA-deficient mutant of *Arabidopsis thaliana*. This plant is one of the most genetically defined angiosperms available and one in which a wide variety of mutants have been isolated. The mutant is characterized by a reduced level of ABA in developing fruits and seeds and by symptoms of wilting. Moreover, in germination tests with immature seeds, embryos of mutants germinate precociously, whereas those of the wild type continue in the embryogenic pathway and lapse into dormancy. Germination responses of immature seeds from reciprocal crosses between the wild type and mutant have indicated that embryogenic development and precocious germination are probably modulated by ABA gene in the embryo and not in the maternal part of the plant.

In an effort to gain further insight into the role of ABA in the precocious germination of embryos, some investigators have attempted to determine the levels of ABA in reproductive tissues of plants and relate them to the patterns of growth of excised embryos in culture. The following account derives from studies of this kind carried out on soybean (Quebedeaux, Sweetser and Rowell, 1976; Ackerson, 1984a) and wheat (King, 1976). In soybean, ABA concentration remains high during the period of active growth of the seed but decreases to low levels at seed maturity. Consistent with this observation it has been found that embryo axes excised from mature seeds germinate soon after culture, while those from immature seeds do not germinate or require a long lag period before germination. According to Ackerson (1984b), when the endogenous pool of ABA is removed from immature soybean embryos by washing them in water or by drying them in the pod, embryos exhibit precocious germination. In a similar way, embryos excised from immature wheat grains containing high concentrations of ABA germinate infrequently as compared to the high germination percentages of mature embryos excised from dry grains containing low levels of ABA. These results suggest that ABA synthesized in the developing seed regulates embryogenesis by inhibiting precocious germination, permitting the immature embryo to play out its full developmental program. High levels of ABA have been found in embryos of *Phaseolus vulgaris* when they are just phased out of the stage vulnerable to precocious germination and again at seed maturity as they become quiescent (Morris, 1978; Hsu, 1979).

In certain plants, mature embryos normally skip an intervening period of developmental arrest and germinate, growing out of the seed coats and fruit while the latter is still attached to the parent plant. This phenomenon, known as vivipary or preharvest sprouting, is the exact antithesis of quiescence or dormancy, which terminates embryogenesis. If high concentrations of ABA prevent precocious germination it is likely that its absence or its presence in low concentrations might be conducive to vivipary. According to Sussex (1975), viviparous embryos of *Rhizophora mangle* are characterized by a high water content and are insensitive to ABA inhibition in culture. In maize, where vivipary is genetically controlled, embryos of certain mutants similarly show a decreased sensitivity to ABA inhibition in culture (Robechaud, Wong and Sussex, 1980). In another line of investigation, fluridone, an inhibitor of carotenoid biosynthesis, applied to developing ears of wild type maize is found to induce vivipary (Fong, Smith and Koehler, 1983). Since carotenoids are probable precursors of ABA, an obvious conclusion from this observation is that vivipary may be due to lesions in ABA synthesis caused by genetic mutation or by chemical inhibition. The relationship between low ABA level in the seed and vivipary was tested in an ABA-deficient mutant of *Arabidopsis thaliana* in which mature seeds were induced to germinate in the fruit in an atmosphere of high humidity (Karssen et al., 1983). This is what one would expect of embryos growing in ovules deficient in ABA.

In summary, it seems reasonable to suppose that precocious germination, vivipary and developmental arrest are complex phenomena and that a variety of additional approaches will be required before it will eventually be possible to determine their cause and explain them in a molecular framework. To the extent that water stress in plants is known to cause a rise in the ABA level of cells, it is of interest to note that embryos germinate precociously when their water content is high and that they lapse into quiescence when the water content is low. The general conclusion suggested by these observations is that ABA might regulate cell metabolism of the embryo during the various developmental stages by its effect on the water potential of the cells. This idea is strongly endorsed in a thoughtful essay by Walbot (1978) and the reader is referred to this work for further details.

Genetics of embryogenesis

Every organ of the mature seed embryo has an embryogenic history associated with it and behind this history in turn are the genes that direct the changes constituting embryogenesis. An area of primary interest for the experimental plant embryologist is the mechanism by which gene expression during embryogenesis is regulated. From this point of view, biochemical analyses of genetically defined mutants impaired at different stages of embryogenesis might provide insights into the nature of gene action and help us to pinpoint the effects of specific genes on developmental processes in the embryo. The production of spontaneous and induced embryo lethal mutants in plants has been known for some time now. Several spontaneous lethal mutants causing different degrees of malformation in the endosperm and embryo have been reported in maize since Mangelsdorf first described them in 1926 (Sheridan and Neuffer, 1981, 1982). As a model, a semidominant lethal gene described in *Lycopersicon esculentum* (tomato) appears to illustrate the concept of gene action on embryo development particularly well. The effect of the gene is specific on the embryo without affecting the endosperm and causes a developmental block in embryo growth before the onset of organogenesis (Huang and Paddock, 1962). Another mutation in tomato is generally expressed in the embryo by the absence of cotyledons or by the presence of a single fused cotyledon. Culture of mutant embryos in a medium fortified by diffusate of normal seeds induces an intermediate phenotypic development; this finding is consistent with the view that the mutation is possibly caused by a deficiency of some factors present in the normal seed, although culture of isolated embryos in media containing selected amino acids and hormones does not give any clues regarding the nutritional block (Mathan and Jenkins, 1960, 1962). Among other plants, occurrence of spontaneous mutants defective in embryogenesis has been reported in *Petunia hybrida* (Vallade and Cornu, 1979). That embryos may be altered by mutagens in their ability to differentiate was suggested by the experiments of Müller (1963), who, following X-irradiation of seeds of *Arabidopsis thaliana*, identified several classes of mutants in the progeny, characterized by the presence of aborted seeds and embryos. However, as these studies provided little insight into the possible causes of developmental defects, they have proved of limited value in understanding the genetic program of embryogenesis.

Genetic analysis of mutants to determine the causes of embryo lethality as well as the factors controlling gene activation during embryogenesis is a recent field of study. The best studied embryo lethal mutants are those characterized in *Arabidopsis thaliana* and maize. Because of the rare occurrence of spontaneous mutants, the ease of recognition of aborted seeds and other desirable features, Meinke and Sussex (1979) used *A. thaliana* as a model system for the genetic analysis of mutations. The method described by these investigators consists of mutagenizing seeds with ethyl methane sulfonate (EMS), screening M_1 plants for mutants and characterizing the mutants by analysis of segregation ratios of M_1 plants and M_2 heterozygotes and tests of allelism. In this way, six nonallelic recessive mutants in which embryos are characteristically arrested at preglobular, early globular or globular-heart-shaped stages have been described. That the defect in the embryos is caused by the action of critical genes essential for completion of specific stages of early embryogenesis is suggested by the observation that no further development occurs in the arrested embryos. It is of interest that the suspensor of embryos of one group of arrested mutants contains 15 to 150 cells and is thus considerably larger than the wild-type suspensor of only six to eight cells. This finding would seem to indicate that, at least initially, the missing gene product that impedes embryo growth permits virtually unlimited growth of the suspensor. Among the possible explanations for the limited growth of the suspensor on the wild-type embryo, the most likely one is that the presence of an actively growing embryo inhibits expression of the full developmental potency of the suspensor (Marsden and Meinke, 1984). The observation that aborted seeds in two out of six mutants are preferentially located toward the half of the ovule closer to the stigma has led to the suggestion that genes for the embryo defect in these mutants are expressed not only during embryogenesis but also at some point prior to fertilization – probably during pollen tube growth (Meinke, 1982). These various results do not provide evidence regarding the mechanism of the gametophytic or sporophytic gene expressions during embryogenesis in *Arabidopsis* – only the fact that they do exist.

A somewhat different picture of gene expression is gained from an analysis of mutants in maize impaired in embryo and endosperm development. The experimental data are not yet extensive; however, because of the effect of mutation on both embryo and endosperm, maize mutants have provided an attractive system to study embryo–endosperm interactions during embryogenesis. The basic approach involves the induction of mutations in normal plants by pollinating them with a pollen suspension treated with EMS. The resulting kernels are raised as M_1 generation and mutants selected by screening self-pollinated M_1 plants. The mutant kernels obtained by this method represent a genetically heterogeneous group, although most are single-gene recessive mutants in which both embryo and endosperm are severely affected (Neuffer and Sheridan, 1980). Included in the latter group are both nutritional-type mutants and developmental-type mutants. Several mutants whose embryos grow into normal plants in an enriched culture medium appear to be auxotrophs, although they have not been fully characterized. In one case, similar to a spontaneously occurring mutant, the induced mutation has been traced to a genetic block in the pathway of proline biosynthesis (Racchi, Gavazzi, Monti and Manitto, 1978; Sheridan and Neuffer, 1980). In a number of developmental mutants analyzed, the defect has been attributed to a permanent block at some stage of embryogenesis between the zygote and the appearance of leaf primordium (Sheridan and Neuffer, 1982). While we await biochemical data to complement our knowledge of the mutants, these studies clearly provide a new handle for the genetic analysis of embryogenesis.

The ultimate change in a mutant is referable to alterations in the internal organization of genes and their flanking regions. In a series of experiments on lectin nonproducing lines of soybean embryos, Goldberg et al. (1983) have shown that the inability of the mutant embryos to accumulate lectin is attributable to a virtual absence of lectin mRNA transcripts in their genome. The transcription lesion in the lectinless phenotypes has been traced to an alteration in the lectin gene structure caused by the insertion of a 3.4-kilobase (kb) DNA segment. How the insertion element interferes with transcription of the lectin gene in the embryo has not yet been established.

Concluding comments

One outcome of the review of work presented in this chapter is the realization that seed embryo is a complex system in which there are different kinds of correlations between different organs. Results of studies on *in vitro* culture of excised embryos have indicated that various hormones conceivably intervene in regulating growth of the different parts of the embryo and in releasing the embryo from physiological dormancy. However, in a large number of cases there is incontrovertible evidence that growth hormones alone or in combination cannot substitute for the complement of nutrient substances of the endosperm in inducing growth of the embryo during its heterotrophic phase. Apparently other factors must be involved, but the nature of these factors and the way they integrate with growth hormones to induce growth have not been identified.

It is clear that we have a long way to go to understand how an embryo proceeds through the orderly series of morphological stages during undisturbed normal development. The various investigations relating to the control of precocious germination of embryos provide a new perspective into the influence of the vegetative parts in providing specific hormones for completion of the embryo developmental program. Since completion of embryogenesis is tied to the ensuing developmental arrest of the embryo, it is expected that a study of the control mechanism of normal embryo development will facilitate an understanding of the basis for inhibition of embryo growth resulting in quiescence.

5
Somatic embryogenesis

Thus far our discussion of angiosperm embryogenesis has been confined to the rigorously programmed development of the embryo from the zygote. The final form attained by the embryo is primarily an expression of the complement of genetic information originally present in the egg and further endowed by fusion with the sperm at fertilization. Although daughter cells born out of simple mitotic divisions of the zygote carry essentially the same legacy of template information as the parent cell, there is a progressive decline in the developmental potential of the newly formed cells as they differentiate into highly specialized cell types. As a result, living cells of the adult plant are normally unable to display their innate morphogenetic potential as they fulfill their prescribed role or when they are subject to systemic controls within the plant body. Under certain experimental conditions, however, cells of the angiosperm sporophyte behave like a zygote and replay with a high degree of fidelity a developmental program leading to the production of embryo-like structures. Since embryo-like structures are derived from the sporophytic or somatic cells of the plant, as opposed to the gametophytic or germ cells, the phenomenon is commonly referred to as somatic embryogenesis. To avoid confusion, there is an increasing tendency among investigators to designate the developmental repertoire of the zygote into an embryo as zygotic embryogenesis. Despite the fact that zygotic embryos and embryo-like structures formed from somatic cells without the intervention of sexual fusion are identical in appearance and possess the same morphogenetic potential, to emphasize the divergent pathways through which they have evolved, the noncommittal term embryoid is generally used to refer to the latter. Attempts have been made to apply some criteria in defining an embryoid. For example, it has been suggested that besides the acquisition of early bipolarity, an embryoid should have had its origin in a single cell (Street and Withers, 1974) and should not be connected to a preexisting vascular strand in the mother tissue (Haccius, 1978). However, these criteria continue to be ignored in contemporary publications that describe the origin of embryoids in tissue cultures; we cannot evade the fact that, in many cases in which perfectly respectable bipolar embryo-like structures have been described, they have not been traced to single-celled beginnings.

Totipotency and somatic embryogenesis

The discovery of somatic embryogenesis is ineluctably tied to the demonstration of totipotency of plant cells. As generally applied to biological systems, a cell is defined as totipotent if it can regenerate in full multicellularity, sexuality and structure the phenotype of the organism of which it was a part. Although this definition is applied by some investigators to assert totipotency of plant cells, for purposes of our discussion, totipotency implies that all plant cells, except those that have undergone irreversible differentiation, possess a profound ability to display their full genetic potential and embark upon a developmental pathway similar to that of the zygote leading to the formation of a new plant. Thus, a tangible expression of totipotency of a plant cell is its ability to regenerate a complete plant through a process of simulated zygotic embryogenesis, while keeping innocent of sex.

The pioneering experiments that led to the demonstration of totipotency of plant cells were initiated by Steward and co-workers (Steward, Mapes and Smith, 1958; Steward, Mapes and Mears, 1958) using the cultured secondary phloem of domestic carrot. These investigators found that culture of slabs of carrot tissue in a solidified tissue culture medium that supplies the requirements of essential salts, vitamins and organic nutrients supplemented with coconut milk typically produces a proliferating mass of callus constituted of simple parenchymatous cells. Subsequent growth of the callus with gentle agitation in a liquid medium of the same composition results in a suspension of individual cells and small cell clusters. Although the origin of the cell clusters appeared to be a vexing question at first, careful examination of a range of cells and cell aggregates established that individual cells dissociating from the callus divide and the daughter cells formed remain attached to one another to form cell clusters. Subsequent growth of this totally disorganized population of cells and cell clumps in the liquid medium without subculture leads to lignification of inner cells of the clump, formation of cambium-like derivatives and eventually to the appearance of lateral root primordia. A normal carrot plant is assembled in the culture flask when the rooted aggregate is subsequently grown on a solid medium lacking coconut milk. In these studies, the more ardous task of monitoring a single cell in isolation during its transformation into a plantlet was not attempted; it remained for Vasil and Hildebrandt (1965) to show that a single cell of a hybrid tobacco nurtured in isolation from other cells in a defined medium forms a completely organized plant. This cell, grown in a drop of the liquid medium in a microculture chamber, divides repeatedly to form a callus that is subsequently induced to form roots and shoots on a solid medium.

The ability of a single somatic cell to form a new sporophyte plant is a remarkable developmental feat. Nevertheless, it falls short of our definition of totipotency as the single cell does not precisely duplicate the pathway of embryogenesis normally followed by the zygote. From this point of view, the work of Reinert (1959) who showed that a callus originating from a strain of carrot root, following a long period of culture in a medium containing coconut milk and IAA, differentiates bipolar embryoids upon being transferred to a synthetic medium enriched with an elaborate mixture of amino acids, amides, vitamins, a purine (hypoxanthin) and IAA, is particularly noteworthy. Although this work did not rigorously establish that embryoids originate in single cells of somatic parentage, the observation nonetheless dispels the notion that formation of embryo-like structures is the private domain of the zygote and takes us a step closer to the demonstration of totipotency of plant cells.

A natural consequence of these observations was an intensification of efforts to induce embryogenic type of development in identifiable single somatic cells. The breakthrough occurred when, almost simultaneously, Steward (1963) and Wetherell and Halperin (1963) reported that a suspension culture of carrot cells and cell clusters regenerates an enormous number of embryoids, which are faithful replicas of zygotic embryos. In Steward's original work, embryos isolated from seeds of domestic or wild carrot are allowed to germinate in a medium containing coconut milk. The free cells that slough off from the hypocotyl region of the seedling are a lavish source of totipotent cells that exhibit typical embryogenic type of development. As shown by Steward, Mapes, Kent and Holsten (1964), the effect is even more dramatic when a cell suspension originating from an immature embryo of wild carrot is plated on a nutrient agar plate; here virtually every cell of the suspension yields an embryoid. Wetherell and Halperin (1963) extended these observations to calluses originating from other parts of wild carrot such as the root, petiole and peduncle. A follow-up work (Halperin and Wetherell, 1964) also explicitly showed that coconut milk is not a factor in inducing totipotency in carrot cells and that a callus reared in a medium containing a moderate dose of the synthetic auxin 2,4-D forms embryoids when it is transferred to the same medium containing a reduced level of auxin (Fig. 5.1). This new idea introduced interpretations that led for a period of time to lively disagreements in the literature on somatic embryogenesis. At about the same time that these discoveries came to light, in a note that has often been ignored, Kato and Takeuchi (1963) described a balanced embryogenic sequence of development in single cells of carrot root callus sloughed off into the growth medium. Finally, the transformation into an embryoid of a single cell from a callus culture of carrot nurtured in isolation from other cells has also been demonstrated, reinforcing the conclusion that embryo-like structures observed in cell suspension cultures indeed have their origin in single cells

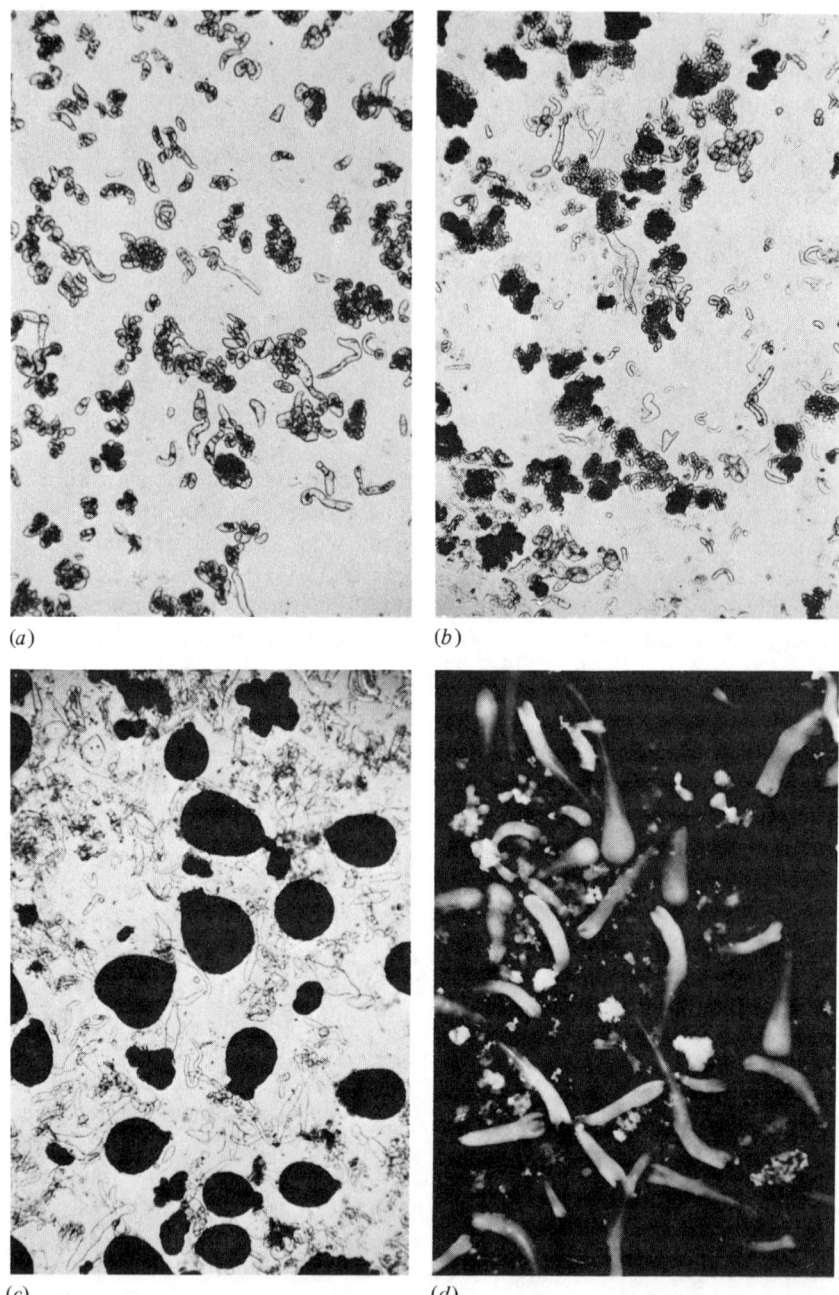

FIGURE 5.1. Somatic embryogenesis in carrot. (*a*) Cell suspension growing in a medium containing 2,4-D for 7 days after subculture. (*b*) Cell clusters representing proembryogenic masses formed upon transfer of the cell suspension to a medium lacking 2,4-D (induction medium). (*c*) Globular embryoids formed in the cell suspension transferred to the induction medium. (*d*) Torpedo-shaped and later stage embryoids formed in the cell suspension transferred to the induction medium. (Photographs made from cultures supplied by C. H. Michler, Jr., The Ohio State University.)

(Backs-Hüsemann and Reinert, 1970). Collectively these studies were of great importance in the practical and intellectual development of the field of somatic embryogenesis because they showed that organized development of free cells and embryogenesis have much that is common and much that is different.

Although several investigators have emphasized a close resemblance among the globular, heart-shaped, torpedo-shaped and cotyledonary stages of zygotic embryos and somatic embryos of carrot, an issue that deserves comment is whether embryos and embryoids follow the same sequence of divisions in their early development. Unraveling this problem is difficult because it has not been possible to identify by any unique cytochemical or ultrastructural features those cells in a suspension culture that are destined to divide in the embryogenic pathway. According to Borthwick (1931), development of the embryo of carrot follows a pattern typified by the Solanad type (see Chapter 2) and begins with a transverse division of the zygote. Both of the resulting cells divide by transverse walls to form a file of four cells. A third round of transverse divisions in each of the four cells gives rise to eight cells held together in a filament. The first longitudinal divisions occur in the three cells farthest from the micropyle whose descendants eventually generate all of the embryo except the root tip. A limited number of divisions in the other five cells produce the root tip and a massive suspensor. However, Halperin (1966) found that in carrot cell suspension culture the presumptive embryoid appears as a globular structure from a disorganized cellular mass without any definite sequence of cell lineage associated with early segmentation. Although a suspensor comparable in stature to that subtending the zygotic embryo is absent, cells of the original aggregate often remain attached to the globular embryoid as a suspensor-like appendage. In the only instance in which a single somatic cell was followed in isolation during embryogenesis it was found that, in contrast to the zygote, which gives rise to a polarized bicellular structure, the isolated cell yields a conglomerate of parenchymatous cells from which the embryoid emerges (Backs-Hüsemann and Reinert, 1970). The most detailed study of early development of somatic embryos of carrot was provided by McWilliam, Smith and Street (1974). These workers showed that the first division of the presumed embryogenic initial in a cell aggregate is in the transverse plane but that the daughter cells born out of this division diverge along different pathways. The terminal cell divides longitudinally to form the embryo proper, while the cell closest to the cellular aggregate divides transversely to form an incipient suspensor-like structure. On the basis of these observations, early division sequences of somatic embryos of carrot appear to have more in common with the Crucifer type than with the Solanad type. Although these results point to a fundamental difference in the early division sequences of zygotic and somatic embryos of carrot, it is important to recognize that both sequences create cell lineages with developmental programs that result in identical structures. The fact that embryo-like structures can be induced in free cells of carrot by simple manipulative techniques seems to suggest that somatic embryogenesis in this plant is not an aberrant form of development displayed by cultured cells but is comparable in many respects to zygotic embryogenesis. Subsequent demonstration of somatic embryogenesis in a variety of plants under different cultural conditions has tended to reinforce this proclivity.

Certain physiological and biochemical evidences also point to a close similarity between zygotic embryos and embryoids. For example, like zygotic embryos, somatic embryos of *Eschscholzia californica* (Kavathekar, Ganapathy and Johri, 1977) and *Vitis vinifera* × *V.rupestris* (grape) (Rajasekaran and Mullins, 1979) exhibit a form of dormancy that is overome by a cold treatment. Cold-induced reversal of dormancy of both seed embryos and somatic embryos of grape is also accompanied by a concomitant decrease in their ABA contents (Rajasekaran, Vine and Mullins, 1982). At the biochemical level, somatic embryos of diverse plants synthesize qualitatively similar fatty acids (Pence, Hasegawa and Janick, 1981a), lipids (Janick, Wright and Hasegawa, 1982), anthocyanins (Pence et al., 1981b), storage proteins (Crouch, 1982) and alkaloids (Schuchmann and Wellmann, 1983) characteristic of maturing zygotic embryos *in vivo*. The level of correspondence in the differentiation of specific tissues in zygotic embryos and somatic embryos has been somewhat equivocal. As noted in *Asclepias syriaca* (Wilson and Mahlberg, 1977), a significant difference between somatic embryo and seed embryo is the presence of laticifers in the latter and their absence in the former; however, laticifers, which are absent in the seed embryo of *Papaver somniferum* (poppy), are nonetheless present in somatic embryos derived from seedling hypocotyl (Nessler, 1982). Clearly, somatic embryos can serve in a limited way as model for the contemporary analysis of embryo development in angiosperms.

Survey of somatic embryogenesis in angiosperms

According to a recent review (Ammirato, 1983a), somatic embryogenesis has been reported in about 71 species of angiosperms, confined to 62 genera distributed within 28 families. From this list and from subsequent reports, it appears that plants belonging to Umbelliferae, Rutaceae, Ranunculaceae and Solanaceae exhibit greater propensity for somatic embryogenesis than members of other families. Embryogenesis has been induced on a variety of explants such as the stem, hypocotyl, root, leaf, pedicel, floral bud, excised seed embryo, endosperm and nucellus. Unfortunately, meristematic cells such as cambium and sites of bud and root initiation are not generally excluded from the cultured explants; it is therefore not established whether embryoids arise from differentiated cells that acquire totipotency or from meristematic cells that have no prior commitments toward differentiation.

It will be recalled that in the successful induction of embryoids from wild carrot, aseptic culture of the tissue explant to produce a callus is the essential first step and that transfer of the callus to a medium of a different composition causes initiation of embryoids (Halperin and Wetherell, 1964). Although this protocol became practicable with many species, in others it is possible to bypass the callus stage and induce embryogenesis directly on the cultured explant. In still other cases, somatic cells enzymatically stripped of their cell walls have been shown to regenerate a new cell wall and pursue an embryogenic type of development. These modes of somatic embryogenesis have no species barriers, since certain organs in the same plant may respond to culture by direct embryogenesis whereas others produce embryoids through the intermediary of a callus – or the same organ may produce embryoids through both pathways. In many cases in which embryoids have been produced by way of naked protoplasts of somatic cells, the latter also yield embryoids easily. The main purpose of the following account is to highlight the developmental aspects of embryogenesis by the three pathways in representative species.

Embryogenesis from callus

The carrot system has been the most comprehensively studied with respect to the developmental aspects of somatic embryogenesis which occurs through the intervention of a callus. For experimental purposes, a callus is initiated from seedling hypocotyl or root segments excised from seeds of domestic varieties of carrot germinated under aseptic conditions and cultured on the surface of a high nitrogen containing mineral salt medium, such as Murashige-Skoog's medium supplemented with sucrose, an organic addendum including *myo*-inositol, a source of iron such as sodium-ferric EDTA (FeEDTA), a cytokinin [kinetin, zeatin or $N^6(\Delta^2$-isopentenyl) adenine (2iP)] and 2,4-D. A suspension culture is initiated by transferring a piece of the callus to a liquid medium of the same composition, but with a reduced level of 2,4-D and is gently agitated on a horizontal shaker. Embryogenesis is induced when an aliquot of the cell suspension is transferred to the basal medium from which 2,4-D is omitted (induction medium). Under these conditions, the first signs of embryogenesis are observed in 4 to 5 days, when the cell aggregates are transformed into globular embryoids. High yields of embryoids of desired stages may be obtained by filtering the suspension successively through glass beads of different sizes followed by centrifugation of the fraction in Ficoll (Warren and Fowler, 1977). According to Wetherell (1984), a 45-minute plasmolytic shock of carrot cells in 1 M sucrose increases the yield of somatic embryos by threefold and produces a high degree of synchrony in the system. Several groups of investigators (Fujimura and Komamine, 1979b; Giuliano, Rosellini and Terzi, 1983; Bradley, El-Fiki and Giles, 1984) have established conditions for obtaining high yields of synchronously developing carrot embryoids starting with a cell suspension. For example, in the method used by Fujimura and Komamine (1979b), cells and cell clusters from a stationary-phase suspension culture retained on a 31 to 47 μm nylon screen are initially subjected to density gradient centrifugation in Ficoll solution. Next, the heaviest fraction from the Ficoll gradient is repeatedly centrifuged at low speed for a short time until most of the contaminating vacuolate cells are removed. When the final fraction is transferred to a medium lacking 2,4-D, the cell clusters exhibit a high degree of synchrony in embryogenesis with more than 90 percent frequency. Serial observations have shown that the most rapid rate of cell division occurs in the cell aggregates in 3 to 4 days after transfer to the embryoid-inducing medium, when presumably the sites of the future root and shoot primordia are determined in the globular embryoid (Fujimura and Komamine, 1980b). It would not be difficult to visualize a different rate of cell multiplication in specific regions of the globular embryoid as it is phased out into the heart-shaped stage with evident cotyledonary primordia.

Carrot has served as a useful model system of somatic embryogenesis in other respects as well. An elegant demonstration of the embryogenic plasticity of carrot was provided by Kato (1968), who observed embryogenic development when epidermal strips accompanied by adjacent cells of the cortex are delicately removed from the seedling hypocotyl and grown in a microculture chamber. That embryoids originate from the cells of the epidermis and not of the cortex is indicated by serial observations of the culture. In some strains of carrot, germinated embryoids regenerate easily a second generation of embryoids directly on the hypocotyl while they are still bathed in the medium (Homès, 1968). Apparently, when a seedling is formed by dedifferentiation of a somatic cell, cells of the seedling are delivered from dependence on a callus intermediary to display totipotency.

Dicots. We shall now consider the requirements for somatic embryogenesis in a few selected dicots. Apparently, stem and embryo explants of several members of the carrot family (Umbelliferae) respond to a culture medium like carrot and no conditions other than nurture of the explant in an auxin-enriched medium to induce a callus, and transfer of the latter to an auxin-impoverished medium would seem to account for somatic embryogenesis in these explants (Steward, Ammirato and Mapes, 1970; Ammirato, 1983a). In some interesting experiments on *Macleaya cordata* (Papaveraceae), isolated mesophyll cells obtained by grinding the leaf in a glass homogenizer are cultured in a medium containing 2,4-D and kinetin. Dedifferentiation of the cell accompanied by mitotic activity results in the formation of a cell cluster resembling a callus. The stimulus for embryogenic development of cells of the callus appears to be none other than omission of 2,4-D from the medium or its replacement with weaker auxins such as IAA or NAA (Lang and Kohlenbach, 1975). Availability of large quantities of embryogenic mesophyll cells will make this system promising for research into the developmental physiology of somatic embryogenesis as it is truly homogeneous with respect to the age and type of cell as well as its physiological condition and genotype.

In several dicots, somatic embryogenesis occurs on explants cultured in a single formulation of a hormone-enriched medium. In evaluating such reports, one fact that stands out prominently is the low frequency of embryogenesis in culture. Such a situation might be expected if developmental blocks exist in the expression of totipotency by cells of the regenerated callus and if a constituent of the medium causes the block. It is important to remember that whereas a callus develops by the division of a previously differentiated cell, an embryoid arises by the division of a meristematic cell of the callus. Sharp, Söndahl, Caldas and Maraffa (1980) focused attention on the consequence of this fact in suggesting that potentially embryogenic cells of the callus are mitotically quiescent in the presence of high concentrations of auxin in the medium. A low-frequency somatic embryogenesis that occurs on explants growing in the presence of high auxin in the medium is apparently caused by the escape of certain embryogenic cells from mitotic arrest. This has led to considerations of experimental systems in which somatic embryogenesis can be induced at variable frequencies by changes in the hormone level of the medium.

In an examination of this question in *Coffea arabica* (coffee), Söndahl and Sharp (1977) found that a leaf-derived callus can be maintained at low-frequency embryogenesis in a medium containing 2,4-D and kinetin. Transfer of the callus to a medium with a reduced concentration of kinetin and auxin at a high ratio leads to a severalfold increase in the number of embryoids formed. However, once arrested cells have entered the mitotic cycle, they do not directly give rise to embryoids; rather, they go through a program involving a renewed growth of the callus. This explains why high-frequency embryogenesis occurs on the surface of a brand new white friable callus that emerges from the old, brownish tissue. In *Apium graveolens* (celery) the problem is not so much with regard to the low frequency of embryoids formed as it is with the failure of embryoids to progress beyond the early stages, when the callus is initiated and maintained in a single formulation of a medium containing 2,4-D and kinetin. Here, progressive reduction in the concentration of auxin and kinetin is necessary to obtain full-term embryoids in culture. These results indicate that somatic embryos, like zygotic embryos, pass through a heterotrophic phase when they are dependent on exogenous nutrients to an independent, autotrophic existence. Limited data on the endogenous auxin and cytokinin levels of developing embryoids of celery suggest that changes in the concentration of cytokinin against a constant level of auxin may have significance in embryogenesis (Williams and Collin, 1976; Al-Abta and Collin, 1979).

Although the first realization of somatic embryogenesis was achieved in a root tissue, roots have not proved to be favorable organs for induction of somatic embryos. Other examples of somatic embryogenesis in root-derived tissues are *Atropa belladonna* (Konar, Thomas and Street, 1972b), *Nigella sativa* (Banerjee and Gupta, 1975), *Crambe maritima* (Bowes, 1976), *Panax ginseng* (ginseng) (Chang and Hsing, 1980a,b), *Ipomoea batatas* (sweet potato) (Liu and Cantliffe, 1984) and *Chicorium intybus* (chicory) (Heirwegh, Banerjee, van Nerum and de Langhe, 1985). In ginseng, embryogenesis occurs when a callus reared in a medium containing 2,4-D (2 mg/liter), kinetin (1 mg/liter) and 2iP (1 mg/liter) is subcultured in the same medium containing a reduced level of 2,4-D but lacking cytokinins. Development of embryoids into plantlets occurs in yet another medium containing half-strength concentration of mineral salts, benzyladenine (BA) (1 mg/liter) and GA (1 mg/liter). An unexpected development observed in this medium is the flowering of embryoids, indicating that they can bypass a vegetative phase in the life cycle before entering the reproductive phase. Whereas many studies with roots have been concerned with the establishment of actively growing cultures of the organ, except in a few cases, the ability of root-derived calluses to grow and undergo morphogenesis has not been critically tested. Consequently, studies on root calluses and totipotency of their cells will have great heuristic value.

The potential for somatic embryogenesis is not confined to the vegetative parts of plants, for they are equally prominent in certain reproductive tissues and organs. Nucellus of *Citrus* is a logical choice for experimental induction of embryoids, as nucellar polyembryony in fertilized ovules is a way of life for many cultivated species and varieties of *Citrus*. Rangaswamy (1961) reported the occurrence of embryo-like structures in a continuously growing culture obtained from the nucellus of fertilized ovules of *Citrus mitis* (*C. microcarpa*). Although the embryo-like structures were referred to as pseudobulbils, they are clearly modified somatic embryos. One could argue that embryogenesis in the nucellar callus from fertilized ovules is an extension of the normal reproductive cycle of the plant. However, as shown in other species of *Citrus,* especially *C. sinensis* (Shamouti orange), it now appears that fertilization of the ovule is not essential to prepare the nucellus to proliferate into a callus and the latter to regenerate embryoids (Kochba and Spiegel-Roy, 1973). Other observations have added further interest to somatic embryogenesis in the nucellar callus of Shamouti

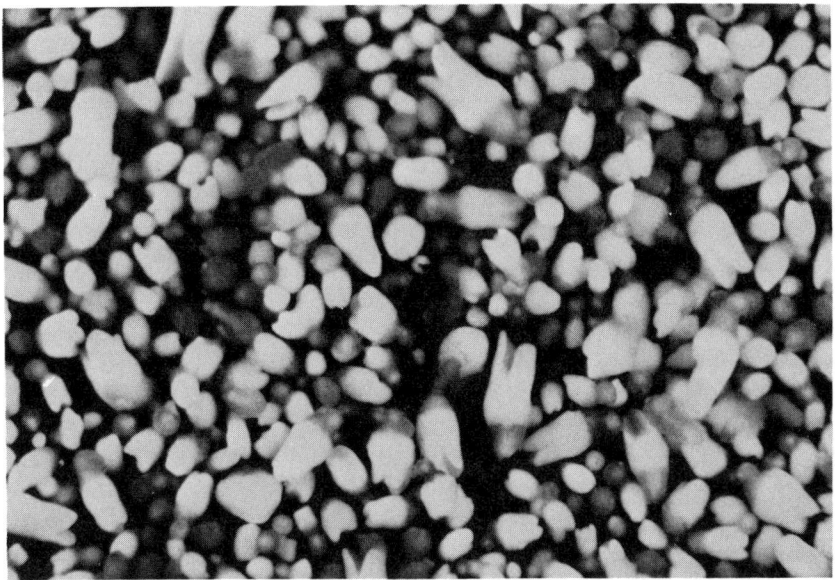

FIGURE 5.2. High-frequency somatic embryo formation in a nucellar callus of grape. Embryoids of different stages of development are seen. (From Srinivasan and Mullins, 1980; photograph supplied by Dr. C. Srinivasan.)

orange. One is that continuous culture of the tissue in a medium containing malt extract and a low concentration of adenine, which is optimal for embryogenesis, leads to its habituation with respect to growth hormone requirements. This finding has resulted in the selection of a number of lines of calluses that are independent of exogenous auxin and cytokinin requirements for growth and that differ greatly in their embryogenic potential. Second, it is possible to increase the embryogenic potential of the callus by aging it in the medium or by withholding sucrose from the medium for one subculture period (Kochba and Button, 1974). Third, single cells separated from the callus by enzyme maceration regenerate embryoids in large numbers (Button and Botha, 1975). These properties offer the choice of a callus that can be manipulated to proliferate or produce embryoids either directly or from single cells. In grape, embryoids are produced in large numbers from a callus of nucellar origin by a sequential treatment of ovules in a medium containing 5 μM 2,4-D or naphthoxyacetic acid (NOA) plus 1 μm BA, followed by a medium supplemented with NOA (10 μM) plus BA (1 μM) and finally in a medium devoid of auxin and cytokinin (Srinivasan and Mullins, 1980) (Fig. 5.2). Somatic embryogenesis has also been reported on nucellus-derived calluses formed on ovules of *Carica papaya* (papaya) (Litz and Conover, 1982) and *Mangifera indica* (Litz, Knight and Gazit, 1984) cultured in the presence of 20 percent coconut milk. When the nucellar callus of papaya is transferred to a medium containing 2,4-D, budding of the existing globular embryoids occurs, leading to a severalfold increase in the frequency of somatic embryogenesis (Litz and Conover, 1983). Placenta (Kavathekar and Ganapathy, 1973; Cheng and Raghavan, 1985), anther wall and connective (Rajasekaran and Mullins, 1979) are other reproductive tissues that regenerate somatic embryos in culture.

From the perspective of somatic embryogenesis in an indefinitely proliferating culture of undifferentiated callus, the endosperm has not lent itself readily to manipulation. The only convincing reports of somatic embryogenesis in endosperm calluses are in *Petroselinum hortense* (parsley) (Masuda, Koda and Okazawa, 1977), *Citrus grandis* (Wang and Chang, 1978), *Santalum album* (sandalwood) (Sita, Ram and Vaidyanathan, 1980) and *Actinidia chinensis* (Gui, Mu and Xu, 1982). Whereas endosperm callus of parsley regenerates embryoids without any exogenous growth hormones, the protocol that secures embryogenesis in sandalwood and *Citrus* endosperm callus cultures is their subculture from a medium containing 2,4-D to one with GA. The general recalcitrance of the endosperm to regenerate embryoids is probably linked to its ploidy status.

Except with reference to leaf callus of coffee, results thus far reviewed have not involved any work with explants of vegetative parts of forest trees and fruit trees. Admittedly, with woody perennials there are no constraints imposed on the use of the same protocols found successful in inducing somatic embryogenesis in herbaceous plants and shrubs. Indeed, it has seemed commercially feasible to exploit tissue culture techniques in inducing adventitious bud formation in calluses obtained from various trees as an alternate to conventional method of vegetative propagation. Nonetheless, only a few instances of generation of embryoids from calluses originating from explants of vegetative parts of trees have been reported. Moreover, these successful cases of somatic embryogenesis have invariably used zygotic or nucellar embryos, embryo segments (Hu and Sussex, 1971; Radojević, 1979; Radojević, Vujičić and Nešković, 1975; Litz, 1984; Kononowicz, Kononowicz and Janick, 1984), and explants from seedlings, young plants or dormant buds (Wilson and Street, 1975; Krul and Worley, 1977; Bapat and Rao, 1979; Gharyal and Maheshwari, 1981). In one case in which a callus was induced from shoot explants of a 20- to 25-year-old sandalwood tree, a continuously growing suspension culture exhibiting somatic embryogenesis at high frequency is obtained (Sita, Shoba and Vaidyanathan, 1980).

Monocots. Despite many experiments conducted by different investigators to induce somatic embryogenesis in monocots, there have been fewer successes than with dicots. This is primarily because of an ingrained feeling that monocots are recalcitrant in tissue culture and the consequent reluctance of investigators to invest in experiments whose conclusions were foregone. Following the well-publicized success with carrot and other dicots, somatic embryogenesis from calluses or suspension cultures was reported in *Asparagus officinalis* (Wilmar and Hellendoorn, 1968; Steward and Mapes, 1971), *Iris* (Reuther, 1977), *Gasteria verrucosa, Haworthia fasciata* (Beyl and Sharma, 1983), *Bellevalia romana* (Lupi, Bennici and Gennai, 1985) and a few palms (Tisserat, 1979; Reynolds and Murashige; 1979; Ahée et al., 1981). In stem segments of *A. officinalis,* the conditions that elicit embryogenesis consist of a complex series of sequential treatments in different media containing synergistic combinations of growth hormones such as induction of callus in White's medium supplemented with coconut milk and NAA, separation of cells and cell clumps in a medium in which NAA is replaced by 2,4-D, induction of embryoids in a high salt medium supplemented with NAA and formation of root and shoot primordia by transfer of embryoids to a medium containing coconut milk and IAA (Steward and Mapes, 1971). These observations imply that progressive loss of capacity to differentiate by events that occur in an orderly sequence may be a necessity for survival of cells of certain plants and that a reversal of the process, as a necessity for further growth, may require another sequential series of steps.

In view of the economic importance of cereals and grasses, it was inevitable that the potential for somatic embryogenesis in members of Gramineae be explored. Norstog (1970) reported that barley embryos cultured in a medium containing kinetin frequently develop embryo-like structures that are complete except for a scutellum. Gamborg, Constabel and Miller (1970) found a suspension culture originating from the mesocotyl of *Bromus inermis* (brome grass) grown in a medium containing 2,4-D for more than 2 years to behave typically like a carrot cell suspension by regenerating embryoids when challenged in a medium lacking 2,4-D. Following these studies, many published reports of plant regeneration from cultured tissues of cereals and grasses would appear to be cases of regeneration through the organization of a multicellular shoot meristem. During the past 5 years, Vasil's group has achieved spectacular success in inducing somatic embryogenesis in a wide range of cereals and grasses by judicious choice of explants, particularly with regard to their physiological age at culture, use of selected hormonal additives and unorthodox manipulation of the medium at critical stages of culture. The example that vividly demonstrates this is *Pennisetum americanum* (Vasil and Vasil, 1981a, 1982b). Culture of immature embryos excised from 10- to 15-day-old caryopses or of segments of unopened inflorescence on a solid medium supplemented with 2.5 mg/liter 2,4-D is found to regenerate a callus. Transfer of the callus to a liquid medium supplemented additionally with 5 percent coconut milk produces a suspension culture consisting predominantly of richly cytoplasmic embryogenic cells that spontaneously develop into globular or heart-shaped structures. Even at this stage, the multicellular entities are not fully committed to the embryogenic pathway; embryoids with characteristic morphology of cereal embryos are obtained when the suspension culture is plated on an agar medium containing ABA but lacking 2,4-D (Fig. 5.3). In an alternate pathway of somatic embryogenesis displayed by cultured immature embryos, some of the superficial cells of the scutellum divide repeatedly to form embryoids or regenerate an embryogenic callus from which embryoids subsequently arise (Vasil and Vasil, 1982a). With some variations in protocol, surgically separated parts of mature embryos of *P. americanum* consisting of the shoot apex and subjacent leaf primordia are also found to yield an embryogenic callus and embryoids in culture (Botti and Vasil, 1983).

The formation of somatic embryos in wheat is another example. Although embryoids formed in tissue cultures of wheat initiated from young inflorescence axes and immature embryos germinate precociously (Ozias-Akins and Vasil, 1982), an embryogenic callus formed in double-strength mineral salt medium readily forms normal full-term embryoids (Ozias-Akins and Vasil, 1983). By subtle manipulation of the medium composition, somatic embryogenesis has also been reported on embryo-derived calluses of *Lolium multiflorum* (Dale, 1980), *Pennisetum americanum* × *P. purpureum* hybrid (Vasil and Vasil, 1981b), *Panicum maximum* (Lu and Vasil, 1982), *Zea mays* (Lu, Vasil and Vasil, 1983), *Dactylis glomerata* (orchardgrass) (McDaniel, Conger and Grahamn, 1982) and *Secale cereale* (Krumbiegel-Schroeren, Finger, Schroeren and Binding, 1984), on calluses derived from leaves and somatic tissues of anthers of *Pennisetum purpureum* (Haydu and Vasil, 1981), on leaf-derived calluses of *Sorghum bicolor* (Wernicke and Brettell, 1980), *Oryza sativa* (Wernicke, Brettell, Wakizuka and Potrykus, 1981), *Panicum maximum* (Lu and Vasil, 1981a) and *D. glomerata* (Hanning and Conger, 1982), on leaf sheath-derived callus of *Echinocloa oryzicola* (Takahashi, Sakuragi, Kamada and Ishizuka, 1984), on calluses originating from immature inflorescences of *Sorghum bicolor* (Brettell, Wernicke and Thomas, 1980), *S. arundinaceum* (Boyes and Vasil, 1984), *L. multiflorum* (Dale, Thomas, Brettell and Wernicke, 1981), *Panicum maximum* (Lu and Vasil, 1982), *P. miliaceum, P. miliare* (Rangan and Vasil, 1983a), *Pennisetum purpureum* (Wang and Vasil, 1982), *Secale cereale* (Krumbiegel-Schroeren et al., 1984), *Echinochloa crusgalli* (Wang and Yan, 1984), *Setaria italica* (Xu, Wang, Yang and Wei, 1984) and *Hordeum vulgare* × *Triticum aestivum* hybrid (Chu et al., 1984) and root callus of *Panicum miliaceum* (Heyser and Nabors, 1982b). Suspension cultures of calluses derived from leaves and apical meristems of *Saccharum*

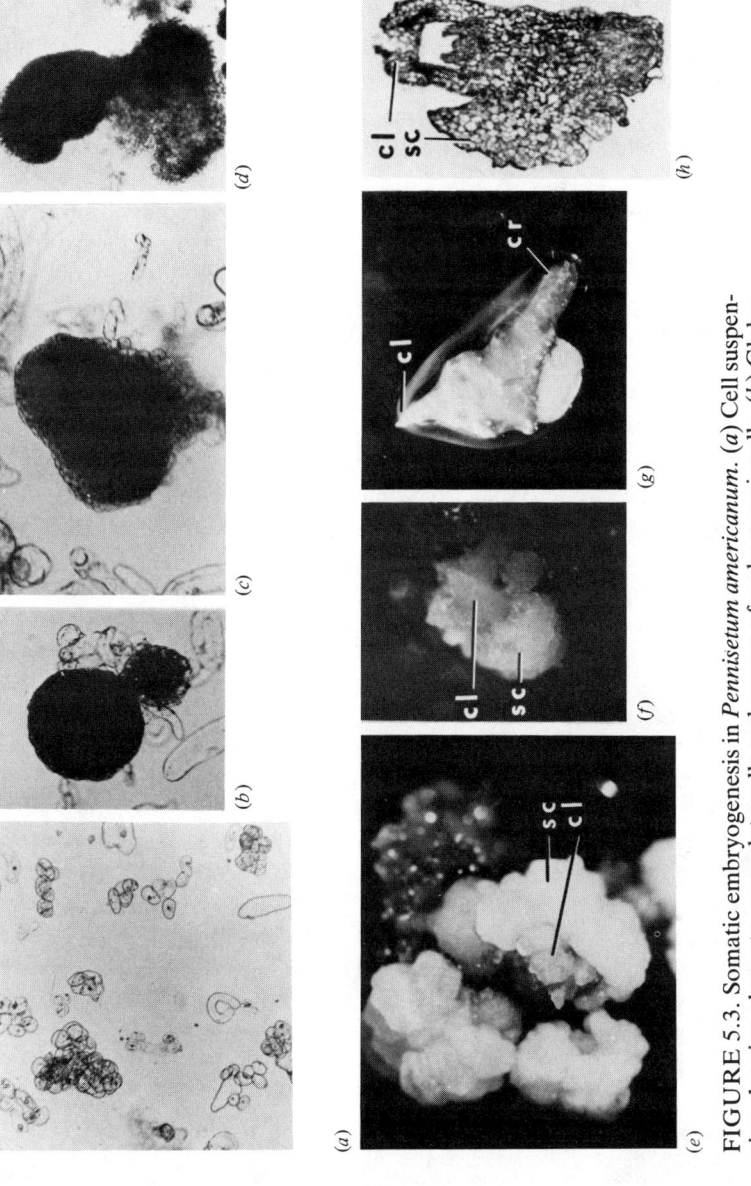

FIGURE 5.3. Somatic embryogenesis in *Pennisetum americanum*. (*a*) Cell suspension showing elongate, vacuolate cells and groups of embryogenic cells. (*b*) Globular embryoid. (*c*) Cotyledonary stage embryoid. (*d*) Formation of the notch in the embryoid, where shoot apex and coleoptile are formed. (*e*) Organization of the embryoid on the callus (sc = scutellum, cl = coleoptile). (*f*) Formation of scutellum (sc) and coleoptile (cl) on the embryoid. (*g*) Embryoid with coleoptile (cl) and coleorhiza (cr) at opposite ends. (*h*) Section of an embryoid (cl = coleoptile, sc = scutellum). (From Vasil and Vasil, 1981a; from negatives supplied by Dr. I. K. Vasil.)

officinarum (sugarcane) (Ho and Vasil, 1983b; Ahloowalia and Maretzki, 1983), immature embryos and young inflorescences of *Panicum maximum* (Lu and Vasil, 1981b), immature embryos of barley (Kott and Kasha, 1984) and *Secale cereale* (Lu, Chandler and Vasil, 1984) and somatic embryos of *Dactylis glomerata* (Gray, Conger and Hanning, 1984) are also a prolific source of somatic embryos. These successful reports of somatic embryogenesis, and others in which somatic embryogenesis was not explicitly demonstrated (Heyser and Nabors, 1982a; Nabors, Heyser, Dykes and DeMott, 1983; Siriwardana and Nabors, 1983), have led to the view that cultured explants of cereals and grasses produce different types of calluses that differ in their regenerative potential and that failure to select the appropriate callus might have accounted in a large measure for the previous lack of success in inducing somatic embryogenesis in members of Gramineae. Two aspects of somatic embryogenesis in gramineam plants that we do not understand at all and that have limited the use of the system in biochemical and genetic investigations are the slow regeneration of the callus and the rapid loss of its embryogenic competence.

Direct somatic embryogenesis

The induction of a callus on the cultured explant and the transfer of the loosely dissociated cells to a medium of a different composition to generate somatic embryos present a complicated situation, particularly in light of the bewildering array of hormonal requirements among different species thus far investigated. From this point of view, the feasibility of somatic embryogenesis directly on the cultured explant might appear to simplify the system and lead to a better understanding of the developmental controls that operate before or as somatic embryos appear. A convincing demonstration of direct somatic embryogenesis was provided by Konar and Nataraja (1965) in the herbaceous annual, *Ranunculus sceleratus*. Virtually any excised organ of this plant can be cultured to yield an embryogenic callus, but in this first report young flower buds were cultured in a medium supplemented with 10 percent coconut milk and 1.0 mg/liter IAA. The dense callus that appears at the cut end subsequently differentiates numerous embryoids without further treatment. This was a remarkable observation, coming as it did in the wake of the unfolding story about somatic embryogenesis in carrot. Equally remarkable was the behavior of embryoids, which germinate precociously into seedlings and bear a second generation of embryoids freely exposed along the entire length of hypocotyl and stem. Other instances of direct embryogenesis on embryoids and seedlings derived originally from embryoids have been described in carrot (Homès, 1968), haploid *Atropa belladonna* (Rashid and Street, 1974a), *Ribes rubrum* (red currant) (Zatykó, Simon and Szabó, 1975), *Ammi ma-*

jus (Grewal, Sachdeva and Atal, 1976), haploid *Brassica napus* (Thomas, Hoffmann, Potrykus and Wenzel, 1976), *Vitis vinifera* (Krul and Worley, 1977), *Eschscholzia californica* (Kavathekar et al., 1977), *Brassica oleracea* (cauliflower) (Pareek and Chandra, 1978) and *Ipomoea batatas* (Liu and Cantliffe, 1984).

Efforts have also been made to demonstrate embryogenesis without the intervention of a callus in leaf explants of coffee (Dublin, 1981), apple (Liu, Sink and Dennis, 1983), *Dactylis glomerata* (Conger, Hanning, Gray and McDaniel, 1983) and coconut (Raju, Kumar, Chandramohan and Iyer, 1984), ovules of *Ribes rubrum* (Zatykó, Simon and Szabó, 1975), *R. nigrum* (black currant) (Zatykó, Kiss and Szalay, 1981) and *Coffea canephora* (Lanaud, 1981), nucellar explants of monoembryonic species of *Citrus* (Rangan, Murashige and Bitters, 1968), *Cyananchum vincetoxicum* (Haccius and Hausner, 1976), apple (Eichholtz, Robitaille and Hasegawa, 1979) and *Mangifera indica* (Litz et al., 1982) and embryos of *Ilex aquifolium* (Hu and Sussex, 1971), *Theobroma cacao* (cacao) (Pence et al., 1980), *Fragaria × Ananassa* (strawberry) (Wang, Wergin and Zimmerman, 1984), *Trifolium repens*, *T. pratense* and *Medicago sativa* (Maheswaran and Williams, 1984). In leaf explants of *D. glomerta* anatomical studies have unequivocally demonstrated that the embryoid arises by the dedifferentiation of a differentiated cell of the explant (Conger et al., 1983), whereas in others such proof is lacking. There is a possibility that in some of the above examples embryoids are formed on an incipient callus regenerated on the cut surface of the explant, as recently shown in punched leaf discs of *Coffea canephora* which almost gives the impression of direct embryogenesis (Pierson, van Lammeren, Schel and Staritsky, 1983).

Cultured leaf segments of *Dactylis glomerata* will serve as an example for comparative evaluation of the underlying basis for direct somatic embryogenesis and embryogenesis through the intervention of a callus (Conger et al., 1983). Segments taken from the more basal parts of the leaf are so dominated by a callus-forming habit that they never form embryoids directly on the explants; by contrast, those excised from the distal portions display direct embryogenesis. Comparison of the response between two leaves of different ages shows that the capacity for formation of embryoids directly on the explant shifts slightly toward the base of the older of the two leaves (Fig. 5.4). In considering these interesting observations, we have to keep in mind the fact that a juvenile grass leaf has an increasing basipetal gradient of meristematic activity and that the meristematic area in a mature leaf is very much restricted to its base. Apparently, direct embryogenesis is perpetuated by a recrudescence of growth in the fully differentiated cells in the more apical part of the leaf, whereas cells that are actively dividing tend to lapse into a callus-forming habit. Thus, the cytological state of the leaf cells at the time of culture is more significant than any special additives

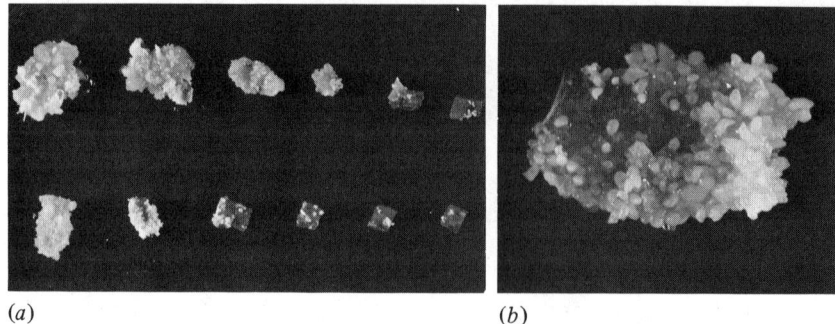

(a) (b)

FIGURE 5.4. Direct embryogenesis from leaf segments of *Dactylis glomerata*. (*a*) Segments from the innermost emerged leaf (*top row*) and next leaf outward (*bottom row*) about 6 weeks after inoculation. Basal segments are to the left and progressively more distal segments to the right. Callus formed on the more basal segments changes to a more direct embryogenic response on distal segments. (*b*) Leaf segment showing the direct origin of numerous embryoids from the explant. (From Conger et al., *Science* 221: 850–851, 1983. Copyright 1983 by the American Association for the Advancement of Science; photographs supplied by Dr. B. V. Conger.)

to the medium in determining the pathways toward morphogenesis. However, there is reason to believe that morphogenetic expression of the leaf cells is also dictated to some degree by genetic, epigenetic and physiological changes in the cultured explant.

Embryogenesis from isolated protoplasts

Since the fertilized egg is only partially covered by a cell wall, an isolated protoplast liberated from the limiting restraint of the wall comes close to resembling a zygote. Thus, the embryogenic episode of a cell divested of its cell wall might shed light on the problems involved in the development of a partially naked zygote enclosed in the embryo sac into an embryo. Unfortunately, this expectation has not been fulfilled, as in all instances thus far investigated, naked protoplasts form embryoids only after they reform a new cell wall. Nonetheless, investigations on the embryogenic development of isolated protoplasts have provided useful insights into the conditions necessary to maximize and stabilize the yield of protoplasts and to promote their transformation into embryoids.

In the first reports of embryogenesis from isolated protoplasts, pieces of carrot root (Kameya and Uchimiya, 1972) and suspension cultures from root and petiole of carrot seedling (Grambow, Kao, Miller and Gamborg, 1972) were used as starting materials. The protoplasts were cleared by filtration and centrifugation and suspended in an isotonic culture medium. They were subsequently cultured in the same medium, the osmolality of which was reduced in a stepwise fashion with progress of culture. Generally, the period leading up to the formation of a new wall was marked by changes in the size and shape of the protoplasts and, irrespective of their origin, the first one or two divisions in the isolated protoplasts were observed about 6 days after culture. Various fates can befall a reformed cell, one of which is repeated divisions to form a cell cluster and transformation of daughter cells into embryo-like structures. Characteristically, the requirements for embryogenic development of cells originating from either cell type by way of isolated protoplasts reflect those of the respective parent cells. Cell clusters originating from protoplasts isolated from suspension cultures regenerate embryoids upon transfer to a medium lacking 2,4-D, whereas those formed from protoplasts of root cells require a medium enriched with coconut milk. Overall, these results show that despite the completely different pressures to which isolated protoplasts are exposed, once they reform a wall their subsequent morphogenesis is as highly precise and predictable as that of a totipotent cell.

As the study of reconstitution of plants from protoplasts was extended to other genera, protoplasts isolated from cell suspension cultures of *Atropa belladonna* (Gosch, Bajaj and Reinert, 1975), *Citrus sinensis* (Vardi, Spiegel-Roy and Galun, (1975), *Pennisetum americanum* (Vasil and Vasil, 1980), *P. purpureum* (Vasil, Wang and Vasil, 1983), *Panicum maximum* (Lu, Vasil and Vasil, 1981) and *Santalum album* (Rao and Ozias-Akin, 1985) and from embryoids of *Brassica napus* (Kohlenbach, Wenzel and Hoffmann, 1982) were also found to undergo the embryogenic type of development and produce embryoids and plantlets with relative ease. The pathways followed by protoplast preparations from these species toward embryogenesis involve cell wall reformation, cell division, formation of cell cluster and regeneration of embryoids. To round out this discussion it should perhaps be pointed out that in such plants as *Asparagus officinalis* (Bui Dang Ha, Norreel and Masset, 1975), *Ranunculus sceleratus* (Dorion, Chupeau and Bourgin, 1975), *Nicotiana tabacum* (Lörz, Potrykus and Thomas, 1977), haploid *Atropa belladonna* (Bajaj et al., 1978), haploid *N. sylvestris* (Facciotti and Pilet, 1979), *Medicago sativa* (Kao and Michayluk, 1980; Lu, Davey and Cocking, 1983), *Lycopersicon peruvianum* (Zapata and Sink, 1981), *Brassica napus* (Li and Kohlenbach, 1982) and *Vigna aconitifolia* (Shekhawat and Galston, 1983), somatic embryogenesis has emerged as a major means of regeneration of mesophyll protoplasts. In *R. sceleratus*, analogous to the behavior of callus-derived embryoids, epidermal cells of embryoids originating from protoplasts also regenerate an abundant crop of secondary embryoids (Dorion, Godin and Bigot, 1984). An explanation of the ease with which embryogenesis is initiated in cells reconstituted from mesophyll protoplasts is not readily available; based on current evidence, it seems fair to say that mesophyll protoplasts could be used in some plants to demonstrate totipotency where the easier route of cell culture has failed.

Cytology of somatic embryogenesis

Few detailed light and electron microscopic studies have been conducted on the cytology of somatic embryogenesis. In many calluses and cell-suspension cultures, it has remained a moot question as to whether the embryoid is formed from a single cell or from a group of cells; in other cases, a critical distinction has not been established between a normal cell and a cell destined to give rise to an embryoid. This uncertainty is attributable partly to the great plasticity of free cells and partly to the difficulty in separating embryogenic cells from others of different lineages in a chaotic jumble of cells. A wide array of observations made on carrot cell suspension cultures grown in the presence of 2,4-D in the culture medium supports this line of thinking. From the earliest studies, for example, it has become clear that certain large vacuolate cells seen in the suspension culture tend to follow a sequence of divisions analogous to zygotic embryos. However, this route to embryogenesis is but rare, as the vacuolate cells, by repeated divisions, end up as clusters of cells designated as proembryogenic masses. Although the cell clusters bear little resemblance to the generalized cell groupings of early-stage embryoids, isolated cells of this inchoate aggregate give rise to embryoids. The combination of increased divisions and decreased differentiation confers a meristematic state on the cell cluster, which is an essential prerequisite for embryogenic competence of its cells (Jones, 1974b). These observations on the origin of embryoids in a single species do not offer a wide base for generalizations applicable to other species in which somatic embryogenesis has been reported; we are left with the feeling that until truly synchronous cultures of single cells of high embryogenic potency are obtained, the origin of embryoids and cytology of somatic embryogenesis in many plants will remain obscure.

In carrot callus growing on a solid medium, embryoids are formed from the large cells of certain meristematic nodules, considered analogous to proembryogenic masses, buried in the callus (Harry, Mestre and Guignard, 1977). In recent studies on somatic embryogenesis in Gramineae, both unicellular (Vasil and Vasil, 1982a; Ho and Vasil, 1983a; Botti and Vasil, 1984) and multicellular (Wernicke, Potrykus and Thomas, 1982) progenitors have been evoked to explain the origin of embryoids from the callus. In the well-illustrated account of the ontogeny of somatic embryos from cultured immature inflorescence of *Pennisetum americanum* (Botti and Vasil, 1984), embryogenic cells formed in a confluent callus are described as being richly cytoplasmic, starch-filled and somewhat thick walled. The last-mentioned property probably places constraints on the embryogenic cell from interacting with the large mass of nonembryogenic cells of the callus, as its developmental program is deflected in the embryogenic pathway. Subsequent cell separation and a sequence of internal segmentations in the embryogenic cell lead to the formation of the embryoid.

The ultrastructural cytology of somatic cells of carrot bathed in an auxin-containing medium has provided a standard of reference for assessing the changes that occur during their transformation into embryoids when confronted with a medium devoid of auxin (Halperin and Jensen, 1967; Wochok, 1973; Wilson, Israel and Steward, 1974; Street and Withers, 1974). In a proembryogenic mass, cells found toward the periphery are considered the progenitors of embryoids, the rest of the cells in the clump probably acting as a nurse tissue to the emerging embryoid. Our minimal ultrastructural image of such a cell has a munificient supply of starch-filled plastids, round, oval to irregular mitochondria, a small number of lipid bodies, free ribosomes, abundant ER studded with polysomes and multivesicular bodies of presumed Golgi origin. Although not seen in carrot cells, a striking organization of rough ER stacked in extensive parallel sheets has been described in the potentially embryogenic cells of *Corylus avellana* (Vujičić, Radojević and Nešković, 1976). The membranes of the ER are associated with a randomly or orderly arrangement of numerous ribosomes. Collectively, these subcellular features in the embryogenically competent cells of carrot and *C. avellana* are indicative of their enhanced metabolic activity preparatory to embryogenesis. The ultrastructural consequence of embryogenic induction in a carrot cell transferred to a medium lacking auxin is implicit in the increase in free ribosomes, decrease in ER, loss of polysomes and increase in the number of Golgi vesicles and lipid bodies. Although random arrays of microtubules predominate the cytoskeletal framework of cells grown in the presence of auxin, microtubules appear to be arranged in multiple layers in the cells of early-stage embryoids (Halperin and Jensen, 1967; Wochok, 1973; Wilson et al., 1974; Street and Withers, 1974).

Button, Kochba and Bornman (1974) described the ontogeny and fine structure of embryoids originating in the habituated nucellar callus of Shamouti orange. Since virtually all cells of the callus are embryogenic, it is relatively easy to characterize the progenitor cells of embryoids. According to these investigators, embryogenesis is usually initiated in single cells located on the periphery of and within an existing proembryoid. Although the ultrastructural image of these cells is undistinguished, the cells are characterized by the presence of a thick wall lacking plasmodesmata. A globular embryoid is formed by divisions of the cell within the confines of this wall, which disintegrates to release the embryoid. The role of the thick wall in inducing divisions of the embryoid mother cell is not known at this stage, whereas the disruption of plasmodesmatal connections indicates that the cell embarking on embryogenesis is physically isolated from its neighbors. Whether physical isolation of the cell is essential for embryogenesis is a debatable issue, and this controversy has been partially covered elsewhere (Raghavan, 1976b).

The involvement of a single cell of the epidermis in embryogenesis has been demonstrated in the hypocotyl of embryoid-derived seedling of *Ranunculus sceleratus* (Konar et al., 1972a). The embryoid mother cells scattered singly or in small groups between regular epidermal cells typically consist of a large central nucleus surrounded by a dense cytoplasm rich in free ribosomes or polysomal clusters, amyloplastids and spherosomes. By contrast, nonembryogenic epidermal cells have a peripheral cytoplasm with sparse organelles. The ultrastructural profile of the embryogenic epidermal cells corresponds closely with that of the superficial cells of an embryogenic callus obtained from flower buds of *R. sceleratus;* it is believed that embryoids regenerating on the callus have their origin in these cells (Thomas, Konar and Street, 1972). What is remarkable about the ultrastructural cytology of somatic embryogenesis on *R. sceleratus* seedlings is that at some stage in its early development, a few epidermal cells of the hypocotyl are reprogrammed as embryoid mother cells, whereas others continue their program of terminal differentiation as epidermal cells. Thus, the same genome is able to display two different phenotypic expressions.

In summary, our knowledge of the cytology of somatic embryogenesis is limited to studies of a few model systems. Even in those cases examined, a general pattern of changes that occur in the somatic cells preparatory to embryogenesis has not emerged. The possibility of subtle ultrastructural changes during the transition of a fully differentiated somatic cell into an embryoid is an attractive concept. As such, it deserves to be explored widely in systems that are uniquely suited to this study.

Physiology of somatic embryogenesis

In the widely investigated carrot cell suspension, such factors as clonal variation, size of the inoculum, conditioning of the medium and environment are known to affect embryogenic development; in other plants, however, the effects of these factors are neither general nor inevitable. The current position is that the critical issues in approaching the physiology of somatic embryogenesis lie with the composition of the nutrient medium and with the role of hormonal substances and unidentified inhibitors. It is to these issues that the discussion in this section is addressed.

Nutritional factors

Nearly all work done on somatic embryogenesis has involved the culture of tissues in a medium which has mineral salts and sucrose as its base. Although these components of the medium, along with amino acids, vitamins and other organic addenda that are also added from time to time undoubtedly control the basal metabolism of the tissues, they are not critical variables that play a unique or specific role in embryogenesis. Nonetheless, a number of observations bear in varying degrees on the importance of the type of nitrogen (N) present in the medium and on the role of potassium (K^+) in somatic embryogenesis. In particular, the extent to which somatic embryogenesis in carrot is controlled by N supplied in the medium has occasioned much discussion between the opposing viewpoints that there is a requirement for NH_4^+ ion and that the form of N supplied is unimportant. Halperin and Wetherell (1965) reported that unorganized proliferation as a rule characterizes the growth pattern of a callus derived from the secondary phloem of wild carrot maintained in a medium containing 60 mmoles NO_3^- as the sole source of N. The tissues display little inclination to form embryoids when challenged in an induction medium, unless the original medium included at least 5 mmoles NH_4^+ ion. It was also found that cells of the petiole-derived callus of carrot grown in a medium containing NO_3^- fail to regenerate embryoids in the induction medium, even though the latter contains NH_4^+ as well. From these results it was suggested that NH_4^+ is essential for the expression of embryogenic competence in carrot cells, whereas NO_3^- probably facilitates the subsequent development of embryoids. The basis for this interpretation was questioned by Reinert, Tazawa and Semenoff (1967), who showed that somatic embryogenesis in cambial explants of domestic carrot occurs with NO_3^- as the sole source of N. The significance of this work is that the actual concentration of N in the medium, and not its form, is decisive in the somatic embryogenesis of carrot. This study was followed by a more extensive investigation of the intracellular levels of N in embryogenic carrot cells; the revealing aspect of this work is that although NH_4^+ in the medium is not essential for embryogenesis, a certain level of intracellular NH_4^+ is prerequisite to the process. Apparently, this level is reached by a moderately high level of NO_3^- in the medium (Tazawa and Reinert, 1969). While reaffirming a role for reduced nitrogen for embryogenesis in carrot cell suspension cultures, Wetherell and Dougall (1976) have suggested that the differences reported from the two laboratories may at least in part reflect differences in experimental protocol and method of data evaluation, exacerbated by the use of wild versus domestic carrot. More likely, somatic embryogenesis in carrot is regulated by factors that extend beyond the direct utilization of NO_3^- or NH_4^+ in protein synthesis.

Despite the uncertainty with respect to a specific role for NH_4^+ ion in somatic embryogenesis in carrot, there is general agreement that some form of reduced N is essential for high-frequency embryogenesis. In carrot cell cultures differing in age, origin and variety, mixtures of amino acids, casein hydrolyzate or individual amino acids such as glutamic acid and alanine and the amide glutamine are found to serve as the sole source of N for embryogenesis (Reinert, Tazawa and Semenoff, 1967; Sussex and Frei, 1968; Wetherell and Dougall, 1976; Kamada and Harada, 1979b). As shown by Dougall and Verma (1978), even NH_4^+ ion can serve as the sole source of N, provided the pH value of the medium is maintained at a constant level during the induction period. Kato and Takeuchi (1966) found that glycine, asparagine, glutamine and arginine are as effective as NH_4^+ in stimulating embryogenesis in hypocotyl segments of carrot.

Another facet of the effect of reduced N on somatic embryogenesis in carrot deserves mention. This is the paradoxical observation (Kamada and Harada, 1979b) that if cells cultured in a medium containing 2,4-D are transferred to an auxin-free medium containing a source of reduced N, embryogenesis ensues; in the absence of reduced N in the induction medium, root formation occurs (Table 5.1). On the basis of these results, the switch from an undifferentiated state that occurs in the presence of 2,4-D to one that overtly expresses embryogenic development seems to require a major infusion of reduced N.

Another link in the chain of evidence that relates somatic embryogenesis to the form of N in the medium is evident in the report that the presence of NH_4^+ in the medium can change the developmental fate of alfalfa callus cultures from root-forming potential to embryo-forming potential. Somatic embryos are formed when the callus is transferred to a medium containing a minimum of 12.5 mM NH_4^+. Although roots are formed from the callus when NH_4^+ ion is absent from the medium, at a fourfold higher concentration of NH_4^+ embryoids are produced from cells exposed to root-inducing conditions (Walker and Sato, 1981). Among a number of amino acids tested, proline, alanine, arginine, lysine, serine and ornithine and the amide asparagine are found to enhance somatic embryogenesis in alfalfa cultures in the presence of a modest level of NH_4^+, whereas glutamine stimulates the process independent of the presence of NH_4^+ in the medium (Stuart and Strickland, 1984a,b). Previous attempts to obtain somatic embryogenesis in soybean were bedeviled by a sporadic occurrence of the event and by the failure of embryoids to grow beyond a certain stage (Beversdorf and Bingham, 1977; Phillips and Collins, 1981). A balanced growth of embryoids was achieved by a scheme of recurrent selection from a callus involving primarily a change from 2,4-D-enriched medium containing 40 mM NH_4^+ to an auxin-free medium with 20 mM NH_4^+ (as ammonium citrate) and 40 mM NO_3^- (Christianson, Warnick and Carlson, 1983). Yet another

Table 5.1. *Embryoid and root formation in carrot cell cultures grown on medium with and without reduced N*

Media used during 2,4-D treatment	Duration of 2,4-D treatment (weeks)	Media used after 2,4-D treatment	Cultures producing embryos (%)	Cultures producing roots (%)
MS	0	W	0.0	0.0
	1	W	14.7	97.1
	2	W	0.0	87.5
	3	W	2.0	74.5
	4	W	11.1	66.7
MS	0	MS	0.0	0.0
	1	MS	82.8	17.2
	2	MS	92.3	10.3
	3	MS	86.4	0.0
	4	MS	100.0	14.3
W	0	W	0.0	0.0
	1	W	0.0	90.9
	2	W	13.3	96.7
	3	W	6.1	84.8
W	0	MS	0.0	0.0
	1	MS	71.0	35.5
	2	MS	97.0	9.1
	3	MS	97.7	4.7

Note: Calluses were transplanted to MS or W medium without 2,4-D after 1–4 weeks of growth on MS or W medium with 2,4-D (MS = medium with reduced N; W = medium without reduced N; 2,4-D = 2,4-dichlorophenoxyacetic acid). Results are shown as percentages of cultures producing embryoids or roots 3 weeks after transfer.
Source: Kamada and Harada (1979b).

case of a specific requirement for reduced N for somatic embryogenesis is *Gossypium klotzschianum,* in which embryoids appear only when the cell suspension is exposed to glutamine (Price and Smith, 1979).

In the unfolding series of investigations on carrot, a new variable has been introduced, with the report that the intracellular amount of K^+ is a limiting factor in somatic embryogenesis in cell-suspension cultures of a domestic variety of carrot (Tazawa and Reinert, 1969). Thus, only a few embryoids appear in suspension cultures grown in a medium that is qualitatively overhauled by removing all KNO_3 and retaining a moderate level of NH_4NO_3. When the K^+ level in the medium is raised by the addition of KH_2PO_4 in an amount equivalent to KNO_3, there is a slight increase in the number of embryoids formed, whereas the same amount of NaH_2PO_4

strongly inhibits embryogenesis. Following this lead, data have been obtained that seem to confirm a role for K^+ in somatic embryogenesis in cell-suspension cultures of wild carrot as well (Brown, Wetherell and Dougall, 1976). The effect of K^+ in this system is particularly striking, since as little as 1 mM K^+ salt supports proliferative growth of cells; yet at least 20 mM is required for optimal embryogenesis. The precise role for K^+ in somatic embryogenesis of carrot cells is not resolved, but the effects of this ion are of considerable interest as a possible developmental model.

Some recent studies have dealt with the comparative effects of carbohydrate sources on somatic embryogenesis. As alluded to earlier, sucrose has been invariably used for this purpose, and rarely has any other sugar been as conspicuously efficient. Wild carrot cell suspensions have been shown to utilize, not only sucrose, but glucose, fructose, mannose, maltose, raffinose and stachyose for growth and embryogenesis (Verma and Dougall, 1977). In this system there is an apparent relationship between *myo*-inositol biosynthesis and carbohydrate supply in the medium, as both glucose and galactose supplied in the medium function as precursors of *myo*-inositol biosynthesis (Verma and Dougall, 1979a,b). A broad range of carbohydrates is also effective in stimulating embryogenesis in nucellar calluses from *Citrus* cultivars, but galactose, lactose and glycerol are superior in this respect (Kochba, Spiegel-Roy, Saad and Neumann, 1978; Kochba, Spiegel-Roy, Neumann and Saad, 1982; Ben-Hayyim and Neumann, 1983). These compounds by no means exhaust the list of possibilities, but they serve to emphasize the unique capabilities of diverse carbohydrates to sustain morphogenesis.

Hormonal factors

Over the past 20 years, a series of reports have provided extensive documentation of the hormonal requirements for induction of somatic embryogenesis in plants. Although an examination of these reports is beyond the scope of this chapter, it can be said that collectively these studies have raised more questions than they have answered, and that they have made us aware that a unified picture of the hormonal control of somatic embryogenesis applicable on a broad level is not likely to emerge in the foreseeable future. Both auxins and cytokinins have been frequently used in media concocted to induce embryogenesis in cultured explants, but it is the use of the former that has aroused considerable interest. As already stated, the standard practice of inducing embryogenesis in a carrot cell suspension reared in a medium containing 2,4-D is to transfer it to a medium lacking auxin or to one containing a reduced concentration of auxin. The controlling influence of 2,4-D, thus clearly indicated, has provided the main focus for much of the physiological investigations on somatic embryogenesis in carrot. Several questions have appeared from time to time in these studies. Is auxin inhibitory to embryogenesis? Is endogenous auxin involved in embryogenesis? What is the role of auxin in inducing embryogenic competence in cells? These difficult questions have ramifications and have tended to make the role of auxin in somatic embryogenesis in carrot somewhat difficult to evaluate.

Because calluses generated from carrot tissue explants continue to increase in mass without producing embryoids when grown in a medium containing 2,4-D, most investigators beginning with Halperin and Wetherell (1964) have emphasized the inhibitory effect of auxin on embryogenesis. The many reports of successful induction of embryoids in calluses and cell suspension cultures of plants grown in an auxin-enriched medium by transferring them to an auxin-free medium or by lowering the concentration of auxin or by substituting weak or less stable auxins are consistent with this interpretation. The existence of a functional connection between a reduced level of 2,4-D in the cells and embryogenesis has also been supported by the demonstration that embryogenic induction in carrot cells, transferred from a 2,4-D medium to one devoid of the auxin, is accompanied by a large-scale release of auxin into the ambient medium (Montague, Enns, Siegel and Jaworski, 1981). However, for a period of time, it remained difficult to accept the role of auxin as an inhibitor of embryogenesis, as there is no assurance that traces of auxin carried over in the inoculum are purged before its transfer to an induction medium. More recent experiments have overcome this objection by extensive washing of cells before transfer and it seems safe to generalize that 2,4-D is a potent inhibitor of somatic embryogenesis in carrot in short-term experiments. But the same cannot be said of other auxins. For example, continuous treatment of carrot hypocotyl segments with NOA, and to some extent with IAA, NAA and indolebutyric acid (IBA) for a 5-week period has been found to favor somatic embryogenesis. Actually it turns out that transfer of tissues to a medium lacking auxin after treatment for a 2-week period with NAA, IBA or IAA also induces somatic embryogenesis to the same extent found with continuous treatment (Kamada and Harada, 1979a). These results raise the question of whether uniformity in the use of the variety of carrot, standardization of protocols followed and duration of treatment would have produced a more generalized picture of the role of auxins in somatic embryogenesis in this system.

There is mounting evidence at this stage to link somatic embryogenesis in carrot cells to changes in the endogenous levels of auxin. Recently, Fujimura and Komamine (1979a) demonstrated the presence of IAA in carrot cells by the classic *Avena* coleoptile curvature method, but there is little change in auxin activity during embryogenesis. In another facet of this study it was found that treatment of a cell suspension of carrot with 2,4-D

Table 5.2. *Effect of treatment of carrot cell suspension with 2,4,6-T or 2,4-D on subsequent embryogenesis*

Culture media		No. of embryoids per tube
First 4 days	Final 7 days	Mean ± SD[d] (n = 10)
BM[a]	BM	445 ± 54
BM	2,4,6-T[b]	349 ± 40
BM	2,4-D[c]	351 ± 63
2,4,6-T	2,4,6-T	11 ± 2
2,4,6-T	BM	46 ± 9
2,4-D	2,4-D	0

Note: Following growth in an auxin-free medium or media containing 2,4-D or 2,4,6-T for 4 days, the cell suspension was transferred again to the auxin-free medium or to media containing 2,4-D or 2,4,6-T for 7 days. The number of embryoids formed was counted 11 days after transfer. The number of embryoids formed in the basal medium 4 days after transfer was 40 ± 5.
[a]Basal medium free of hormones.
[b]10^{-5} M 2,4,6-T.
[c]5×10^{-7} M 2,4-D.
[d]Standard deviation.
Source: Fujimura and Komamine (1979a).

or the antiauxin 2,4,6-trichlorophenoxyacetic acid (2,4-6-T) during the first 4 days inhibits embryogenesis when examined 7 days later, whereas these same compounds do not affect embryogenesis when supplied during the last 7 days of culture (Table 5.2). The sensitivity of carrot cells to auxin and antiauxin suggests that a certain critical level of endogenous auxin may be important during the early period of embryogenesis, although it is unlikely to be a limiting factor. The effect of 2,4-D also does not appear to be caused by an auxin-mediated evolution of ethylene (Tisserat and Murashige, 1977a).

Another experimental approach to relate somatic embryogenesis in carrot to endogenous auxin level has involved the use of variant cell lines. Sung (1979) found that a cell line of wild carrot selected for resistance to 5-methyltryptophan accumulates an excessive amount of IAA and continues to proliferate as a callus in a medium lacking 2,4-D. Moreover, these cells do not fare any better in producing embryoids in the absence of auxin than wild-type cells grown in the presence of auxin, emphasizing again that the high level of endogenous auxin is responsible for suppression of embryogenesis. The use of variant cell lines promises to open new experimental avenues into the underlying mechanism of somatic embryogenesis in cell-suspension cultures initiated by auxin depletion from the medium.

We shall now ponder the role of auxin in the induction of embryogenic competence in cells. The key question is whether a cell is programmed for embryogenesis before it encounters auxin in the medium or after. Although it has been difficult to dissociate the role of auxin in promoting callus growth from its role in conferring embryogenic competence in the cell, the demonstration of direct embryogenesis in cultured epidermal strips of carrot (Kato, 1968) provides a way out of this dilemma. The fact that the epidermal cells are not transformed into embryoids unless 2,4-D is present in the medium suggests that embryogenic competence is attained after treatment of cells with auxin. The situation is somewhat obscured in calluses and cell suspensions of carrot where auxin increases the quantity of embryogenic cells by promoting division of vacuolate cells to form proembryogenic masses as well as by preventing the initiation of organized growth in them. Single cells in culture are rarely transformed directly into embryoids; a hallmark of attainment of embryogenic competence in a single cell is its division to form a cluster of cells constituting a proembryogenic mass. In line with this argument, Fujimura and Komamine (1980a) showed that a carrot cell suspension consisting predominantly (98%) of single cells does not undergo extensive embryogenesis in an auxin-free medium, but shows more than a twofold increase in the number of embryoids formed when cultured in a medium containing 2,4-D for 4 days. Apparently single cells are deprived of the head start given to cell clusters in initiating embryogenesis, and 2,4-D treatment remedies this.

Studies on the habituated callus of Shamouti orange, which grows in the absence of auxin and cytokinin in the medium, have also contributed to a more explicit understanding of the role of auxin in somatic embryogenesis. The addition of even a low concentration of IAA or NAA significantly reduces or inhibits embryogenesis whereas the addition of inhibitors of auxin biosynthesis greatly stimulates the process (Kochba and Spiegel-Roy, 1977b). A series of experiments carried out on the effects of γ-irradiation of the callus on embryogenesis are also consistent with the assumption that auxin level is one of the factors controlling embryogenesis in this tissue. Radiation is known to inactivate the endogenous auxin of plants which explains why embryogenesis is stimulated when the callus is irradiated prior to subculture. Interestingly, depending on the radiation dosage, exogenous IAA modifies the propensity for somatic embryogenesis of the tissues (Fig. 5.5). Thus, at low doses of radiation, IAA inhibits embryogenesis, but the inhibition is gradually overcome by increasing the radiation dosage (Kochba and Spiegel-Roy, 1977a). These results underscore the point that transformation of a cell into an embryoid proceeds by the activation of biosynthetic systems, resulting in a decreased auxin con-

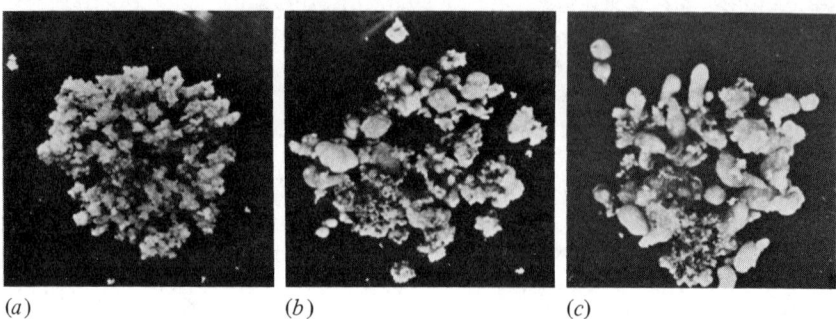

(a) (b) (c)

FIGURE 5.5. Effect of γ-radiation dose and indoleacetic acid (IAA) on somatic embryogenesis in Shamouti orange callus. (a) Nonirradiated callus grown in a medium containing 0.1 mg/liter IAA. (b) Callus exposed to 16 kR γ-radiation, grown in the basal medium. (c) Callus irradiated as in (b), grown in a medium containing 0.1 mg/liter IAA. A number of embryoids are seen. (From Kochba and Spiegel-Roy, 1977a.)

centration. Comparative studies of auxin metabolism of embryogenic and nonembryogenic strains of Shamouti orange callus have shown that the former has a very streamlined system for removal of auxin by conjugation with aspartic acid (Epstein, Kochba and Neumann, 1977) and by inactivation by peroxidases (Kochba, Lavee and Spiegel-Roy, 1977). In this way an optimum level of auxin favorable for low-frequency embryogenesis is maintained in the habituated state of the callus.

In most species investigated thus far, little is known about auxin specificity for somatic embryogenesis. Since a variety of auxins has been shown to induce embryogenesis, one could argue that the process is comparatively nonspecific in the effective range of auxins and that it can be induced to some degree with substances exhibiting auxin activity. Thus, a well-documented case of auxin specificity recently described in leaf explants of *Solanum melongena* (eggplant) can be considered atypical of that in many other plants (Gleddie, Keller and Setterfield, 1983). In this system, auxins such as 2,4-D and 2,4,5-T tested at a range of concentrations induce prolific callus growth without accompanying embryogenesis, whereas IAA or phenylacetic acid give little or no response. Only addition of NAA favors embryogenesis, although the transformation into plantlets requires the removal of auxin from the medium. Specificity of auxin in somatic embryogenesis is displayed in a different way by cell cultures derived from shoots of *Musa* sp. (plantain). In contrast to the usual situation in which a reduction in the concentration of exogenous auxin is necessary for somatic embryogenesis to occur, here the differentiation of proembryoids into full-term embryoids occurs only in the presence of the specific auxin, 2,4,5-T

(Cronauer and Krikorian, 1983). In these cases, the chemical nature of the auxin appears to determine the reprogramming of cells in the embryogenic pathway.

Other plant hormones play diverse roles as morphogenetic cues in the initiation and development of somatic embryos. Cytokinins have generally found their way into culture media used to induce embryogenesis in explanted tissues and organs; both inhibitory and promotory effects have been assigned to them. Since the publication of the report by Halperin and Wetherell (1964), a cytokinin has been an invariable component of the standard medium used for monitoring embryogenesis in carrot cell suspension cultures. Wochok and Wetherell (1972) have claimed that the decline in embryogenic potential experienced by carrot cells during prolonged culture can be alleviated to some extent by the addition of low concentrations of kinetin to the medium. A clear-cut role for cytokinins in somatic embryogenesis in carrot is seen in the work of Sung, Smith and Horowitz (1979), who showed that 2iP promotes embryogenesis in wild-type carrot cells grown in the presence of 2,4-D in the medium, as well as in 5-methyltryptophan-resistant cell lines, in which embryogenesis is impaired by high levels of endogenous IAA. A role for cytokinins in promoting embryogenesis by stimulating divisions in the proembryogenic masses of cells, implied in this work, is also evident in the investigations of Fujimura and Komamine (1980a), who found that the capacity of cell clusters to undergo embryogenesis is enhanced when zeatin is continuously present in the induction medium. If the hormone is added at different times after transfer of the cell suspension to the induction medium, high-frequency embryogenesis results in cells exposed to zeatin during the first 3 to 4 days after transfer. A logical construction of the data is the assumption that zeatin enhances cell division, which generally occurs in the proembryogenic masses during the first 3 to 4 days in culture (Fujimura and Komamine, 1980b). Results leading to the same conclusion have been obtained when 2iP or isopentenyladenosine are added to embryogenic cultures of *Pimpinella anisum* (anise) (Ernst and Oesterhelt, 1984). An increase in cell number probably accounts for a kinetin-induced enhancement in embryogenesis in celery cell suspension cultures, as the embryoids are phased out from the globular torpedo-shaped to torpedo-shaped stage (Al-Abta and Collin, 1978). Since compounds with cytokinin activity have been identified in carrot cell cultures (Salem, Linstedt and Reinert, 1979) and celery embryoids (Al-Abta and Collin, 1979), it will indeed be surprising if they do not exercise a regulatory role in somatic embryogenesis.

Because of the apparent ease of embryoid formation in many cultured plant tissues in the absence of GA in the medium, this hormone has been written off as inconsequential as a controlling factor in the process. Moreover, a critical evaluation of the role of GA in somatic embryogenesis has not been helped by the conflicting reports of its effects on embryogenesis. For example, in carrot cell cultures (Fujimura and Komamine, 1975) as well as in nucellar tissues of *Citrus* (Tisserat and Murashige, 1977c; Kochba, Spiegel-Roy, Neumann and Saad, 1978), GA appears to inhibit embryogenesis by reducing the number of embryoids formed, while in a callus strain originating from cotyledons of cacao it stimulates embryogenesis (Kononowicz and Janick, 1984). However, a new twist has been added to the story by the discovery that somatic cells of carrot contain several endogenous gibberellins that exhibit characteristic transformations during embryogenic development (Noma, Huber, Ernst and Pharis, 1982). In particular, in place of a high level of endogenous polar GA (probably GA_1) present in cells grown in a medium containing 2,4-D, embryoids formed in the induction medium have severalfold higher levels of nonpolar GA (probably GA_4, GA_7). A similar tendency is noted in the presence of free GA-like substances during somatic embryogenesis in grape (Takeno et al., 1983). In view of these results, it may not be misleading to suggest that removal of 2,4-D from the medium may be targeted at an aspect of GA metabolism, reducing it to a level permitting embryogenesis.

The evolution of an embryoid with its precise regularity of form from a homogeneous mass of cells probably involves several hormone-controlled processes, any one of which may be rendered limiting; occasionally, the hormonal control mechanisms may interact with environmental factors to foster normal development of the embryoid. By supplying the limiting substance exogenously or by compensating for the environmental factor, it is possible to induce normal development of the embryoid. The reality of an interaction of hormonal and environmental factors in somatic embryogenesis has figured prominently in recent work. This relates to the occurrence of aberrant forms of embryoids when a cell suspension culture of *Carum carvi* (caraway) reared in a medium containing auxin is transferred to an induction medium lacking auxin in light. A considerable restoration of the original form of embryoids in the population is achieved if the cells are subcultured in a medium containing ABA in the dark (Ammirato, 1974); in the presence of ABA in light, the restoration is less complete. This interesting result led to studies involving the use of zeatin and GA on the restoration of form in the embryoids (Ammirato, 1977). Zeatin antagonizes the effects of ABA in both light and dark conditions and reinstates the developmental anomalies in the embryoids; residence of the cell clusters in a medium containing GA and ABA in the dark promotes normal development of embryoids. When the cell clusters are grown in equimolar concentrations of ABA, zeatin and GA in the dark, GA negates the disruptive effects of zeatin and ensures ABA-induced restoration of the normal form of embryoids. Clearly, the formation of normal embryoids in tissue cultures of caraway can be controlled by a balance among ABA, GA and zeatin and by modifications in the environmental conditions of culture. Similar effects of ABA in normalizing the growth of embryoids have also been observed during somatic embryogenesis in carrot (Kamada and Harada, 1981; Ammirato, 1983b). As we saw in Chapter 4, exogenous ABA is involved in inhibiting vivipary during zygotic embryogenesis; it will be surprising if an analogous hormonal control mechanism ensuring normal growth of embryoids is not found.

Inhibitors of embryogenesis

With the discussion of the role of hormones in somatic embryogenesis behind us, we can conclude this section with a look at substances in the cell that function as inhibitors of embryogenesis. By repeated subculture of carrot cell suspension, a falling off in the number of embryoids formed has frequently been observed. A convincing demonstration that this is caused by an accumulation of inhibitors is a formidable task and has not been attempted; however, experiments in which addition of activated charcoal to the medium reinstates normal level of embryogenesis in aging cultures suggest the presence of inhibitory substances (Fridborg, Pedersén, Landström and Eriksson, 1978; Drew, 1979; Warren and Fowler, 1981). More clear-cut evidence for the presence of inhibitors of embryogenesis has come from experiments with monoembryonic (*Citrus medica*) and polyembryonic (*C. reticulata*) species of *Citrus*. Thus, when nucellar tissues of *C. medica* are cocultured with those of *C. reticulata* or with carrot callus, a decrease is noted in the number of embryoids formed in the latter two. As the effects of the live nucellar tissues of *C. medica* are duplicated to some extent by the addition of ethanol, as well as 2,4-D, ABA and a source of ethylene, it appears that volatile and nonvolatile substances act in concert to inhibit somatic embryogenesis (Tisserat and Murashige, 1977b,c). It is hoped that extraction and identification of inhibitors from the nucellus of *C. medica* will provide further insight into their role in somatic embryogenesis – insight that is not readily available from studies on other systems.

Genetics of somatic embryogenesis

The development of a mature embryoid from its single-celled beginning is conditioned by the genetic and epigenetic mechanisms of the cell. The workings of the former are implicit in the precise fashion by which tissues and organs are laid down in the embryoid. Epigenetic changes against a constant cellular genome are conspicuous in the regularity by which form changes occur in the developing embryoid. But these facts have been overlooked in most of the published studies and consequently, until recently, genetic concepts have not been employed to gain insight into the basis of somatic embryogenesis. Even now, information available on the genetics of somatic embryogenesis is limited to studies on the effects of the genotype and on the isolation and characterization of variant cell lines impaired in embryogenesis.

Genotypic differences in the potential for somatic embryogenesis of cultured explants are manifest in different ways, especially in species that are highly heterogeneous as a result of open pollination. In *Medicago sativa*, shoot tip explants from different plants of the same variety exhibit a requirement for different levels of auxin and mineral salts for somatic embryogenesis (Kao and Michayluk, 1981). Under similar conditions of culture, cell suspensions originating from hypocotyl or epicotyl explants of different cultivars of soybean give variable yields of embryoids; genotype effects are also seen in the frequency of occurrence of embryoids with abnormal cotyledons (Phillips and Collins, 1981). Keyes, Collins and Taylor (1980) evaluated the potential for somatic embryogenesis of hypocotyl explants of *Trifolium pratense* (red clover) obtained by crosses among randomly selected clones and found that an additive genetic effect is a significant source of variability in the number of embryoids formed on calluses. The frequency of embryoids regenerating on leaf explants of *Solanum melongena* is also influenced by the donor genotype, although all cultivars tested exhibit some degree of embryogenesis (Gleddie et al., 1983). Overall, these results suggest that the ability of somatic cells to form embryoids may be an inherited character and that alleles permitting high-frequency somatic embryogenesis in some of our cultivated plants may be found by screening a number of genetic lines. Alternatively, highly embryogenic genotypes can be effectively procured by breeding and genetic selection from standard cultivars.

For genetic studies of somatic embryogenesis, mutants isolated for modifications of specific biochemical functions such as auxotrophy, deficiency of critical enzymes, and resistance to drugs or those in which progression of embryogenesis is blocked at specific stages are of particular interest. Use of the wild-type cell as a standard of reference permits gene expression at different stages of embryogenesis to be monitored in the mutant cell lines impaired for specific functions. Despite these obvious advantages, no true genetic mutants have been obtained from embryogenic cell-suspension cultures, but a few variant cell lines isolated and characterized are worthy of comment. Although carrot cells have been exclusively used in this work, other embryogenic cell lines, such as those of *Citrus* (Kochba et al., 1982) and members of Gramineae (Rangan and Vasil, 1983b), are potentially useful as well.

In the previous section, reference was made to 5-methyltryptophan-resistant cell lines of carrot, which present a striking contrast to the wild-type cells in their inability to regenerate embryoids. How can this effect on morphogenesis be explained by biochemical alterations at which genetic or epigenetic episodes must act? This appears to occur because in the variant

Genetics of somatic embryogenesis 149

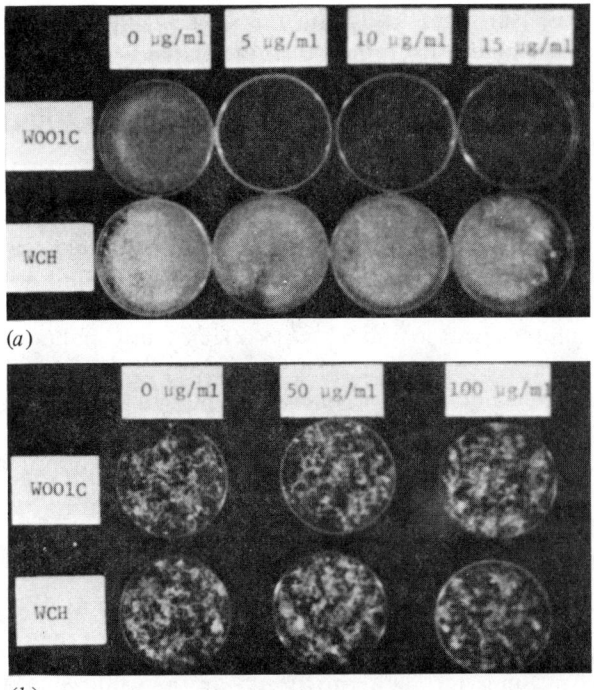

FIGURE 5.6. Callus growth and somatic embryogenesis in wild-type (W001C) and cycloheximide-resistant (WCH) cell lines of carrot grown in increasing concentrations of cycloheximide. (a) Cell suspensions grown in a medium containing 2,4-D favoring callus growth, supplemented with various concentrations of cycloheximide; photographed 4 weeks after inoculation. Callus proliferation in the wild-type cells, but not in the mutant, is sensitive to the drug. (b) Three-week-old embryoids grown in the induction medium supplemented with cycloheximide for 2 weeks. Somatic embryogenesis in both wild-type and mutant cells is insensitive to the drug. (From Sung et al., 1981.)

system, the enzyme anthranilate synthetase, which catalyzes the synthesis of tryptophan is resistant to feedback inhibition by the end product tryptophan. This allows an accumulation of tryptophan, in turn leading to an increase in the endogenous IAA level in the cells to a point at which embryogenesis is impaired (Sung, 1979). Some indication of the alteration in cellular properties during embryogenesis as a function of gene regulation is seen in the responses to cycloheximide of wild-type cells and cycloheximide-resistant variant cell line of carrot. In the former, callus proliferation but not somatic embryogenesis, is sensitive to the drug. In the variant cell line both functions are insensitive to the drug (Fig. 5.6). It has been shown that cycloheximide resistance, which is expressed in the embryoids of the wild type as well as in the callus of the variant system, is caused by internal cellular mechanisms that inactivate the drug (Sung, Lazar and Dudits, 1981). Indeed, similarities in the basis for cycloheximide resistance in the wild-type cells and variant cell lines suggest that the general form of organization of the gene coding for this function is the same but that its expression is regulated differently in the two cell types. Recently, a haploid cell line of carrot was used for the recovery of temperature-sensitive variants impaired at specific stages of embryogenesis and displaying altered phenotypes. There is no firm indication as to whether temperature sensitivity can be transmitted through sexual crosses, but the possibility cannot be excluded as some impaired embryogenic traits are retained in plants regenerated from calluses (Breton and Sung, 1982). Because temperature sensitivity indicates that a gene product that is sensitive to the temperature is not made for completion of embryogenesis, it is possible to identify the protein mediating at specific stages of embryogenesis.

The equivalence between gene and proteins implies that impairment of developmental processes by genetic or epigenetic mechanisms must alter the synthesis of proteins. Biochemical observations on wild type and variant cell lines of carrot are consistent with this view. For example, synthesis of new proteins connected with embryogenesis (embryo-specific proteins) is initiated within 4 hours after wild-type cells are transferred to an induction medium, and cell lines incapable of embryogenesis do not synthesize these proteins (Sung and Okimoto, 1981). In the wild-type cells, the synthesis of new proteins that specify callus proliferation (i.e., callus-specific proteins) is a sequel to growth in a medium containing 2,4-D; surprisingly, the cycloheximide-resistant strain that grows as a callus in the presence of 2,4-D synthesizes only embryo-specific proteins (Sung and Okimoto, 1983). The corollary to these findings is that callus growth can proceed even in the absence of synthesis of callus-specific proteins, whereas synthesis of embryo-specific proteins does not necessarily induce embryogenesis. The evidence based on these studies supports the concept that callus trait and embryogenic trait are coordinately expressed by a common mechanism, the activation of one function being accompanied by the elimination of the other. If this interpretation is valid, it might appear that in the somatic cells of carrot, the action of a single gene provides the signal either for continued proliferative growth or for embryogenic development.

Concluding comments

The development of methods for the culture of cells, tissues and organs of plants has made it possible to expose the cryptic potentiality of cells to go through stages of simulated zygotic embryogenesis. Despite the large number of examples left out of this survey, it is clear that certain common principles are involved in the induction of embryogenic development in somatic cells of angiosperms. What primarily distinguishes the requirements for somatic embryogenesis in the various species is the type of explants used and the composition of the medium for the induction process.

It is hardly necessary to point out that much remains to be done to describe somatic embryogenesis in terms of cellular, biochemical and molecular events. Although carrot has retained preeminence as the workhorse in much of the research already done in this area, information is needed on the molecular aspects of somatic embryogenesis of most other species. Only then we can claim to possess what would appear to be the nearest approach to a central theme to elucidate the mechanism controlling the morphogenetic change of a differentiated cell in a specific way.

6
Pollen embryogenesis

The pollen grain in angiosperms is the product of a reduction division of the pollen mother cell in the anther and is the basic entity from which the male gametophyte evolves. The essential features of male gametogenesis in angiosperms have long been understood in a gross morphological and cytological sense. Briefly, these are the asymmetrical division of the pollen grain into a vegetative cell and a generative cell, germination of the pollen grain, growth of the pollen tube and division of the generative cell into two sperm. Although pollen grains of angiosperms are thus programmed for terminal differentiation into gametes, a small number of pollen grains have been found to divide in an essentially immortal way when anthers of certain plants are cultured at an appropriate stage of development in a relatively simple mineral salt medium. The multicellular pollen grains thus formed go through typical stages simulating zygotic embryogenesis to form embryoids and plantlets with the haploid or gametic set of chromosomes. This phenomenon is known as androgenesis, haploid embryogenesis or pollen embryogenesis, but the last-mentioned term is preferred for use in this book. The occult potentiality of pollen grains of cultured anthers to go through an embryogenic type of development was first demonstrated by Guha and Maheshwari (1964) in a basic discovery. These investigators developed a procedure for culturing excised anthers of *Datura innoxia* (Solanaceae) at various stages of development under aseptic conditions. It was found that when anthers at the pollen grain stage were cultured in a mineral salt medium supplemented with casein hydrolyzate, IAA and kinetin or with coconut milk, grape juice or plum juice, embryo-like structures surfaced from the sides of the anther in about 6 to 7 weeks. Although it was initially suspected that embryoids might have originated in the somatic tissues of the anther, later work confirmed their origin from pollen grains and consequently their haploid nature. A brief ontogenetic study also showed that in anthers cultured in a medium containing coconut milk, a variable but substantial number of pollen grains begin to enlarge and divide repeatedly, forming multicellular units. Later, the exine gives way to free the cellular masses, which organize into typical bipolar embryoids (Guha and Maheshwari, 1966). This study represents a milestone in plant tissue culture investigations and its significance has been heightened by the realization that it opened up the way to produce haploid plants in quantity for genetic and breeding experiments – a hopeless pursuit until a few years ago but now a promising reality. Viewed from a developmental standpoint, the observations offer tangible evidence to show that differentiation of the angiosperm pollen grain is not accompanied by an irreversible change in the genome; rather, the same genome is reprogrammed to evoke totipotent development.

Conceptually, the sporophytic types of growth assumed by pollen grains of cultured anthers can, for convenience, be assigned to two basic modes. The first is exemplified by plants like *Datura innoxia,* in which the pollen grain directly goes through stages reminiscent of zygotic embryogenesis to form embryoids and plantlets. The other group includes plants in which the pollen grain initially forms a multicellular callus, which subsequently regenerates plantlets by somatic embryogenesis or organogenesis. The leading work in this field was done by Niizeki and Oono (1968), who showed that when anthers of rice are cultured at the mature pollen grain stage in a medium supplemented with IAA, 2,4-D and kinetin, multicellular bodies appear from within the anther in about 4 to 8 weeks. These later yield a dense callus. Cytological examination of the callus confirmed its haploid nature and hence origin from pollen grains. Transfer of the callus to a medium containing IAA and kinetin elicits regeneration of shoot and root systems (organogenesis).

Following these investigations, embryogenic or callus mode of growth has been induced in pollen grains of cultured anthers of a number of plants. According to recent surveys, pollen embryogenesis, including callus type of growth, has been documented in more than 170 species including some hybrids, distributed within 68 genera and 28 families of angiosperms, and the list continues to grow (Maheshwari, Rashid and Tyagi, 1982; Bajaj, 1983). The experimental value of pollen embryogenesis as a developmental system has also been extended by the formulation of techniques for the induction of embryogenic or callus type of growth in isolated pollen grains. The stages in pollen embryogenesis in cultured anthers of *Hyoscyamus niger* (henbane) presented in Figure 6.1, are representative of a broad spectrum of plants investigated.

Survey of pollen embryogenesis in angiosperms

The salient features associated with induction of sporophytic type of growth in pollen grains of representative angiosperms are described in this section. Some aspects of the culture conditions permitting specific developmental patterns of growth in pollen grains of cultured anthers are woven into this account as well.

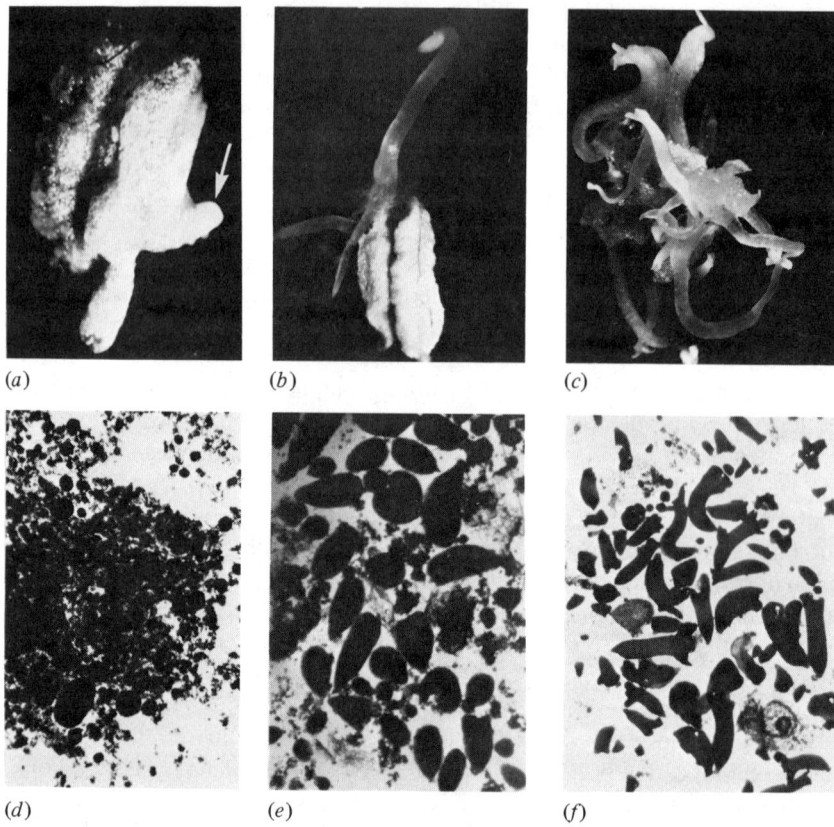

FIGURE 6.1. Pollen embryogenesis in cultured anthers of *Hyoscyamus niger*. (*a*) Anther 18 days after culture showing the appearance of an embryoid (*arrow*) outside. (*b*) Anther 22 days after culture showing emergence of embryoids and plantlets. (*c*) Anther 28 days after culture showing several plantlets. (*d*) Globular stage embryoids. Disintegrating pollen grains are also seen in large numbers in the background. (*e*) Heart-shaped and torpedo-shaped embryoids. (*f*) Cotyledonary-stage embryoids. (*d–f*) Almost the entire yield of embryoids from selected anthers is shown.

Dicotyledons

Among dicots, plants belonging to Solanaceae have consistently proved a good source of anthers for induction of the embryogenic type of development in pollen grains. Great success has been achieved with anthers of several species of *Datura, Hyoscyamus, Lycopersicon, Nicotiana, Petunia, Scopolia* and *Solanum, Atropa belladonna, Capsicum annuum, C. frutes-*

cens, Lycium halimifolium, L. barbarum (Vasil, 1980; Chu, 1982), *Physalis minima, P. ixocarpa* and *Withania somnifera* (Maheshwari, Tyagi, Malhotra and Sopory, 1980; Bapat and Wenzel, 1982). Although the pioneering study on anthers of *Datura innoxia* included hormones and undefined natural plant extracts in the medium, later work showed that perfectly normal embryoids of pollen grain origin appear in anthers of this species cultured in a mineral salt–sucrose medium (Nitsch, 1972). It is also clear that hormonal additives are not essential ingredients of the medium for inducing pollen embryogenesis in cultured anthers of several species of *Nicotiana* (Sunderland, 1971; Collins and Sunderland, 1974), despite the fact that in some early studies hormones were invariable constituents of the medium. Since the first report of pollen embryogenesis in cultured anthers of *N. tabacum* (Nitsch and Nitsch, 1969), this plant has been one of the most extensively studied with respect to the factors inducing embryogenic development of pollen grains. As a result of these investigations, some of which are referred to in later sections of this chapter, *N. tabacum* is considered to be a model system for research on pollen embryogenesis. In *Solanum,* pollen embryogenesis has been induced in a truly impressive list of species, for most of which the addition of IAA, 2,4-D, *p*-chlorophenoxyacetic acid (CPA), NAA, kinetin, benzylaminopurine (BAP), activated charcoal, coconut milk and potato extract alone or in combination appears necessary (Irikura, 1975; Anand and Arekal, 1979; Sopory, 1979; Sopory and Rogan, 1976; Sopory, Jacobsen and Wenzel, 1978; Reynolds, 1984b). Whereas the presence of hormones and complex additives in the medium promotes the growth of a callus, in all but a few species structures resembling zygotic embryos originating from pollen grains have been identified.

Both embryogenic and callus types of growth have been reported to originate in pollen grains of cultured anthers of plants belonging to Cruciferae. Typical bipolar embryoids appear on cultured anthers of *Brassica campestris* (Keller, Rajhathy and Lacapra, 1975), *B. napus* (Thomas and Wenzel, 1975), *B. pekinensis* (Yang, Pan and Liu, 1979), *B. juncea* (George and Rao, 1982) and *B. hirta* (*Sinapis alba*) (Klimaszewska and Keller, 1983; Leelavathi, Reddy and Sen, 1984), whereas anthers of several species of *Arabidopsis* (Amos and Scholl, 1978), *B. chinensis* (Zhong, Ren and Dai, 1978), *Iberis amara* (Babbar, Mittal and Gupta, 1980) and *Lobularia maritima* (Goel, Mudgal and Gupta, 1981) yield only calluses. For *B. campestris,* concentration of sucrose in the medium appears to be critical, since in the presence of less than 6 percent sucrose no embryogenic development of pollen grains occurs; the optimum concentration of sucrose is 10 percent (Keller et al., 1975). A systematic study of the effect of temperature on the frequency of pollen embryogenesis in *B. campestris* and *B. napus* showed that an elevated temperature of 30°C to 35°C during the early period of culture of anthers increases the yield of embryoids appreciably (Keller and Armstrong, 1978, 1979). As recently shown by Dunwell, Cornish and de Courcel (1985), pretreatment of cultured anthers at 35°C in some genotypes of *B. napus* spp. *oleifera* is an absolute necessity for embryogenic induction of pollen grains. It is also worth mentioning that pollen embryoids of *B. campestris* and *B. napus* contain biochemical markers of the respective zygotic embryos like the fatty acid erucic acid and 12S storage protein, cruciferin (de la Roche and Keller, 1977; Crouch, 1982). From this we can conclude that similarities between zygotic embryos and pollen embryoids transcend the morphological to the biochemical level.

Success has been sporadic in inducing embryogenic or callus type of growth from pollen grains of members of Rosaceae and Ranunculaceae. Multicellular units that could be traced to pollen grains are generally observed in anther cultures of *Helleborus foetidus, Paeonia lutea, P. suffruticosa, Prunus avium* (Jordan, 1974; Zenkteler, Misiura and Ponitka, 1975) and apple (Kubicki, Telezynska and Milewska-Pawliczuk, 1975) while anthers of *Prunus amygdalus, P. persica* (Michellon, Hugard and Jonard, 1974; Hammerschlag, 1983), *Paeonia lactiflora* (Ono and Tsukida, 1978), *Rosa damascena* and *R. hybrida* (Tabaeezadeh and Khosh-Khui, 1981) yield mostly haploid calluses. Thus, in these latter plants, the possibility that some pollen grains divide to form calluses cannot be discounted, even though the evidence is not ontogenetic. In a related vein, it may be noted that cultured anthers of certain species of *Anemone* and *Clematis* yield typical bipolar embryoids, although they are not traced to pollen grains (Johansson and Eriksson, 1977; Johansson, Andersson and Eriksson, 1982). The only indubitable proof that octoploid plants arising from cultured anthers of octoploid cultivars of strawberry have their origin in pollen grains is based on genetic segregation of the offspring (Rosati, Devreux and Laneri, 1975).

Compared with the limited number of cases in which embryogenic or callus type of growth is induced routinely and reproducibly in pollen grains of cultured anthers, there are a large number of plants in which induction of pollen embryogenesis has not been attempted; in those in which it has been attempted, the method has not met with the expected degree of success. One of the most important family of plants included in the latter category is the Leguminosae. From a review of the published reports it appears that the particular problems associated with legumes center on the failure of multicellular pollen grains to complete embryogenic development and the difficulty of regenerating plants from pollen-derived calluses. Cultured anthers of *Phaseolus vulgaris* (Peters et al., 1977), *Trifolium alexandrinum* (Mokhtarzadeh and Constantin, 1978) and *Vicia faba* (Hesemann, 1980) generally yield calluses characterized by extreme tardiness in growth. The presence of cell groups with the gametic number of chromosomes in the calluses suggests their origin from pollen grains; however, the haploid callus is generally so intermingled with calluses of other ploidy levels that it is virtually impossible to disentangle it and culture it separately. Multicellular pollen grains and calluses have been described in anthers of *Pisum sativum* (Gupta, 1976), *Arachis correntina, A. villosa* (Mroginski and Fernandez, 1979), *A. hypogaea, A. glabrata* (Bajaj, Labana and Dhanju, 1980), *Phaseolus aureus* (Bajaj and Singh, 1980), *Cajanus cajan* (Bajaj, Singh and Gosal, 1980), *Crotalaria pallida* (Debata and Patnaik, 1983), *Cicer arietinum* (Khan and Ghosh, 1983), *Psophocarpus tetragonolobus* (Pal, 1983), *Cassia siamea* (Gharyal, Rashid and Maheshwari, 1983a) and *Albizzia lebbeck* (Gharyal et al., 1983b) cultured in media supplemented with auxins and cytokinins. The dilemma in regenerating plants from calluses appears closer to resolution with the demonstration that the key to shoot and root formation from pollen calluses of *Arachis hypogaea, A. correntina, A. villosa* (Mroginski and Fernandez, 1979; Bajaj, Ram, Labana and Singh, 1981), *Trifolium alexandrinum* (Mokhtarzadeh and Constantin, 1978) and *Albizzia lebbeck* (Gharyal et al., 1983b) is in the manipulation of the hormonal components of the medium.

No real evidence has been found to justify the view that requirements for pollen embryogenesis in plants belonging to other dicot families differ materially from those considered earlier in this section; yet, only isolated genera from this group have been shown to initiate sporophytic type of growth from pollen grains. Among the species included in this list, pollen grains of cultured anthers of *Aesculus hippocastanum* (Sapindaceae: Radojević, 1978), *Luffa cylindrica, L. echinata* (Cucurbitaceae: Sinha, Jha and Roy, 1978), *Poncirus trifoliata* (Rutaceae: Hidaka, Yamada and Shichijo, 1979), *Vitis vinifera* (Vitaceae: Zou and Li, 1981), *Primula obconica* (Primulaceae: Bajaj, 1981a), *Sambucus nigra* (Caprifoliaceae: Sehgal, Arora and Narang, 1982), *Hevea brasiliensis* (Euphorbiaceae: Chen et al., 1982) and *Carica papaya* (Caricaceae: Tsay and Su, 1985) undergo a complete or partial embryogenic type of development to yield embryoids. On the other hand, haploid and other ploidy level calluses or multicellular pollen grains are formed in cultured anthers of *Pelargonium hortorum* (Geraniaceae: Abo El-Nil and Hildebrandt, 1973), *Coffea arabica* (Rubiaceae: Sharp et al., 1973), *Citrus limon* (Rutaceae: Drira and Benbadis, 1975), *Saintpaulia ionantha* (Gesneriaceae: Hughes, Bell and Caponetti, 1975), *Digitalis purpurea* (Scrophulariaceae: Corduan and Spix, 1975), *Pharbitis nil* (Sangwan and Norreel, 1975a), *Ipomoea batatas* (Convolvulaceae: Tsay and Tseng, 1979), *Oenothera coronifera* (Onagraceae: Jean et al., 1976), *Gerbera jamesonii* (Compositae: Preil, Huhnke, Engelhardt and Hoffmann, 1977), *Gossypium hirsutum* (Malvaceae: Barrow, Katterman and Williams, 1978), *Ulmus americana* (Ulmaceae: Redenbaugh, Westfall and Karnosky, 1981), several species of *Populus* (Salicaceae: Chu, 1982),

Coriandrum sativum (Umbelliferae: Egorova and Reznikova, 1982), *Papaver setigerum* and *P. radicatum* (Papaveraceae: Johansson et al., 1982). A systematic analysis of the nutrient requirements for initiation of sporophytic type of growth of pollen grains in these plants shows that a combination of an auxin and a cytokinin necessary for induction of callus type of growth from pollen grains is equally essential for pollen embryogenesis in cultured anthers; in some cases, the spectrum of hormonal combinations is even more complex. Any increase in the requirements for N or sucrose in the medium, if it occurs at all, appears to be nominal. A change in the medium composition can induce shoot and root formation in the calluses; rarely, as in *Pharbitis nil* (Sangwan and Norreel, 1975a) and *Ipomoea batatas* (Tsay and Tseng, 1979), the calluses regenerate embryoids.

Monocotyledons

We now turn to the monocots and to a review of the work done on the initiation of embryogenic or callus type of growth in pollen grains of this subclass of plants. The regeneration of a callus from pollen grains of cultured anthers of rice described earlier spawned much of the later experimental approaches to induce sporophytic type of growth in pollen grains of monocots. There is a long list of plants belonging to Gramineae, including several grasses and major crop plants in which pollen grains are induced to form callus and, by manipulation of the hormonal ingredients of the medium, plant regeneration by organogenesis is accomplished. Inherent in these studies is the hope of raising haploid plants of cereals for genetic and breeding experiments but this expectation has not been fully realized due to the low yield of haploids and the troublesome endopolyploidy (repeated doubling of chromosomes within the nuclear envelope by curtailment of spindle formation) of cells of the callus.

Of the numerous cultivated varieties of rice so far studied, a genotype of *Oryza sativa* var. *indica* is the only one in which pollen grains respond respectably to reprogramming cues by the formation of embryoids rather than calluses. This occurs when anthers are cultured in a medium containing coconut milk, yeast extract, IAA, 2,4-D and kinetin (Guha-Mukherjee, 1973); in other cases, the pollen grain initially differentiates into a callus. From all indications it appears that a great interest in the anther culture of rice lies in the modifications of culture conditions to obtain high yield of calluses. From this point of view, differences in genotype (Guha-Mukherjee, 1973; Cornejo-Martin and Primo-Millo, 1981; Schaeffer, 1982), physiological and developmental stage of donor plants (Chen and Lin, 1976), variables in the composition of medium such as choice of hormones (Guha, Iyer, Gupta and Swaminathan, 1970), the ratio of NH_4^+ to NO_3^- (Chu et al., 1975), sucrose concentration (Chen, 1978) and osmolality (Wang, Sun and Chu, 1974) and administration of a cold stress to anthers (Genovesi and Magill, 1979; Chaleff and Stolarz, 1981) have been found to play important roles in the frequency of callus induction and the number of green versus albino plants regenerated.

Chu et al. (1973), Ouyang, Hu, Chuang and Tseng (1973) and Picard and de Buyser (1973) reported the first successful induction of pollen plants from cultured anthers of *Triticum aestivum* and *T. vulgare* (wheat). The basic technique followed in both species is culture of anthers in a medium supplemented with 2,4-D or NAA to induce proliferation of a pollen callus, and transfer of the callus to a medium containing IAA and kinetin to induce plantlet formation by organogenesis. Direct transformation of pollen grains into embryoids has also been documented when anthers of certain varieties of wheat are cultured in media containing high concentrations of IAA and kinetin (Pan and Kao, 1978). Similar to the work on rice anther culture, the direction of future studies on the induction of pollen callus in wheat is seriously hampered by variations in the frequency of callus formation among genotypes (Schaeffer, Baenziger and Worley, 1979; Liang, Sangduen, Heyne and Sears, 1982).

Pollen morphogenesis in cultured anthers of *Hordeum vulgare* has been the subject of a number of studies notably by Clapham (1971), Grunewaldt and Malepszy (1975), Foroughi-Wehr, Mix, Gaul and Wilson, (1976) and Gonzalez-Medina and Bouharmont (1978). The behavior of barley pollen grains in culture is not much different from that described in wheat and essentially involves the formation of a callus as an obligate stepping stone toward formation of plantlets by organogenesis. Major obstacles in the anther culture of barley are the low yields of callus, variability in yield between cultivars and the abysmally low proportion of green plants to albinos regenerated from the callus.

Attempts to improve the yield of calluses from cultured barley anthers have produced a considerable body of information that holds potential significance not only in future studies on this crop, but also in relationship to the whole problem of low yields that continue to plague cereal anther culture. In *H. vulgare* var. Dissa, Wilson (1977) used a combination of pretreatment of tillers in water at laboratory temperature for 1 to 2 days and culture of the entire spike in an agitated liquid medium containing a critical combination of 10 percent sucrose and 1 mg/liter each of 2,4-D, IAA and BAP to stimulate high-frequency formation of pollen calluses. Under these conditions, the gradient properties of the spike are sufficiently perturbed to stimulate callus formation from a good number of pollen grains in the anther and from several anthers simultaneously. Another variety, *H. vulgare* var. Sabarlis has proved an almost ideal plant in which to analyze the effectiveness of various pretreatments on pollen efficiency, since in the absence of any pretreatments virtually no calluses are formed

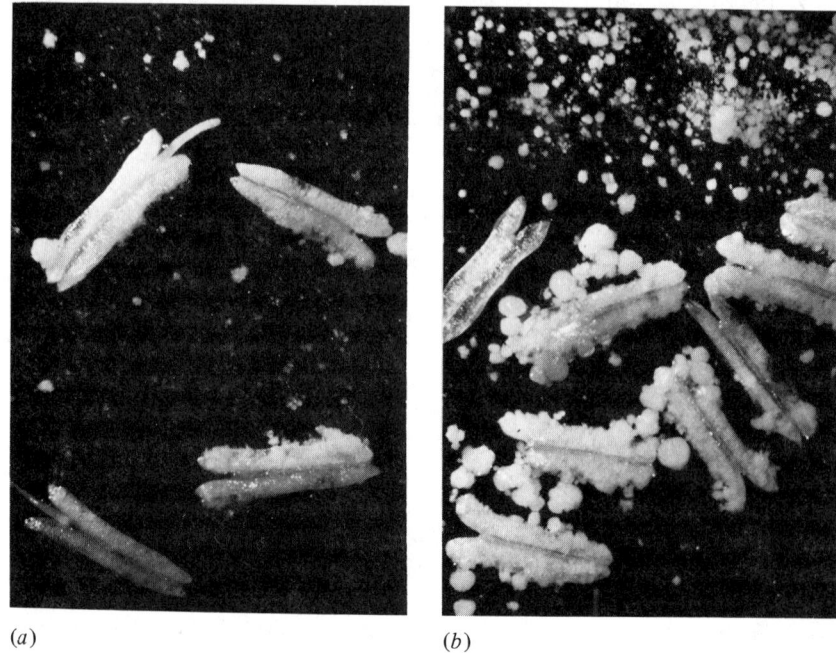

(a)   (b)

FIGURE 6.2. Yield of pollen calluses in barley Sabarlis anthers. (*a*) Anthers cultured in liquid medium containing 2,4-D and kinetin. (*b*) Anthers cultured in the same medium previously conditioned by the growth of anthers. (From Xu et al., 1981; photographs supplied by Dr. N. Sunderland.)

in cultured anthers. In this cultivar, pretreatment of excised spikes is more effective than pretreatment of excised tillers; moreover, callus yield in cultured anthers is higher following pretreatment of spikes or tillers at 4°C than at 25°C (Huang and Sunderland, 1982). High-yielding cultures of Sabarlis barley are routinely obtained by incubating cold-stressed anthers in a medium previously conditioned by the growth of anthers or ovaries of barley (Xu, Huang and Sunderland, 1981), by supplementing the conditioned medium further with *myo*-inositol (Xu and Sunderland, 1981) or by increasing anther inoculum density (Xu and Huang, 1984) (Fig. 6.2). The basis for these remarkable effects of temperature and conditioned medium on pollen efficiency in cultured anthers of barley has not been satisfactorily explained.

Progress in the culture of anthers of maize was disappointingly slow until 1975, when the first reports from China appeared. In retrospect, one primary reason for this appears to be the extreme genotypic specificity of maize anthers, as shown by an array of observations. According to Miao et al. (1978), anthers of only nine out of 159 genotypes of maize tested produced calluses or embryoids. Successful reports (Brettell, Thomas and Wernicke, 1981; Ting, Yu and Zheng, 1981; Nitsch et al., 1982) of induction of embryogenic or callus type of growth in maize anthers from outside China have invariably used Chinese or Chinese–U.S. hybrid germ plasm. In one investigation using 12 indigenous genotypes, anthers of only four responded positively to culture (Genovesi and Collins, 1982). A second reason for the recalcitrance of pollen grains of maize is that the culture medium that finally induced morphogenesis turns out to be capricious with regard to certain critical ingredients. The improvements to the medium that secured successful induction of sporophytic type of growth include increase in sucrose concentration to an optimum of 12 to 15 percent and addition of activated charcoal, casein hydrolyzate, organic acids, and a spectrum of growth hormones (Ku et al., 1978; Miao et al., 1978). Although one can make some sense of this medium in terms of its complexity, it is difficult to say how a complex medium can override the expression of gametophytic program ingrained in the pollen and induce sporophytic type of growth. Nonetheless, it is now clear that a small proportion of pollen grains of maize anthers cultured in a complex medium yields embryoids and calluses, although the conditions that favor one or the other type of growth have not been identified. Another strategy employed to enhance anther response in maize is to subject the tassel to a pretreatment regimen of several days at a low temperature before culture of anthers (Nitsch et al., 1982; Genovesi and Collins, 1982). Among other genera of Gramineae, embryogenic or callus type of growth has been induced in pollen grains of cultured anthers of *Aegilops, Agropyron, Coix, Festuca, Lolium, Secale, Setaria, Triticale* (Vasil, 1980; Chu, 1982), *Avena* (Rines, 1983), *Pennisetum* (Bui Dang Ha and Pernes, 1982) and *Saccharum* (Fitch and Moore, 1983).

In summary, it is clear that Gramineae shares with Solanaceae a commanding role as suitable material for research on pollen embryogenesis. The varied protocols employed to induce morphogenesis in pollen grains of cultured anthers of graminean genera, while imaginative and interesting, cannot really be considered as adequate and established. Unfortunately, other monocotyledonous genera have not fared as well as graminean genera; in this category are included only *Asparagus* (Pelletier, Raquin and Simon, 1972), *Lilium* (Gu and Cheng, 1982) and *Cocos* (Thanh-Tuyen and de Guzman, 1983). The limited success obtained with non-graminean monocots would appear to show that each group of closely allied plants has its own methods to be devised and its own problems to be overcome before sporophytic type of growth is induced in pollen grains.

Embryogenesis in isolated pollen grains

One problem with much of the work discussed above is that pollen grains are encased within the anther locule during the crucial period in which the initial divisions in the embryogenic pathway occur. This makes it difficult to dissociate any influence of the anther wall and tapetum on embryogenic divisions from the more obvious effects of the components of the medium. For an analysis of embryogenic development of pollen grains in the absence of influence from the somatic tissues of the anther, a significant development has been the elaboration of methods for inducing embryogenesis in isolated pollen grains. In some of the early work along these lines, the objective was not easily attained within the limitations imposed by the methods employed; consequently, until recently, attempts to induce embryogenesis in isolated pollen grains have involved a series of pretreatments to anthers before pollen grains are isolated and cultured. The ability of isolated pollen grains to undergo embryogenic development was first demonstrated in *Datura innoxia* by Nitsch and Norreel (1973). In this work, pollen grains from anthers of flower buds pretreated at 3°C for 48 hours are released into an isotonic medium and cleaned by filtration and centrifugation. Aliquots of the pollen suspension are cultured in a liquid medium conditioned by the addition of an extract of cultured anthers of *D. innoxia*. Under these conditions, embryogenic pollen grains in the aliquot divide and form structures resembling zygotic embryos. Embryogenesis is also observed when pollen grains isolated from cold-stressed anthers of *Lycopersicon esculentum* and *L. pimpinellifolium* are nurtured in the conditioned medium enriched with *Datura* anther extract (Debergh and Nitsch, 1973).

The need to determine the factors in *Datura* anther extract that promote embryogenesis in isolated pollen grains is apparent from these studies. Toward this end, a culture medium was developed in which the anther extract was replaced by defined organic substances. In a key study in this area, anthers excised from cold-treated flower buds of *Nicotiana tabacum* var. Coulo are grown in a mineral salt medium for 8 days. Pollen grains extracted from anthers after this period are cultured in an embryogenesis-inducing medium containing high concentrations of glutamine, *myo*-inositol and serine, and their fate is followed for the next several days. The result of this procedure is the transformation of a small number of pollen grains directly into embryoids (Nitsch, 1974). With modifications, this method has been adapted for initiating pollen cultures in other varieties of tobacco (Reinert, Bajaj and Heberle, 1975; Horner and Street, 1978a; Lu, 1978; Wernicke, Harms, Lörz and Thomas, 1978), rye (Wenzel, Hoffmann, Potrykus and Thomas, 1975), *Hyoscyamus niger* (Wernicke and Kohlenbach, 1977), *Datura innoxia* (Tyagi, Rashid and Maheshwari,

1979), *Solanum tuberosum* (Weatherhead and Henshaw, 1979), rice (Cornejo-Martin and Primo-Millo, 1981), *Brassica napus* (Lichter, 1982) and *Saccharum spontaneum* (Hinchee and Fitch, 1984).

Such studies provide information about the morphogenetic potency of isolated pollen grains but little about the inductive processes that occur during periods of cold treatment or preculture of anthers. Another offshoot of research on this topic has been directed toward inducing embryogenic divisions in pollen grains cultured directly in a medium in a single-step method. Although varying degrees of success have been reported in inducing embryoid or callus formation in isolated pollen grains of *Petunia hybrida* (Sangwan and Norreel, 1975b), dihaploid *Solanum tuberosum* (Sopory, 1977), *Paeonia lactiflora* (Ono and Harashima, 1981), *Nicotiana tabacum* var. Samsun, *N. rustica* (Imamura, Okabe, Kyo and Harada, 1982) and *Hyoscyamus niger* (Nagmani and Raghavan, 1983), it is in *N. tabacum* var. Badischer Burley that the procedures have been streamlined to ensure high-frequency *ab initio* pollen embryogenesis. In tracing the modifications of the procedures, the observation that it is possible to secure an enriched fraction of embryogenic pollen from a heterogeneous suspension is critical (Herberle-Bors and Reinert, 1980). This is accomplished by washing the pollen suspension in 0.25 M sucrose by repeated centrifugation, followed by density-gradient centrifugation in sucrose–Percoll mixture, when most of the embryogenic pollen is layered at the top of the gradient. Starting from this point, Rashid and Reinert (1983) have induced high-frequency pollen embryogenesis by selecting anthers from plants induced to flower in short days at 15°C, followed by an additional cold treatment at 10°C for 10 days in the dark and isolating embryogenic pollen by gradient centrifugation in sucrose (4 percent)–Percoll. Following these treatments, pollen grains are primed to undergo embryogenic division which is easily accomplished in a mineral salt–sucrose medium. By faithfully following this protocol, as high as 30 percent of the cultured pollen grains regenerates embryoids (Fig. 6.3). The critical step that determines the life course of the pollen, and therefore its destiny, is the cold treatment administered to flower buds before excision of anthers. Actual counts have shown that pollen grains extracted from anthers of plants grown in short days at 18°C and subjected to a cold stress at 10°C for 10 days in culture form embryoids at a higher frequency than do pollen cultures that are not subjected to the temperature stress (Rashid and Reinert, 1981). By administering cold treatment at 10°C for 15 days to flower buds and culturing isolated pollen in the mineral salt–sucrose medium, moderately efficient embryogenesis has been obtained in the less responsive *N. sylvestris* (Rashid, 1982). The underlying basis for the increased frequency of embryogenesis in pollen cultures following low-temperature treatment is totally obscure but presumably depends on changes in the microenviron-

164 *Pollen embryogenesis*

FIGURE 6.3. Embryogenesis in isolated pollen cultures of *Nicotiana tabacum* var. Badischer Burley. Embryoids are mostly in the globular stage. (From Rashid and Reinert, 1983.)

ment of the pollen grains. Experiments designed to determine whether this is related to auxin level in the anther have yielded equivocal results. For example, anthers collected from plants grown in short days that yield pollen embryoids in high frequency have higher levels of IAA than do their long-day counterparts, indicating a role for endogenous auxin in embryogenesis. The incidence of pollen embryogenesis in such anthers is also enhanced by the addition of an inhibitor of auxin biosynthesis or an antiauxin, reinforcing the opposite view that reduction in auxin level is important in embryogenesis (Dollmantel and Reinert, 1980).

A procedure for generating high-efficiency cultures of embryogenic pollen grains devised by Sunderland and Roberts (1977) relies on the fact that in some plants anthers dehisce soon after culture, allowing all or some of the enclosed pollen to escape. If anthers are cultured in a liquid medium, the pollen shed into the medium can be handled for all intents and purposes as a typical pollen culture. By serially transferring cultured anthers at intervals to fresh media, embryogenic pollen grains and embryoids of different ages are obtained. Such pollen cultures, designated shed pollen cultures, have been obtained in tobacco (Sunderland and Roberts, 1977), *Datura innoxia* (Tyagi, Rashid and Maheshwari, 1979), rice (Chen et al., 1980), wheat (Henry and de Buyser, 1981) and barley (Sunderland and Xu, 1982). Like isolated pollen cultures, a low-temperature stress administered to anthers in the flower bud or inflorescence leads to appreciable yields of embryoids and calluses in shed pollen cultures (Sunderland and Roberts, 1979; Chen et al., 1980; Sunderland and Xu, 1982). The use of shed pollen culture to study the biochemical aspects of pollen embryogenesis leaves much to be desired since, in many cases, pollen grains are shed only after the initial events of embryogenic induction are completed inside the anther. In terms of specific nutritional requirements of pollen grains to differentiate into embryoids, the role of unknown substances leaching out from the anther into the medium cannot be ignored.

Pathways of pollen embryogenesis

A pollen grain poised on the threshold of sporophytic mode of development will give us the impression that completion of the first haploid mitosis is almost certainly a prerequisite for this event. Based on the general observation that a large vegetative cell and a small generative cell are born out of this division, models to account for the origin of embryoids and calluses from pollen grains can be considered in three general classes. One is to suppose that the vegetative cell in which DNA synthesis has been turned off enters a phase of renewed cell divisions to give rise to an embryoid or a callus. A more likely alternative is that the generative cell that is already primed for DNA synthesis in preparation to form sperm continues to produce new cells in a rapid burst of mitotic activity. The third possibility is that both vegetative and generative cells divide and contribute to the formation of the sporophytic outgrowth from the pollen grain. With this preamble we shall now review studies that have traced the origin of pollen embryoids and calluses to their single-celled beginning. We shall keep in mind that a successful model should be flexible enough to accommodate any variations from the norm and that it should be cytologically sound to explain the ploidy level of the final product.

The repeated division of the vegetative cell to form an embryoid is an odd kind of event, but it is one of the most common pathways of pollen embryogenesis and one that has been thoroughly investigated. We begin with a typical bicellular pollen grain enclosing a vegetative cell and a generative cell. The former loses its morphogenetic individuality and is initially partitioned by a series of equal divisions without intervening growth until a mass of cells typical of somatic cell size is produced. Unable to contain the burgeoning growth within, the exine breaks open, liberating the cellular mass into the anther locule, where it differentiates into facsimiles of zygotic embryos. The culmination of this developmental feat is the germination of the embryoid, which appears outside the anther wall as a small plantlet. As the initial divisions of the vegetative cell are under way, the generative cell either disintegrates or undergoes but a few divisions. Without contributing to the formation of the embryoid, these cells may simply wilt away. The essential cytological details of the origin of embryoids by the division of the vegetative cell were first traced by Sunderland and Wicks (1971) in *Nicotiana tabacum* var. White Burley (Fig. 6.4). Since then, this route to embryogenesis, designated as the A pathway (Sunderland, 1973), has been described in other cultivars of *N. tabacum,* as well as in other species of *Nicotiana* and in *Capsicum annuum, Datura innoxia, D. metel, Paeonia* (Sunderland and Dunwell, 1977; Vasil, 1980), *Solanum surattense* and *Luffa cylindrica* (Sinha et al., 1978).

Clapham (1971) showed that the vegetative cell is instrumental in giving rise to a pollen callus in cultured anthers of *Hordeum vulgare* var. Sabarlis. Upon release from the exine, the cellular mass follows a pattern of disorganized growth typical of a callus. Later investigations by other workers have supported the occurrence of the A pathway of pollen callus formation in cultured anthers of several cultivars of *H. vulgare* (Wilson, Mix and Foroughi-Wehr, 1978; Sun, 1978; Zhou and Yang, 1980), *Triticum aestivum* (Wang et al., 1973; Pan, Gao and Ban, 1983), *Oryza sativa* (Chen, 1977; Sun, 1978; Yang and Zhou, 1979), *Zea mays* (Miao et al., 1978) and *Secale cereale* (Sun, 1978).

Whereas most pollen grains exhibiting the A pathway follow the division sequences infallibly, in others irregularities may arise during or after division. Such situations have been described during pollen callus formation in *Triticum aestivum* (Zhu, Sun and Wang, 1978) and *Hordeum vulgare* (Zhou and Yang, 1980; Sunderland and Evans, 1980). In *H. vulgare* var. Sabarlis in which detailed studies have been made, one common variation of the A pathway is an intermediate free nuclear stage during the division

Pathways of pollen embryogenesis 167

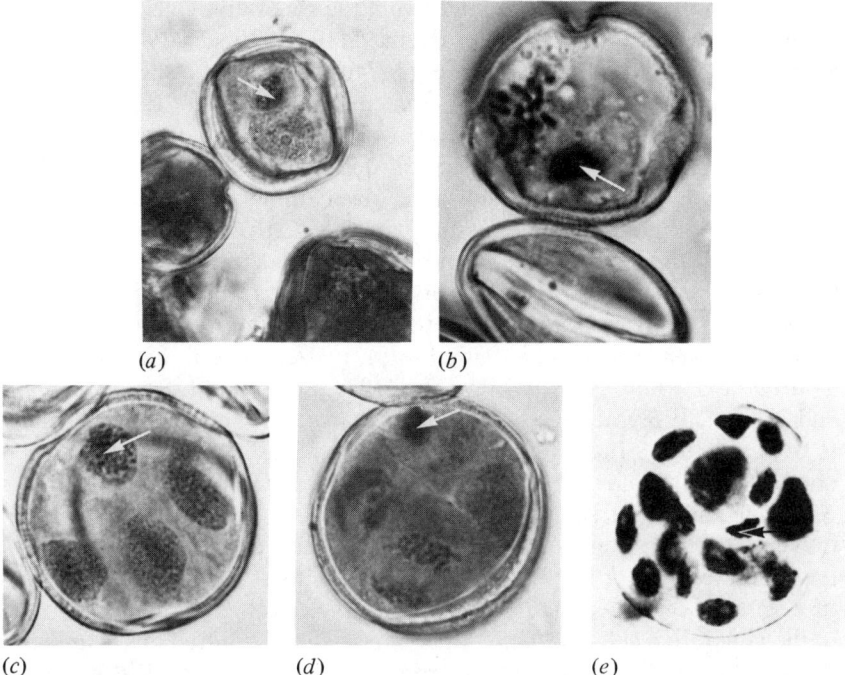

FIGURE 6.4. Early stages in the transformation of pollen grains of cultured anthers of *Nicotiana tabacum* var. White Burley into embryoids by the A pathway. (*a*) Bicellular pollen grain 6 days after culture. (*b*) Embryogenic pollen with vegetative cell nucleus in division. (*c*) Four-celled pollen grain with three cells formed from the vegetative cell. (*d*) Multicellular pollen grain; eight cells formed from the vegetative cell. (*e*) A pollen embryoid composed of about 20 cells, still enclosed in the exine, formed from the vegetative cell; the generative cell has divided once. In all cases the arrow points to the generative cell or its division products. (From Sunderland and Wicks, 1971; photographs supplied by Dr. N. Sunderland.)

of the vegetative cell, when fusion between two or three nuclei occurs; in another case, the generative cell nucleus is incorporated into a derivative of the vegetative cell by fusion. This means that a callus emerging from a pollen grain following the A pathway may consist of cells derived wholly from the vegetative cell, from a fused product of vegetative cell derivatives or from a fused product between a vegetative cell derivative and the generative cell. These events are believed to account for the appearance of cells with different ploidy levels in a callus of single pollen grain origin (Sunderland and Evans, 1980). A few additional variations of the A pathway have been described in tobacco as well (Misoo, Yoshida and Mastubayashi, 1979).

In tracing the origin of embryoids from the generative cell of a bicellular pollen grain, the investigations of Bernard (1971), Nitsch (1972), Iyer and Raina (1972) and Rashid and Street (1974b) appear to be suggestive. The experiments in question showed that in cultured anthers of various plants, as the embryoid is formed from the vegetative cell, the generative cell divides at a slow pace to produce cell derivatives. Although the fate of these latter cells was not established, observations indicate that embryogenic development of some pollen grains is directly geared to the division of the generative cell. The concept of an independent role for the generative cell in pollen embryogenesis or pollen callus formation has received strong support from cytological observations on cultured anthers of *Hyoscyamus niger* (Raghavan, 1976a, 1978), maize (Miao et al., 1978), rye, *Triticale,* rice (Sun, 1978), barley (Sun, 1978; Sunderland, Roberts, Evans and Wildon, 1979), tobacco (Anand, Arekal and Swamy, 1980) and wheat (Pan et al., 1983). In *H. niger,* in which this was first demonstrated, a good number of embryoids are formed by the repeated division of the generative cell, which apparently functions as the embryo mother cell. Here the vegetative cell does not divide, or undergoes but a few divisions; in both cases, the vegetative cell or its division products constitute a suspensor-like structure on the organogenetic part of the embryoid formed by the derivatives of the generative cell (Fig. 6.5). In pollen grains of cultured maize anthers, the vegetative cell divides only once or twice and then gradually disintegrates, the callus being formed entirely from the generative cell (Miao et al., 1978). In *H. niger* (Raghavan, 1978), tobacco (Anand et al., 1980) and several cereals (Bouharmont, 1977; Miao et al., 1978; Sun, 1978; Sunderland et al., 1979), embryoids are also formed by the repeated division of both vegetative and generative cells; the contribution of neither cell can be considered more important than the other. The resulting structures are chimeras at the cellular level, that is, amalgams of unlike parts. Less obvious types of changes also occur in the vegetative and generative cell derivatives before their incorporation into the embryoid or callus, such as a free nuclear phase in the vegetative cell and its polyploidization as in barley (Sunderland et al., 1979) or a free nuclear phase in the generative cell as in barley and rye (Sun, 1978). We shall refer to the pathway of embryogenesis involving the generative cell or both generative and vegetative cells as the E pathway (Sun, 1978).

In *Datura innoxia,* which has been a favored material for the study of pollen embryogenesis, a remarkable situation exists. In cultured anthers, under conditions which are as yet imperfectly understood, the vegetative cell and generative cell, rather than embark on pathways of independent divisions, fuse with one another. The fusion might occur between one or two haploid vegetative nuclei and a haploid or an endopolyploid generative nucleus so that nuclei with nonhaploid modal chromosome numbers are

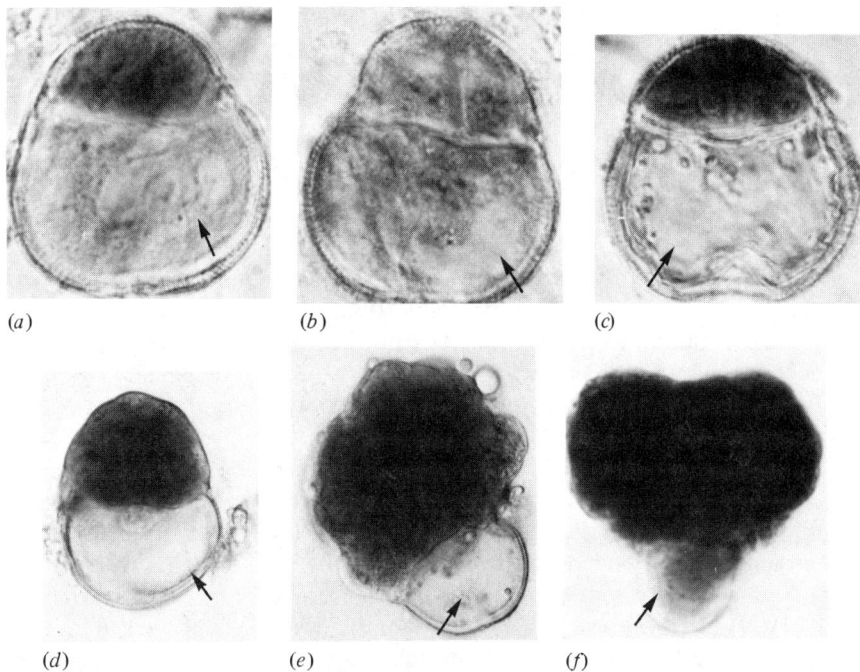

FIGURE 6.5. Stages in the transformation of pollen grains of cultured anther segments of *Hyoscyamus niger* into embryoids by the E pathway. (*a*) Formation of generative cell and vegetative cell. (*b*) Division of the generative cell. (*c–e*) Formation of the organogenetic part of the embryoid by the division of the generative cell; the vegetative cell is undivided. (*f*) Heart-shaped embryoid with undivided vegetative cell. In all cases the arrow points to the vegetative cell. (From Raghavan, 1978.)

formed. Repeated divisions of the fusion products account for the formation of embryoids with different ploidy levels from the same anther (Sunderland, Collins and Dunwell, 1974). This route to embryogenesis, designated the C pathway, has also been described, but with much less detail, in barley (Sunderland and Evans, 1980).

In all the above pathways of pollen embryogenesis, division programs set at the beginning of embryogenic development result in the formation of two asymmetrical cells or nuclei. The distinctive feature of the B pathway of embryogenesis (Sunderland, 1973) to be described now is the appearance of two symmetrical cells or nuclei after the first haploid mitosis. The subsequent behavior of these cells to form the embryoid diverges from this common plan later on. Most frequently, as shown in *Anemone coronaria*, *Atropa belladonna*, *Datura innoxia*, *Nicotiana sylvestris*, *N. tabacum*,

Oryza sativa, Triticum aestivum (Vasil, 1980) and *Hordeum vulgare* (Zhou and Yang, 1980) both cells contribute derivatives to the formation of the embryoid. Convincing evidence for the participation of only one cell from an identical pair in pollen callus formation in *Triticale* has been provided by Sun, Wang and Chu (1974). Here a series of free nuclei generated from the nonparticipating cell are extruded into the anther locule and do not contribute to callus formation. Another scenario, which is not yet fully documented, occurs in *H. vulgare* var. Sabarlis (Sunderland and Evans, 1980); here the two nuclei initially fuse before they participate in embryoid or callus formation. In yet another variation, described in *Triticum aestivum*, two identical nuclei divide repeatedly to form a series of free nuclei, but their differentiation into a callus has not been followed; this division sequence is designated as the D pathway (Zhu, Sun and Wang, 1978). The dispositions of cells and nuclei in the main pathways of pollen embryogenesis described here are illustrated in Figure 6.6.

In cultured anthers, pollen embryoids have multiple origins. The relative abundance of embryoids formed by a specific division sequence varies not only in anthers of different flower buds of the same plant, but in anthers of the same flower bud as well. What prompts pollen grains to choose a particular division sequence is also not clear. According to Sunderland et al. (1979) in *Hordeum vulgare* var. Sabarlis, the phase of the cell cycle of the first haploid mitosis in which pollen grains are held at the time of excision and culture of anthers is important in determining the specific division sequence, while administration of a temperature stress enhances the frequency of occurrence of certain division pathways. However, the possibility that the determining factor in pollen division sequence may differ among different plants and that multiple mechanisms may come into play needs be explored.

Cytology of pollen embryogenesis

As one might expect, several ultrastructural, cytochemical and biochemical changes should be occurring during the transformation of a pollen grain into an embryoid or a callus. In analyzing these changes, a variety of questions are to be considered. Is the gametophytic cytoplasm of the vegetative and generative cells recruited for the initiation of sporophytic type of growth? Or is the cytoplasm of these cells replaced by a new cytoplasm that carries information for sporophytic growth? What are the distinct ultrastructural features of the constituent cells of the pollen that augur embryogenic development? What is the pattern of RNA and protein synthesis in the embryogenic pollen grain and how is it different from that occurring in the normal pollen? Although the pollen grain holds great promise for research into these questions, not much effort has been mounted on the ultrastructural and biochemical cytology of pollen embryogenesis. At the

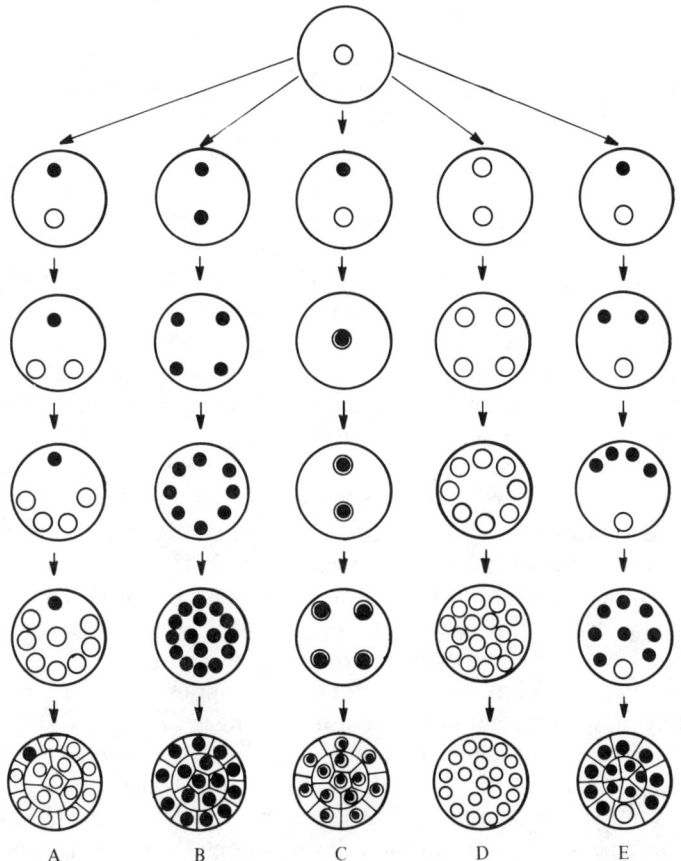

FIGURE 6.6. Fates of nuclei arising out of the first haploid pollen mitosis, in the different pathways of embryogenesis. In pathways A, C and E, solid circles represent nuclei of the generative cell or its division products; hollow circles represent nuclei of the vegetative cell or its division products. In C, solid circles enclosed in hollow circles indicate fusion between the nuclei of vegetative and generative cells. In B and D, circles represent symmetrical nuclei born out of the first haploid mitosis and their division products. Variations in each pathway described in the text are not shown here.

ultrastructural level, only a few model systems have been studied; in the area of biochemical cytology, even much less work has been done.

Ultrastructural cytology

Rather than present a comparative discussion of the subcellular changes in the embryogenic pollen grains of different species, we shall first concentrate on *Datura innoxia* and consider other examples to illustrate the variations. In *D. innoxia,* there are no fundamental differences between the ultrastructural changes that occur in the pollen grain during the first haploid mitosis in culture and that *in vivo.* In either case, the result is the formation of a vegetative cell packed with ribosomes and highly electron-dense mitochondria and plastids. It is this basic subcellular organization that is partitioned repeatedly as embryogenic division is initiated. Apparently, synthesis of new cytoplasmic organelles associated with the altered program of the vegetative cell does not occur until after its first division in the embryogenic pathway (Dunwell and Sunderland, 1976; Sangwan-Norreel, 1978). In searching for specific ultrastructural features associated with embryogenesis, Sangwan and Camefort (1982) noted in pollen embryoids of *D. metel* a transient appearance of several cytoplasmic structures which have the aspect of dense groupings of polysomes and rough ER. They appear as early as the two- to four-celled stage but disappear as globular embryoids are phased out into the heart-shaped stage. Another distinctive ultrastructural feature of pollen embryogenesis in this plant is the acquisition of a tannin-like deposit by the tonoplast of embryogenic pollen grains. This marker is absent in normal pollen grains and persists only up to the stage of globular embryoids (Sangwan and Camefort, 1983). These considerations appear to imply that the tonoplast change is a necessary feature in the initial transformation of the normal pollen to the embryogenic type.

In *Nicotiana tabacum,* contradictory observations have been presented depending upon whether ultrastructural changes are followed in pollen grains of cultured anthers or in pollen grains in *ab initio* cultures. According to Dunwell and Sunderland (1974a), during the first 5 days of culture of anthers, embryogenic pollen grains reveal no structural attributes suggestive of an impending switch in their developmental pattern. This period in culture may well be critical in establishing that gametophytic differentiation in the pollen will proceed only so much and then will proceed no more, before sporophytic type of differentiation sets in. Ultrastructural changes noted in the vegetative cell of embryogenic pollen grains at later periods of culture (7 to 8 days) are assignable to a structural regression of its cytoplasmic organization resulting in the elimination of much of the gametophytic influence. Signs of regression of the cytoplasm seen in the electron microscope are highly characteristic and include appearance of zones of multivesiculate bodies analogous to lysosomes, elimination of ribosomes and degradation of other organelles (Dunwell and Sunderland, 1974b). During the redifferentiation phase that follows regression, the vegetative cell divides in the embryogenic pathway; the newly formed daughter cells synthesize fresh cytoplasm complete with an array of ribosomes arranged in polysomal profiles (Dunwell and Sunderland, 1975). These observations are in conformity with the view that embryogenic pollen differentiates from the gametophytic pollen by way of a programmed destruction of the cytoplasm of the vegetative cell, thereby depleting it of a specific phenotype, followed by repopulation of the cell by a new set of organelles.

In *ab initio* pollen cultures originating from cold-treated anthers of *Nicotiana tabacum*, embryogenic pollen grains do not exhibit any ultrastructural features that could be construed as organelle regression; rather, they appear in a repressed state characterized by an attenuated cytoplasm, with condensed mitochondria and sparsity of ribosomes (Rashid, Siddiqui and Reinert, 1981). Embryogenic division of the vegetative cell is associated with the laying down of a convoluted fibrillar wall between the plasmalemma and the inner layer of intine, thus insulating the entire pollen cytoplasm. At this time, there is also an increase in the ribosome population in the cytoplasm in sufficient abundance to crowd out other organelles. Symptomatic of the generally heightened activity of the cell is the attainment of a normal configuration by the mitochondria (Rashid, Siddiqui and Reinert, 1982). Observations on pollen grains of *Aesculus hippocastanum* (Radojević, Zylberberg and Kovoor, 1980) and *Hordeum vulgare* (Idzikowska, 1981) during induction of sporophytic type of growth also point to a role for increased ribosomes and complex mitochondria for meeting the metabolic requirement during this period.

Although the ultrastructural changes observed in the vegetative cell of embryogenic pollen grains of tobacco before and after division are striking, a direct analogy of pollen from cold-treated plants with those from cultured anthers may be a facile comparison. It does not take into account any effect of temperature stress in preventing organelle regression or in causing other ultrastructural anomalies in the cytoplasm, as shown in cold-stressed pollen grains of *Datura metel* (Camefort and Sangwan, 1979). Thus, during culture of anthers, the possibility of a regression of gametophytic cytoplasm in pollen grains destined to yield embryoids cannot be ruled out. Although low temperature is conducive to the depolymerization of microtubules (see Hepler and Palevitz, 1974, for review), this fact does not appear to have entered into discussions about the consequences of cold stress on pollen grains.

In *Nicotiana tabacum*, the generative cell is undistinguished cytologically. As the vegetative cell divides in the embryogenic pathway, the generative cell remains ultrastructurally unchanged until it eventually disintegrates (Dunwell and Sunderland, 1975). Ultrastructural and stereological lines of evidence have converged in indicating that in *Hyoscyamus niger* the generative cell that functions as the embryo mother cell shows changes suggestive of high metabolic activity as it prepares to divide in the embryogenic pathway. Very impressive are the quantitative changes in the volume of granular and fibrillar zones in the nucleolus and in the amount of decondensed chromatin in the uninucleate, embryogenic pollen grains, seen as early as 6 hours after culture of anthers and carried over unchanged into the generative cell. There are no signs of vulnerability of the gametophytic cytoplasm of the generative cell to regression; rather the generative cell appears to undergo the full gametophytic development before cytoplasmic changes concerned with embryogenic growth are launched (Reynolds, 1984a).

Biochemical cytology

Our present understanding of the biochemical cytology of pollen embryogenesis is the confluence of two streams of inquiry: (1) analysis by cytochemical and cytophotometric methods of the changes in nucleic acid contents of cells and (2) autoradiographic analyses of the patterns of nucleic acid and protein synthesis. Since changes occurring in pollen grains during embryogenic transformation are profoundly related to those occurring during normal pollen development, we shall take a brief look at the biochemical cytology of nuclear differentiation that occurs during pollen development. It is now well established that after the first haploid mitosis, RNA and protein synthesis occurs rapidly in the vegetative cell; the nucleus of this cell also enlarges, although its DNA content remains unchanged. On the other hand, the nucleus of the generative cell undergoes DNA replication, but the cytoplasm of this cell synthesizes only insignificant amounts of RNA and proteins (La Cour, 1949; Woodard, 1958; Sauter, 1969; Sangwan-Norreel, 1979; Reynolds and Raghavan, 1981; Raghavan, 1984). Development of the pollen grain thus involves switching off DNA synthesis in one cell without interfering with transcription; in the other cell, transcription is slowed down without interfering with DNA synthesis.

DNA synthesis is directly linked to the initiation of embryogenic divisions in pollen grains; one factor accentuating the importance of DNA replication is that we are dealing with initiation of divisions in one cell which has essentially stopped division and in another cell which is just fated to go through one more round of division before terminal differentiation. Microspectrophotometric determination of DNA content of pollen grains of cultured *Datura innoxia* anthers showed that only a few bicellular pollen grains housing two identical nuclei double their DNA content preparatory to division (Sangwan-Norreel, 1983). Embryoids formed from these pollen grains also have abnormally high levels of DNA in their cell nuclei (Sangwan-Norreel, 1981). Somewhat similar observations have been made in embryogenic pollen grains of cultured anthers of wheat and *Petunia* (Raquin et al., 1982). These results suggest that fusion between nuclei is rampant in the evolution of pollen embryoids in these plants. In tobacco pollen, in which the embryoid arises by the division of the vegetative cell, microspectrophotometric measurements have revealed that DNA doubling occurs in this cell before any overt signs of dedifferentiation; in contrast, during normal pollen ontogeny, DNA content of this cell remains at 1C level (Chu, Liu and Du, 1982). In confirmation of the origin of embryoids in *Hyoscyamus niger* from the generative cell, DNA synthetic activity studied by ^{3}H-thymidine incorporation showed that dedifferentiation of the pollen grain into an embryoid is accompanied by DNA synthesis in the generative cell nucleus, while the nucleus of the vegetative cell synthesizes no DNA after it is cut off or synthesizes DNA only during a limited number of cell cycles (Raghavan, 1977a).

The accumulation and synthesis of RNA and proteins in the pollen grains are also responsive to culture conditions leading to embryogenic divisions. In cultured anthers of tobacco, depending on the temperature of incubation, there is an induction period of 6 to 12 days before the vegetative cell divides in the embryogenic pathway. In pollen grains that continue the gametophytic program in culture and become nonembryogenic, cytochemically detectable RNA and proteins register a four- to six-fold increase during the induction period, whereas both components decline in the embryogenic pollen. On the basis of these results, and those based on ultrastructural studies discussed in the previous section (Dunwell and Sunderland, 1974a,b), it has been suggested that the initial event of embryogenic induction is a suppression of the gametophytic program concerned with accumulation of RNA and proteins (Bhojwani, Dunwell and Sunderland, 1973). This would ensure that genes coding for proteins involved in embryogenic divisions could be expressed fully without being masked by the simultaneous expression of genes for pollen maturation and germination. Cytochemical study of pollen embryogenesis in *Datura innoxia* (Sangwan-Norreel, 1978) has painted a different picture of RNA and protein changes, however; here, compared with nonembryogenic pollen grains, embryogenic pollen grains exhibit an increased stainability for cytoplasmic RNA preparatory to the first haploid mitosis. In the embryogenic binucleate pollen, the nucleus of the generative cell stains more densely for histones than the nucleus of the vegetative cell; this might focus attention on histones in preventing the division of the generative cell in the embryogenic pathway.

The correspondence between the overall division patterns of pollen grains in cultured anthers of *Hyoscyamus niger* and incorporation of ^{3}H-uridine has suggested that synthesis of rRNA might be an important facet of the biochemical cytology of pollen embryogenesis. In this plant, embryogenic divisions are initiated in pollen grains in 6 to 12 hours after culture of anthers. Consistent with this observation, embryogenic pollen grains were found to synthesize RNA as early as the first hour of culture of anthers. During normal pollen development, the generative cell of the bicellular pollen grain does not incorporate any appreciable amount of

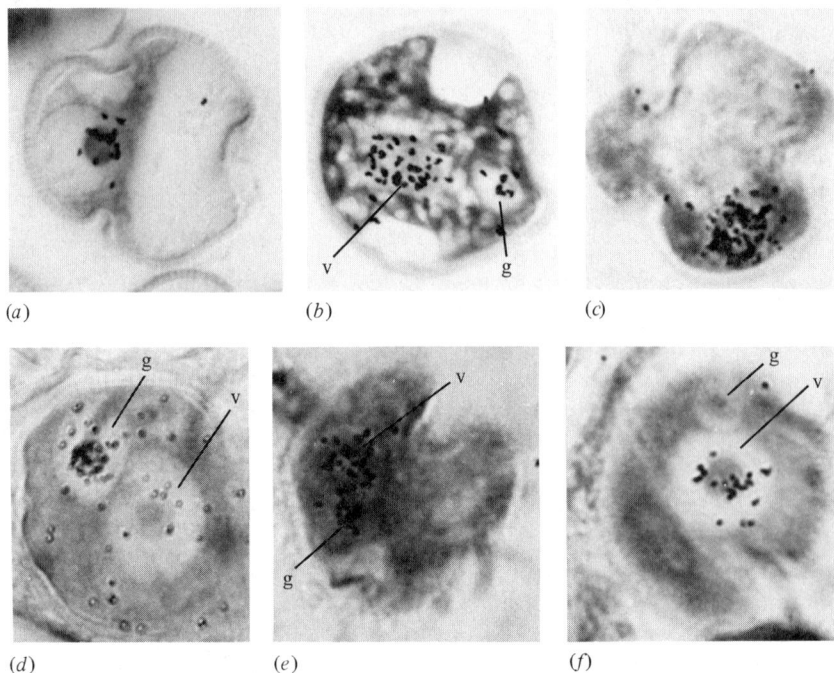

FIGURE 6.7. Autoradiographs showing ^{3}H-uridine incorporation in uninucleate and bicellular pollen grains of *Hyoscyamus niger* during normal ontogeny and during embryogenic development. (*a*) Uninucleate pollen grain from intact anthers; silver grains are seen over the nucleus. (*b*) Bicellular pollen grain from intact anthers; vegetative cell nucleus is heavily labeled. (*c*) Uninucleate pollen grain from a cultured anther segment; the nucleus is heavily labeled. (*d*) Bicellular pollen grain from a cultured anther segment showing the nucleus of the generative cell heavily labeled. (*e*) Bicellular pollen from a cultured anther segment showing nuclei of both generative and vegetative cells labeled. (*f*) Bicellular pollen grain from a cultured anther segment showing only the nucleus of the vegetative cell labeled. g = nucleus of the generative cell, v = nucleus of the vegetative cell. (Parts *a* and *b* are from Reynolds and Raghavan, 1982. Parts *c–f* are from Raghavan, 1979b.)

^{3}H-uridine into RNA. By contrast, in the potentially embryogenic bicellular pollen grains there is an accelerated incorporation of ^{3}H-uridine into the generative cell (Fig. 6.7). Analysis of ^{3}H-uridine incorporation in the embryogenic bicellular pollen grains during progressive embryogenesis has indicated that only those pollen grains in which RNA synthesis occurs in the generative cell, or in both generative and vegetative cells, divide further in the embryogenic pathway, whereas those in which RNA synthesis occurs exclusively in the vegetative cell become starch filled and nonembryogenic (Raghavan, 1979a,b; Reynolds and Raghavan, 1982). Thus, transcription in the generative cell is a more important prerequisite for embryogenic divisions of pollen grains of *H. niger* than is transcription in the vegetative nucleus; put another way, a constraint that prevents the generative cell from embarking on an embryogenic pathway might be its inability to synthesize the appropriate RNA.

The regulation of gene expression requires that the input of genetic information into the protein synthetic machinery also be regulated at different stages of transformation of the pollen grain into an embryoid. A recent analysis of incorporation of ^{3}H-arginine, ^{3}H-leucine, ^{3}H-lysine and ^{3}H-tryptophan into pollen grains of *Hyoscyamus niger* correlates well with the findings from studies of ^{3}H-uridine incorporation. This work has shown that while uninucleate pollen grains of intact anthers do not incorporate any ^{3}H-tryptophan and ^{3}H-leucine into proteins, culture of anthers is associated with the incorporation of these amino acids as well as with an accelerated incorporation of ^{3}H-arginine and ^{3}H-lysine into uninucleate embryogenic pollen grains. The division of the pollen grain in the embryogenic pathway or the continuation of the gametophytic program of the pollen grain is correlated with the synthesis of proteins in the generative cell or the vegetative cell, respectively. Generally, those pollen grains which incorporate ^{3}H-arginine, ^{3}H-leucine and ^{3}H-lysine into proteins of the generative cell or of both generative and vegetative cells become embryogenic, whereas those in which the vegetative cell alones incorporates these amino acids into proteins become starch filled and nonembryogenic. These results imply that the synthesis of certain general and basic proteins in the uninucleate pollen grains and subsequently in the generative cell of the bicellular pollen grains is linked to the switching on of their embryo developmental program (Raghavan, 1984).

In summary, these observations of different embryogenic systems indicate that much further work is necessary before a comprehensive picture of the involvement of nucleic acid and protein synthesis in the pollen cells during their transformation into embryoids can be obtained.

Physiology of pollen embryogenesis

A simple list of the factors that affect the formation of embryoids in anther and pollen cultures is admittedly impressive. In addition to genotypic effects, this includes the physiological state and conditions of growth of the donor plant, plant age, stage of pollen development, pretreatment of flower buds, temperature of culture, photoperiodic conditions of culture, solid versus liquid medium, composition of the culture vessel atmosphere, effect of nutrient and non-nutrient substances in the culture medium and effect of somatic tissues of the anther (Dunwell, 1979). It is not yet possible to assign specific morphogenetic significance to any of these factors, as they do not appear essential for the selective induction of pollen grains to become embryoids, but merely cause an increase in the number of embryoids formed from pollen grains already diverted in the embryogenic route. For our purposes we shall consider those factors that affect most importantly the physiology of pollen grains causing them to form embryoids or calluses in a relatively efficient way. The critical division of the pollen grain in the embryogenic pathway is a complex process and, when one considers the numerous factors that influence it, studies that do not take them into account are less satisfactory.

Stage of pollen development

To be able to produce embryoids, pollen grains must be at an appropriate cytological stage of development at the time of culture. For each species studied, this function can be achieved by initially staging anthers based on the length of flower buds and the frequency of embryogenic development and using anthers from flower buds as close as possible to the propitious stage for subsequent culture. The work of Sunderland and Wicks (1971) established that *Nicotiana tabacum* anthers are responsive only if they are cultured within a certain period of development, beginning with the liberation of pollen grains from the tetrad and ending with bicellular pollen grains. Embryoids are readily formed from anthers cultured at the uninucleate stage or as the pollen nucleus enters mitosis. In many other plants as well, pollen embryogenesis is best achieved in anthers cultured at the uninucleate stage of development, and rarely have anthers containing pollen at earlier or later stages of development been as bountiful. Both for shed pollen cultures (Sunderland and Roberts, 1977) and isolated pollen cultures (Heberle-Bors and Reinert, 1979) of tobacco, pollen grains at the binucleate stage of development yield embryoids in high frequency. Undoubtedly, these stages of pollen development at which embryogenic induction occurs easily are related to a vulnerable stage of the mitotic cycle of the first haploid mitosis at which the trigger is perceived. However, as no studies on this question have been undertaken, the relationship of the stage of cell division cycle of the pollen grain to its embryogenic development must be regarded as an open question.

Physiological stress of flower buds

We have referred in passing to the effects of a pretreatment of flower buds at low temperatures in enhancing embryogenic response in cultured anthers of several cereals and in isolated pollen cultures of *Nicotiana tabacum* and *Datura innoxia*. Other studies have shown that a low-temperature stress administered to flower buds of *D. innoxia* (Tyagi, et al., 1979), *D.*

metel (Gupta and Babbar, 1980), *Hyoscyamus niger* (Sunderland and Wildon, 1979), *Lycium* (Gu, Gui and Xu, 1984) and *Brassica napus* (Dunwell et al., 1985) increases the pollen response in anthers during subsequent culture. In *D. innoxia* a more successful experimental design to increase anther productivity has involved pretreatment of flower buds at a low temperature combined with centrifugation of excised anthers at a low speed (Sangwan-Norreel, 1977). The most vivid effect of low temperature is seen in cold-stressed flower buds of *Petunia hybrida*. Here, not only is anther productivity increased upon culture of cold-treated anthers (Malhotra and Maheshwari, 1977), but simply incubating cold-stressed buds at a higher temperature for 24 hours without any culture medium is sufficient to induce embryogenic divisions in pollen grains (Babbar and Gupta, 1980). In investigations dealing with the effect of temperature pretreatments of flower buds of *N. tabacum*, it has become clear that the stage of development of the anther and duration of cold treatment are crucial in maximizing the yield of embryoids. Generally, temperature stress is effective if given to flower buds before, during or after the first haploid mitosis in pollen grains encased in the anthers. With regard to specific temperatures, pretreatment at 7°C to 9°C is more effective than 5°C, and the younger the stage of pollen development, the longer is the duration of treatment required (Sunderland and Roberts, 1979). There is a dearth of complete data of this kind for thermal stress of flower buds of other species.

There are speculations on the manner in which a thermal trauma increases the number of responding anthers and their embryogenic productivity. Nitsch and Norreel (1973) suggested that the course of the highly polarized first haploid mitosis of the pollen grain is deranged by a thermal shock. This results in the formation of pollen grains with two symmetrical cells that are more favorably disposed to enter the embryogenic pathway than are pollen grains with asymmetrical cells. However, this can be true only in a special and limited sense – special because of the relative ease with which pollen grains of tobacco respond to cold stress and limited because pollen grains of tobacco and other plants dividing asymmetrically become embryogenic in anthers cultured without a cold stress. Moreover, it is doubtful whether there is an actual increase in the number of pollen grains with symmetrical cells following cold treatment of anthers. For example, Duncan and Heberle (1976), who subjected tobacco anthers to a cold stress, observed an increase in the number of viable pollen grains but not in the number of pollen with symmetrical cells. Sunderland and Roberts (1979) have suggested that low-temperature treatment of anthers might impede the senescence of somatic anther tissues and thus ensure survival of a greater number of pollen grains for subsequent embryogenic divisions. This idea is attractive and will become much more so if it is possible to establish a relationship between the production of senescence factor and pollen degradation.

Subjecting flower buds of tobacco to water stress (Imamura and Harada, 1980a), to reduced atmospheric pressure (Imamura and Harada, 1980b), to anaerobic conditions (Imamura and Harada, 1981) and to a water-saturated atmosphere (Dunwell, 1981a) has also been shown to promote pollen embryogenesis in cultured anthers. Although their meanings are unclear, the observations have prompted somewhat vague proposals about the mode of action of these physiological stresses on anthers.

Composition of the culture medium

In plants such as tobacco, *Datura, Saintpaulia, Hyoscyamus, Atropa* and *Solanum*, pollen embryogenesis is induced in anthers cultured in a medium that includes only macronutrient salts, micronutrient salts, vitamins, a sugar alcohol (*myo*-inositol) and sucrose. Using this limited diet as a base, variations in the frequency and pattern of pollen embryogenesis in cultured anthers may be detected by changing the form of iron present, concentration and quality of sugars supplied in the medium and hormonal components of the medium and by supplementing the medium with a variety of other substances.

Among the mineral constituents of the medium, the importance of iron and its participation in pollen morphogenesis have been the subject of some studies. In most of the anther and pollen culture investigations, iron is generally supplied in a chelated form as FeEDTA or as ferric ethylenediamine-di-O-hydroxyphenylacetic acid (FeEDDHA). In some particularly favorable systems like *Nicotiana, Datura* and *Atropa,* iron-free medium can support embryogenic development of pollen grains in cultured anthers up to the globular stage, but further development of the embryoid is completed only in the presence of iron (Heberle-Bors, 1980; Vagera and Havránek, 1982). This means that embryogenic induction does not require an external source of iron, except what is available as contaminants in the major salts used.

Although sucrose is generally provided in the medium at a concentration of 2 percent, anthers of some plants are exacting in their sucrose requirements, and concentrations in the range of 6 to 20 percent have been used in anther cultures of *Triticale, Hordeum vulgare, Triticum aestivum, Brassica campestris, Solanum tuberosum, Zea mays* and *Saccharum officinarum* (Chu, 1982). It appears that in *B. campestris* (Keller, Rajhathy and Lacapra, 1975) and *S. tuberosum* (Sopory, 1979) high levels of sucrose play a role in triggering embryogenic divisions in pollen grains but not in the growth of embryoids, since the latter do not adapt easily to high-sucrose-containing medium. Although sucrose supplied in the medium far in excess of the amount generally metabolized might be construed to function as an osmoticum, this does not appear to be the case for anther cultures of *S.*

tuberosum; here a mixture of sucrose and mannitol does not match the effectiveness of high levels of sucrose in inducing pollen embryogenesis (Sopory, 1979).

Based on a survey of published reports, it has been surmised that successful induction of embryogenic or callus type of growth in pollen grains of cultured anthers of the vast majority of plants requires either an auxin or a cytokinin or a combination of both in the medium. By contrast, cultured anthers of only a handful of plants form embryoids without the intervention of exogenous hormones (Maheshwari et al., 1982). Evidence tabulated in recent reviews (Reinert and Bajaj, 1977; Clapham, 1977) also seems quite clear in indicating that there are wide variations in the apparent concentrations of hormones necessary for embryogenic or callus type of growth in cultured anthers of various plants. It would be outside the scope of this chapter to review this literature; for present purposes, it is sufficient to point out that the role of auxins and cytokinins in inducing the normally recalcitrant pollen grains to embark on a sporophytic type of growth is nowhere better illustrated than in cultured anthers of cereals, which, as alluded to earlier, have the peculiar virtue of forming pollen-derived calluses. Subsequent regeneration of plantlets by organogenesis occurs when the callus is transferred to the basal medium alone, to a medium containing cytokinin with a low level of auxin or to a medium containing a weak auxin (Clapham, 1977). Admittedly, when pollen grains fail to embark upon sporophytic type of growth in the absence of hormones, the latter perform a morphogenetic function. The current dilemma in evaluating these results is to determine the extent to which positive responses of pollen grains to specific hormones result from cellular changes and from increase in the endogenous content of hormones required to mediate in growth and cell division.

In anthers of *Datura innoxia,* the role of hormones is one of enhancing the production of pollen embryoids. Although auxins such as IAA, NAA, and IBA and cytokinins such as zeatin, kinetin and BAP are effective in this role, the best response is obtained with zeatin and kinetin. These observations have formed the basis for the suggestion that embryogenic divisions are related to the maintenance of a certain level of endogenous hormones in the pollen grain (Sopory and Maheshwari, 1976). Cultured anthers of *Hyoscyamus niger* respond to high concentrations of 2,4-D (more than 2.0 mg/liter) by regenerating calluses rather than bipolar embryoids from pollen grains. Compared with anthers cultured in the basal medium, auxin does not affect the number of pollen grains dividing in the sporophytic pathway or the patterns of early division sequences involving the vegetative cell and generative cell. However, after its release from the exine, undifferentiated growth prevails in the cellular mass, resulting in a small nodule. By continued production of such nodules, accompanied by coalescence of the erstwhile separate nodules, a callus is formed (Corduan, 1975; Raghavan, 1978). It requires only the removal of 2,4-D from the medium to permit regeneration of numerous embryoids from the callus (Raghavan and Nagmani, 1983). The behavior of the pollen callus augments the general impression that the action of 2,4-D here is very similar to its role in somatic embryogenesis in carrot. Significance of the effects of other hormones like ethylene (Horner, McComb, McComb and Street, 1977; Dunwell, 1979) on pollen embryogenesis is impossible to assess, as are the various observations that assign embryoid-inducing effects to undefined substances such as coconut milk, potato extract, yeast extract, malt extract and casein hydrolyzate (Chu, 1982) or to substances emanating from cultured leaves and anthers or calluses (Sharp, Raskin and Sommer, 1972; Pelletier and Durran, 1972; Zenkteler and Stefaniak, 1982).

Among other additives to the culture medium, the beneficial effects of activated charcoal on pollen embryogenesis seem to be well established. The most convincing explanation of these effects stems from the current assumptions that charcoal might adsorb inhibitors released from the anther, or those present in the agar, or those arising from degradation of other metabolities present in the medium (Kohlenbach and Wernicke, 1978; Horner et al., 1977; Weatherhead, Burdon and Henshaw, 1978). Phenolic substances appear to be prime candidates for the role of anther-derived inhibitors. In anther culture of *Anemone canadensis,* the concentration of phenolic substances is found to be high in the culture medium lacking activated charcoal. The addition of activated charcoal, which reduces the concentration of phenolics to less than 20 percent of the original value, also enhances embryogenesis in cultured anthers (Johansson, 1983). It is likely that an increase in the frequency of pollen embryogenesis reported in *Datura innoxia* anthers by polyvinylpolypyrrolidone is attributable to the well-known action of this substance in adsorbing phenolic compounds (Tyagi et al., 1981b). Whether this compound can be used on a wider level like activated carbon to enhance anther productivity remains to be seen.

Role of anther wall and tapetum

A typical angiosperm anther enclosing uninucleate pollen grains consists of a well-defined layer of epidermis, followed on the inside by a layer of endothelium, one to three middle layers (all constituting the anther wall) and a layer of radially elongate cells of the tapetum. The facts so far presented have emphasized that pollen embryogenesis occurs most readily in intact anthers of responsive species cultured at an appropriate stage of development in a suitable medium. It is therefore not surprising to find that, except in a few cases, the tendency for embryogenic or callus type of growth tends to vanish in isolated pollen grains cultured *in vitro*. Apparently, in the majority of plants studied, pollen embryogenesis, beyond the favorable conditions supplied in a nutrient medium, is also favored by conditions in which pollen grains are set up in the close association of an intact anther. This point is important enough in its implications for the physiology of pollen embryogenesis to merit some consideration.

Nevertheless, the available evidence in support of a causal relationship between the somatic tissues of the anther and embryogenic development of pollen grains is rather indirect. The idea that a conditioning effect of the somatic tissues of the anther is a typical feature of pollen embryogenesis may be traced to the work of Pelletier and Ilami (1972). These workers showed by pollen transplantation experiments that temporary residence of pollen grains of *Nicotiana tabacum* in anthers of the same species, or in anthers of the related *N. glutinosa* or of *Petunia hybrida,* is essential for embryogenic development. It will be recalled that in the first successful induction of embryogenesis in isolated pollen grains of *Datura innoxia* (Nitsch and Norreel, 1973), the regular mineral salt medium was supplemented with an extract of cultured anthers of the same species. Efforts to improve the yield of calluses in cultured barley anthers have shown a medium conditioned by the growth of anthers of barley to be beneficial; this finding is consistent with the view that substances released from the anther into the medium provide the key to the induction of some phase of sporophytic type of growth in the pollen grains (Xu et al., 1981). In cultured anthers of *Hyoscyamus niger* (Raghavan, 1978) and barley (Idzikowska, Ponitka and Młodzianowski, 1982), pollen grains found in the vicinity of the tapetum are found to differentiate into embryoids and calluses, respectively. An additional observation on the role of tapetum is related to the prerequisite for a cold stress of anthers for pollen embryogenesis. Transmission electron microscopy of anthers of *Datura innoxia* (Cadic and Sangwan-Norreel, 1983) and scanning electron microscopy of anthers of barley (Sunderland, Huang and Hills, 1984) have shown that chilling treatment is associated with dispersal and loss of tapetal cells. Although the loss of tapetal integrity can influence pollen embryogenesis in many possible ways, this observation reinforces the view that a positive role of the tapetum in pollen embryogenesis should not be dismissed hastily.

We are far from understanding the mechanism by which somatic tissues of the anther influence embryogenic development of pollen grains. One of the basic aspects of the problem concerns the nature and timing of appearance of specific regulatory substances in the tissues. Amino acid analyses have shown that culture of anthers of *Nicotiana tabacum* is associated with a rapid increase in the level of serine (Nitsch, 1974) and glutamine (Horner and Pratt, 1979); it is interesting to note that supplementation of the medium with high concentrations of these compounds led to the first successful culture of isolated tobacco pollen in a defined medium (Nitsch, 1974). Extensive changes in the levels of free and bound amino acids, particularly of threonine, serine, glutamic acid, proline and γ-aminobutyric acid have been detected during critical phases of culture of anthers of *Datura metel* (Sangwan, 1978). Although amino acid analyses were carried out on whole anthers including enclosed pollen grains, it is quite possible that some of the amino acids are synthesized by the somatic tissues of the anther and supplied to the pollen grains. More recently, it has been shown (Sangwan, 1983) that certain amino acids when added individually to growth media containing an inorganic source of nitrogen are highly favorable for pollen embryogenesis in anther cultures of *D. metel*.

Some experimental results obtained in cultured anthers of tobacco have been interpreted to indicate that as the somatic tissues of the anther begin to senesce, they exert an inhibitory effect on pollen embryogenesis. Depending upon the time after culture of anthers when senescence begins, the effect of the senescing tissues may be manifest either in the degeneration of a large number of pollen grains or in the retardation of growth of embryoids formed (Pelletier and Ilami, 1972; Mii, 1976, 1980). In *Datura innoxia* serial transfer of anthers to fresh liquid medium results in an increase in the number of embryoids harvested, compared with continuous culture of anthers in the same medium (Tyagi et al., 1979). These results, together with other evidence, have been taken to imply the production by the anthers of an inhibitor whose concentration is apparently diluted by the transfer of anthers to fresh medium. From these data it is clear that in cultured anthers of such widely investigated systems as tobacco and *Datura* pollen embryogenesis is delicately balanced between the actions of promotory and inhibitory substances; other conditions being equal, we do not know what tips the balance one way or the other.

Embryogenic competence of pollen grains

Our discussion thus far has assumed that pollen grains enclosed in the anther locule are programmed to undergo gametophytic differentiation but are deflected in the embryogenic pathway as a result of the trauma of excision and culture of anthers. The first signs of change conducive to embryogenesis are observed in developmental bias without overt signs in pollen grains of anthers sampled during early periods of culture. For example, in *Hyoscyamus niger* we could identify within the first few hours of culture of anthers two types of uninucleate pollen grains – a small number of densely stained nonvacuolate pollen confined to the periphery of the anther locule and a large number of lightly stained pollen in varying stages of vacuolation found toward the center of the anther locule. Embryoids are formed from the nonvacuolate, densely stained pollen (Raghavan, 1979a).

About 8 days after culture of anthers of *Nicotiana tabacum,* binucleate embryogenic pollen grains are identified by their low frequency and light-staining nature from the large number of densely staining gametophytic pollen (Sunderland and Wicks, 1971). Thus, viewed in terms of individual pollen grains and what goes on within them, diversity rather than similarity marks their development in cultured anthers; some become embryogenically competent, whereas others mark their time as gametophytic pollen or disintegrate.

Some investigators have gone a step further and have claimed that pollen grains become competent to form embryoids during their development in the anther locule and that subsequent culture of the anther provides the environment for expression of this predetermined potential. This claim is based on the presence in the same anther, of a type of variant pollen grains which differ from the main population mainly with respect to their size and staining properties, a phenomenon known as dimorphism. Although dimorphic pollen grains occur in many plants, Dale (1975) first called attention to a correlation between the number of small pollen grains present in anthers of *Hordeum vulgare* var. Akka and the number of calluses formed in cultured anthers. According to Horner and Street (1978b), a population of weakly staining, small, binucleate pollen grains which occurs in low frequency in tobacco anthers is clearly distinguished from normal, binucleate, starch-filled pollen grains that occur in high frequency. Here also, the frequency of distribution of the small pollen grains and embryo frequency in cultured anthers are similar (Horner and Mott, 1979). In wheat anther cultures, a case for the origin of embryoids from small pollen grains that occur in low frequency is supported by obvious similarities in the development of these anomalous pollen grains and early stages of embryoid formation (Zhou, 1980). These studies suggest the possible origin of embryoids from the small pollen grains, which are considered embryogenically competent at the time of culture of anthers.

What are the factors responsible for the evolution of embryogenically competent pollen grains in a developing anther? This question has been studied in tobacco, where conditions under which plants are grown play an important role in the frequency of embryogenesis in isolated pollen culture. According to Heberle-Bors and Reinert (1981), in plants raised under short days and low temperature, there is a direct relationship between the yield of embryoids in pollen culture and the number of embryogenically competent pollen present in the anther, suggesting an environmental control of dimorphism. This could not have occurred if the frequencies of formation of embryogenic and gametophytic pollen grains were not in some way linked to the development of sexuality in flowers. Some information on this has come from the finding that embryogenically competent pollen grains are produced under conditions that induce high pollen sterility in the anther and that thus favor a shift in sex balance of the flower toward femaleness (Heberle-Bors, 1982a). It has been proposed that when the sex balance of the flower is tipped toward femaleness, there is also a tendency toward large-scale pollen sterility and that embryogenically competent pollen grains are in reality sterile pollen grains that are still viable (Heberle-Bors, 1982b).

These ideas on the evolution of embryogenic competence in pollen grains are provocative; under the impact of these ideas, some early reports of occurrence of pollen dimorphism and multicellular pollen grains in anthers of several plants (Maheshwari, 1950; Sunderland, 1982) might appear in a new light. There are also some reports on the spontaneous formation of embryoid-like pollen grains in anthers of a hybrid *Solanum* (Ramanna, 1974; Ramanna and Hermsen, 1974), *Narcissus biflorus* (Koul and Karihaloo, 1977) and *Paeonia* (Li, 1982). To these must be added the induction of multiple pollen grains in plants by treatment with chemicals, such as in wheat by ethylene-releasing substances (Bennett and Hughes, 1972). These examples represent cases in which embryogenically competent pollen grains formed in the anther have apparently escaped from the intracellular control system and expressed incipient sporophytic features. Although a strong case can be made for the embryogenic predetermination of pollen grains in such plants as tobacco, evidence for such a conclusion is scanty in other plants. In still others, such as *Datura* and *Hyoscyamus*, there is reason to believe that pollen grains become embryogenically competent only upon culture of anthers. For the moment we are unable to generalize beyond saying that this is a case for further analysis, with the caveat that we may not achieve a consensus.

Genetics of pollen embryogenesis

Varietal differences generally observed in the regenerative ability of pollen grains of cultured anthers suggest that the process is largely determined by the intrinsic hereditary properties of the individual plant, which at some point impinge on the process of pollen dedifferentiation. References were made in a previous section to several cereals in which pollen callus formation is controlled genotypically. In addition to these examples, there is a long list of plants in which genotypic differences in the regenerative capacity of pollen grains have been fully documented. This is not to say that all cases of failure to obtain pollen embryogenesis are referrable to genotypic effects, but it focuses attention on a problem that has to be reckoned with in anther culture studies of highly heterogeneous populations of plants.

Demonstrating the level at which genotype exerts its effects on the functional properties of the anther and pollen grains is not an easy task. Based on an analysis of the frequency of embryogenesis in cultured anthers and isolated pollen grains of different varieties of tobacco, it has been claimed that the genotype determines the frequency of embryoids formed by its influence on the number of embryogenically competent pollen grains differentiated and the metabolic state of the cultured anthers (Heberle-Bors, 1984). It will be instructive to learn how a particular genetic combination of the donor plant facilitates pollen grains to become embryogenically competent.

In many ways the current central problem in the genetics of pollen embryogenesis is whether genotypic differences in anther response to *in vitro* culture are heritable. Experiments undertaken in wheat (Picard and de Buyser, 1977; Lazar, Baenziger and Schaeffer, 1984), potato (Wenzel and Uhrig, 1981) and barley (Foroughi-Wehr, Friedt and Wenzel, 1982), involving culture of anthers of F_1 and their reciprocal crosses made between genotypes differing in their ability to form embryoids or calluses, have indicated that it is possible to transfer regenerative propensity of pollen grains of a parent to its F_1 hybrid in a cross with a parent in which this propensity is negligible or nonexistent. In another line of investigations carried out in potato, it was also possible by crossing progenies to recombine and accumulate genes favoring embryoid production in anthers of weakly responding clones (Jacobsen and Sopory, 1978). A recent study of the effect of different temperature regimens on pollen callus formation in anthers of F_1 hybrids of wheat has clearly demonstrated the influence of the parents, indicating that temperature sensitivity of pollen grains is also heritable (Ouyang, Zhou and Jia, 1983).

Frequently discussed in connection with the genetics of pollen embryogenesis is the fact that embryoids and plants formed directly from pollen grains or through the intermediary of a callus do not always display haploidy, but constitute a mixture of different levels of ploidy. Such chromosomal patterns have been regularly observed in several species *Nicotiana*, *Datura*, and *Petunia*, in *Brassica campestris*, *Atropa belladonna*, *Solanum nigrum*, *Digitalis purpurea*, *Hyoscyamus niger*, *Arachis hypogaea*, *Cajanus cajan*, *Cicer arietinum*, and cereals such as rice, barley and rye (Sunderland and Dunwell, 1977; Chu, 1982; Bajaj, 1983). The frequencies with which nonhaploids are represented in anther cultures vary in different species. For example, in *B. campestris* and rye no haploids have been reported, whereas other genera show a preponderance of nonhaploids over haploids (Sunderland and Dunwell, 1977; Bajaj, 1983). On the other hand, even secondary embryoids produced in cultured anthers of *B. napus* spp. *oleifera* retain essentially the gametic chromosome number for more than a year of subculture (Loh and Ingram, 1983). In *N. sylvestris*, nearly 70 percent of plants formed from cultured anthers are haploid (McComb and McComb, 1977).

As the origin of calluses and embryoids from the somatic tissues of the anther has been rigorously excluded in these studies, the reasons for the increased chromosomal complement of cells of calluses and embryoids obviously lie in the pollen grains or in the products of their division. It has been shown that in plants such as *Datura innoxia* the proportion of non-haploids regenerated from pollen grains varies with the age of anthers at the time of culture, haploids being predominant in anthers enclosing early-stage pollen grains but decreasing with later stages of pollen with a concomitant appearance of nonhaploids of higher levels of ploidy (Engvild, Linde-Laursen and Lundqvist, 1972; Sunderland et al., 1974). There is little or no evidence that these pollen grains at the time of culture are anything but haploid, thereby strengthening the conclusion from cytological studies that nonhaploid embryoids might have arisen by nuclear fusion between division products of the pollen cells during embryogenic development (Sunderland et al., 1974).

Endopolyploidy appears to be the primary mechanism by which nonhaploid calluses are formed in anther cultures. Even though prolonged growth of the callus seems to increase the ploidy level of cells, this does not rule out the possibility that the initial growth of the callus is accompanied by chromosomal changes of a less extreme form. Not infrequently, calluses and plants regenerated from calluses appear to be mixoploids, constituted of cells of different ploidy levels. In *Lycopersicon peruvianum* (Sree Ramulu, Devreux, Ancora and Laneri, 1976) and *Hordeum vulgare* (Mix, Wilson and Foroughi-Wehr, 1978), in which the origin of mixoploids has been analyzed in some detail, evidence seems to indicate that more than one type of pollen cell is involved in their formation. The observation in barley is in accord with the cytological analysis of the early division sequences in pollen grains of cultured anthers (Sunderland and Evans, 1980).

The isolation and characterization of mutant cell lines originating from haploid plants has become of late a major area of study. This effort stems from the ease with which suspension cultures of haploid cells are initiated from haploid plants by routine tissue-culture techniques. From the perspective of embryogenesis, the ability of pollen grains to reconstitute an embryo-like structure is potentially useful for selection of haploid mutant embryoids impaired in embryogenesis, for nutritional deficiency or for drug resistance. Although efforts toward this end are most likely in progress in some laboratories, success has not yet been achieved.

Concluding comments

The analysis of pollen embryogenesis discussed in the preceding pages leaves us with the clear impression that the vegetative and generative cells of the angiosperm pollen grain are not only specialized reproductive cells, but that they also possess the innate potential to develop into whole new organisms of which they were integral parts by a process of simulated embryogenesis. Thus, totipotency of the germ cells has been demonstrated by anther culture experiments. From the demonstration of totipotency of an undivided generative cell, it is only one step further to demonstrate totipotency of the sperm, the ultimate unit of the short-lived male gametophyte in angiosperms. Undoubtedly a great deal of manipulative skill will be required before the sperm is brought within the confines of a culture vessel to express its full genomic potential.

The great riddle facing investigators in this field is to know what turns on the pollen grain in the embryogenic pathway. Knowledge concerning the mechanism of pollen embryogenesis is limited, based on cytological and histochemical studies and not on modern biochemical methods. A major disadvantage of anther and pollen culture systems is the lack of uniformity in response and the relatively small percentage of pollen grains that become embryogenic. In order to apply molecular and biochemical techniques to study the mechanism of pollen embryogenesis priority should be given to the formulation of methods of inducing high-frequency embryogenesis in pollen grains routinely and reproducibly. Results obtained with model systems like tobacco are encouraging, but much more needs be done to simplify the system before it can be fully exploited.

7
Regulation of gene activity during embryogenesis

It is clear by now that in angiosperms the embryo has its origin in the fertilized egg, whereas embryo-like structures or embryoids are generated from somatic cells and pollen grains. From the moment of fertilization or from the time the somatic cell or pollen grain enters the embryogenic pathway, development results from the temporal expression of genes in the progenitor cells. Obviously, the strategies for marshaling genes into action differ widely in the different cell types as they evolve into embryos or embryoids. The aim of this chapter is to explore how genetic information is parceled out to control differentiation of these different types of embryo mother cells. In line with this pursuit is an investigation of the mechanism controlling transcriptional and translational processes during embryogenesis. Some of these studies have been reviewed in Chapter 3.

The unique differences between cells that give rise to the embryo and embryoid require that concepts and experiments relating to the programming of developmental information during zygotic, somatic and pollen embryogenesis be discussed separately. Each cell type presents its own special advantages and disadvantages to the study of information flow during embryogenesis. For example, it is impossible to investigate the molecular biology of fertilization and the early division phase of embryos because these are accomplished in the relatively inaccessible environment of the embryo sac. On the other hand, suspensions of single cells and cell clusters possessing the potential to form embryoids offer a convenient experimental system that can be used to study the early events of somatic embryogenesis. But gene activation during the transformation of a somatic cell into an embryoid is quite different from the information processing that occurs after genetic recombination at fertilization, the basic mechanism of which can scarcely be established by studying anything but fertilization itself. Pollen grains enclosed within anthers are intermediate between the fertilized egg and somatic cell suspensions as objects for study. Although pollen grains do not present vexatious problems in experimental handling, they do not develop synchronously in the embryogenic pathway in mass culture.

It is interesting to note that out of more than 1,000 publications on zygotic, somatic and pollen embryogenesis that have appeared during the past quarter-century, only a few are devoted to the question of regulation of gene activity.

Zygotic embryogenesis

This section begins with a search for maternal mRNA during angiosperm embryogenesis and is followed by a discussion of the developmental regulation of gene action during embryogenesis in soybean and cotton in which it has been studied in great detail.

Maternal mRNA during embryogenesis

The concept of maternal message, which has its roots in animal embryology, assumes that the mature egg contains large stores of maternal mRNAs which become available after fertilization to code for the first proteins of the zygote. Three considerations appear to have provided the impetus for much of the experimental work done to test the validity of this concept. One is the development of cross-species hybrids such as between sea urchin species that follow the pattern of the maternal partner until a few rounds of divisions are completed; subsequently, paternal or hybrid characters appear. A second consideration is the well-known demonstration that parthenogenetically activated anucleate halves of certain animal eggs are not prevented from undergoing a few rounds of divisions, suggesting that protein synthesis necessary for the execution of these divisions may be occurring in the absence of concomitant DNA-dependent RNA synthesis. Third is the finding that very little RNA is synthesized in the egg immediately after fertilization to account for the vigorous protein synthesis observed. On the basis of recent and extensive experiments on sea urchins and amphibians, the existence in the eggs of template information stored outside the nuclear genome used to direct protein synthesis immediately after fertilization is now widely accepted.

Given the paucity of data on RNA and protein synthetic activities of the egg before and after fertilization, it is not surprising that there is no hard evidence for the existence of a legacy of template information in the angiosperm egg. However, in one study (Nagato, 1979), it has been shown that when rice and barley florets are cultured in a medium containing ^{3}H-uridine, no autoradiographically detectable incorporation of the label occurs in developing embryos until they have produced more than about 100 cells and are phased out of the globular stage. At the same time there is no restraint to the incorporation of ^{3}H-leucine, a precursor of protein synthesis, into embryos of all ages. Knowing that eggs are fertilized after the florets are placed in culture and assuming that cells of the globular embryo do not possess a large uridine pool, it has been reasoned that, as in animal embryogenesis, the first proteins of the embryo are coded on stored maternal mRNAs.

The many reported cases of parthenogenetic activation of the egg in angiosperms seem to offer indirect evidence for an independent role for the maternal genome in directing the early embryo developmental program. In a number of cases listed by Battaglia (1963), the haploid egg apparently divides in the total absence of a stimulus from the male gametophyte; in other examples tabulated by Lacadena (1974), several chemical and physical agents are able to hinder the growth and penetration of the pollen tube, thereby stimulating the egg to divide using its own stored information. A remarkable demonstration of the autonomy of the egg is seen in diplosporous apomicts in which the megaspore mother cell does not undergo reduction division and the embryo is formed from the diploid egg in the absence of fertilization. Particularly convincing is the demonstration that when unpollinated female flowers of *Aerva javanica* are raised in an *in vitro* culture, viable seeds are produced in great numbers; this could not have happened unless the nuclear genome of the egg was functionally active throughout embryogenesis (Puri, 1964). Haploids that arise by chromosome elimination in interspecific crosses between *Hordeum vulgare* (and other species of *Hordeum*) and *H. bulbosum* (Kasha and Kao, 1970; Subrahmanyam 1977, 1979, 1980) also seem to offer a basis for the assumption that maternal templates alone can support embryo development, albeit under somewhat abnormal conditions. Since the male chromosomes disappear even before they are integrated into the genome of the egg, the haploid embryo essentially develops by transcribing the stored maternal information.

Another area of research that might substantiate a role for the maternal program in the early division of the embryo relates to the recent reports of origin of embryo sac-based embryos from unfertilized ovules or ovaries containing unfertilized ovules (see San Noeum, 1976; Yang and Zhou, 1982, for review). It remains to be firmly established whether, in these instances, embryos have their origin in the fertilized egg or in the accessory cells of the embryo sac.

In the examples considered above, partially or fully developed embryos arise in the absence of fertilization because of the ability of the genome of the egg to support its continued divisions. Thus, it is likely that, as in animal embryos and in angiosperms also, following fertilization, maternal genomes regulate the initial developmental program of the zygote. Since the potential consequences of this for gene expression during embryogenesis are profound, this is an area of research that begs resolution.

Structural gene transcription during embryogenesis

For purposes of this discussion, a structural gene is defined as the unit of genetic information coding for a specific protein and is represented at the molecular level by DNA sequences on which mRNA is transcribed. Little qualification is needed to assert that a growing embryo should be synthesizing certain housekeeping proteins and a complement of enzymes for normal cellular metabolism, but cell specialization associated with organ initiation requires the synthesis of specific sets of proteins in excess of the production of housekeeping proteins. This implies that marked changes in gene activity should be occurring at certain stages of embryogenesis to direct the transcription of structural genes for organogenesis.

Many investigations on gene regulation during embryogenesis have concerned protein synthesis. As discussed in Chapter 3, the qualitative and quantitative details of the synthesis of nutritional storage proteins in certain angiosperm embryos provide a paradigm of gene activity during embryogenesis. Here the cotyledons become committed to the synthesis of specialized proteins at a certain stage of their development. As shown by polysomal analysis, cDNA–RNA hybridization and hybridization using cloned messages for storage proteins, the synthesis and accumulation of the latter are associated with the predominance of mRNAs in the cotyledons almost to the exclusion of other parts of the embryo. We have also seen that genes for storage proteins are transcribed in the cotyledons of small embryos long before they begin to accumulate proteins and that these genes gradually decay after protein accumulation has reached peak levels. However, study of storage protein accumulation gives only a partial picture of gene activity during embryogenesis because the appearance of gene products is not associated with organ differentiation in the embryo. Moreover, despite the fact that form change in the embryo is accompanied by differentiation of the cotyledons, hypocotyl and root, there is no evidence to link the synthesis of particular proteins among specific cell lineages with organogenesis in the embryo.

Goldberg et al. (1981b) recently studied the regulation of gene action during embryogenesis by determining the complexity, abundance classes and their distribution as well as differences among sequences of mRNA populations in soybean embryos of different ages. As an intermediate between the structural gene and the protein it codes for, the mRNA populations in embryos should provide an overview of the variety of proteins synthesized during embryogenesis. Two protocols have been employed to gain information on the distribution of mRNAs and their sequence complexity. In one method, cDNA prepared from poly(A)RNA of cotyledonary-stage and mid-maturation-stage embryos by reverse transcription of mRNA is reacted with an excess of homologous mRNAs. All the mRNA sequences should be represented in cDNA with a frequency corresponding to their abundance in mRNA. The second protocol has used a set of single-copy DNA tracer prepared from soybean leaf to hybridize with mRNA preparations from cotyledonary-stage embryos and from embryo axes and cotyledons of mid-maturation-stage embryos. Messengers that react with single-copy DNA account for most of the sequence complexity of RNA. Since they constitute only a small fraction of the total mRNA mass, the messenger sequence in this class is represented only by a few copies. The results of these experiments have led to the delineation of superabundant (>19,000 molecules per cell per sequence), moderately prevalent (550 to 800 molecules per cell per sequence) and rare (3 to 17 molecules per cell per sequence) classes of mRNAs in soybean embryo cells. In terms of the number of different mRNA sequences present in each class, a general conclusion is that change in form in the embryo is associated with an increase in mRNA complexity in the rare class. The fact that mid-maturation-stage embryos have a very high representation of superabundant class of messages, which is absent from the cotyledonary-stage embryos, attests to the diversity in the distribution of different classes of mRNA. There is an important qualification here: Although the superabundant mRNAs account for nearly 50 percent of the mRNA mass of mid-maturation-stage embryos, this class contains only 6 to 10 diverse mRNAs. Despite the fact that morphological complexity of the embryo increases with development, a surprising finding is that cotyledonary-stage embryos and mid-maturation-stage embryo axes and cotyledons contain approximately the same total number (14,000 to 18,000) of diverse mRNAs. This signifies that in terms of the absolute number of structural genes expressed there are no detectable changes in gene expression during morphological differentiation of the embryo and within morphologically distinct parts of the embryo. The basis for derivation of these data and conclusions lies outside the scope of this chapter. The interested reader is referred to the paper by Goldberg et al. (1981b) and, for a developmental biologist's point of view, to Davidson's (1976) book.

Another type of developmental analysis of soybean embryo mRNAs is accomplished by hybridization of kinetic fractions of mid-maturation-stage embryo abundant and rare cDNA (selected by repeated cycles of hybridization with homologous mRNA, followed by hydroxyapatite fractionation) to excess of cotyledonary-stage and mid-maturation-stage embryo mRNA (Goldberg et al., 1981b). Comparison of the mRNA sequences present in the embryos of two developmental stages shows that almost the entire sequence comprising the mass and diversity of mid-maturation-stage embryo is contained in the early-stage embryo. Although these comparisons are sensitive only to changes in at least hundreds of rare transcripts, we can conclude that mRNAs of embryos of different ages tend to be qualitatively similar. It appears likely that in soybean the entire set of structural genes that code for proteins of embryogenesis are transcribed at a very early stage regardless of whether they are translated. Finally, when the fate of the embryogenic mRNA sequences is determined in the mature seed embryo by hybridization with cDNA from cotyledonary-stage and

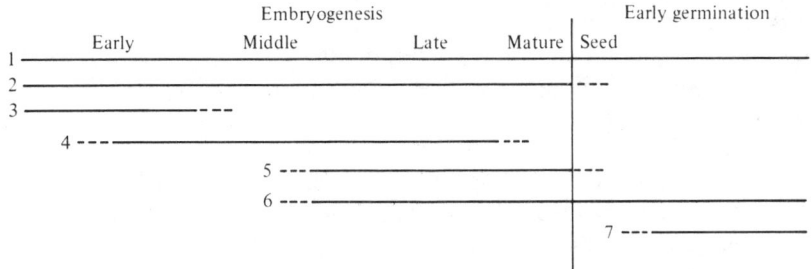

FIGURE 7.1. Diagrammatic representation of the time span of appearance and disappearance of mRNA subsets coding for different sets of proteins during embryogenesis and germination in cotton. (From Chlan and Dure, 1983.)

mid-maturation-stage embryos, a surprising result emerges. The cells of the quiescent seed embryo appear to conserve the majority of cotyledonary-stage and mid-maturation-stage embryo mRNA sequences, although not in the same abundance as in the early-stage embryos. This finding is in contrast to the changes that occur during embryogenesis in cellular concentrations of the superabundant mRNAs, which code for storage proteins (Chapter 3).

A great number of biochemical analyses have been undertaken on cotton embryos by Dure and associates from which a fairly complete picture of the complex pattern of gene regulation during embryogenesis can be deduced. These analyses have been mainly concerned with extant proteins, proteins synthesized *in vivo* and *in vitro* and the changing mRNA populations in cotyledons of embryos of different ages. Two major questions are raised in this work: Does the population of extant proteins in the young embryo include some whose synthesis is not required for its continued differentiation? Does a mature embryo synthesize proteins which are not needed for the survival of the young embryo? These questions were addressed by an analysis of two-dimensional gels of labeled proteins from young (50 mg fresh weight), maturation-stage (100 mg fresh weight) and mature embryos (seed embryos) (Dure, Greenway and Galau, 1981; Chlan and Dure, 1983). When the soluble fractions of proteins from cotyledons of embryos of different ages are analyzed, substantial modulations are found to occur in several sets of proteins whose synthesis is developmentally regulated; the corresponding changes in mRNA subsets of these proteins are illustrated in Figure 7.1. For example, one set of proteins (set 1) comprising an impressively large number of members is synthesized throughout embryogenesis whereas a small set (set 2) of six members is synthesized in the young and maturation-stage embryos, but not in the mature ones. Comparison of the products of cell-free translation of RNAs isolated from embryos of different ages with proteins synthesized *in vivo* has demonstrated that many translatable mRNAs of the first set of proteins are detectable only when their corresponding proteins are synthesized *in vivo*. This observation indicates that gene activity with regard to the synthesis of these proteins is regulated at the transcriptional level. However, proteins of the second set are synthesized *in vitro* by embryos of all ages. It has been suggested that these proteins, which are rapidly hydrolyzed during the first hours of seed germination, suffer degradation during incubation of the mature embryo in the isotope, a process analogous to imbibition during germination. A preliminary characterization of *in vitro* translation products of *Raphanus sativus* (radish) embryos of different ages has indicated similar changes in the coding capacity of the prevalent mRNAs (Laroche-Raynal, Aspart, Delseny and Penon, 1984).

In cotton embryos, another set of proteins, numbering about four (set 3), accumulates extensively in embryos of all stages of development. They are synthesized only in young and maturation-stage embryos, and their mRNAs are not active either in *in vivo* or *in vitro* protein synthesis of mature embryos. The behavior of these proteins is comparable to that of storage proteins (Chapter 3). Two other proteins that become abundant during late embryogenesis differ in one striking respect. One set (set 5) disappears during germination but can be induced prematurely by incubating excised embryos in ABA. This is just as expected of a group of proteins whose synthesis in the embryo is regulated by endogenous ABA. The other set of proteins (set 6) persists during normal germination as well as during precocious germination. The most extensively changing proteins of the embryo are those enriched for storage proteins (set 4), described in Chapter 3. Overall, this analysis shows that in cotton, as in soybean, proteins needed at a given stage of embryogenesis may be synthesized over a long period and that post-transcriptional events may be of profound importance in assembling and maintaining the appropriate level of gene products during embryogenesis.

Changes in the protein sets during embryogenesis in cotton described above are transcriptionally mediated, being dependent on the recruitment of mRNAs into polysomes. This has been verified by comparison of mRNA populations at different stages of embryogenesis by DNA-RNA hybridization technology (Galau and Dure, 1981). Particularly informative have been the comparisons of reassociation kinetics between cDNAs from cotyledons of all three embryo developmental stages and excess of homologous mRNAs. These have essentially revealed that the different protein sets detectable in embryos of different ages are represented by corresponding mRNA sets. The mRNAs belong to several abundance classes that show more than a twofold increase in their sequences in the mature embryo. Another type of analysis involves hybridization of each cDNA to mRNA isolated from the other embryo stages. Such heterologous reassociations have shown that embryos of all ages contain most of the mRNA sequences, although sequences of one stage may be present at a reduced level in other stages. They have also focused on some major differences in concentration of the more abundant mRNAs between embryo developmental stages, the most dramatic being a stark reduction in the presumptive storage protein mRNAs in the mature embryo.

Embryogenesis in other systems is not nearly as well understood at the molecular level as it is in soybean and cotton. Investigations show that both species apply the same solutions to the informational demands of developing embryos. The only important difference is one of timing. More information is needed to define the functional meaning of several interesting proteins that are likely to be involved in the developmental commitment of specific cell lineages of early embryos. An important but neglected aspect of the molecular biology of embryogenesis concerns the spatial organization of developmental information. Data relating to this are likely to provide new insights into the role of maternal mRNA and the timing of gene activity in the early division phase of embryos.

Somatic embryogenesis

Our model system for studying gene expression during somatic embryogenesis is the carrot cell suspension. In Chapter 5 we saw that a cell suspension of carrot consisting of single cells and cell clumps can be maintained in an undifferentiated state in a medium containing 2,4-D and that removal of auxin from the medium triggers embryogenic development. Although the simplicity with which carrot cell suspension can be manipulated makes it ideally suited to studies monitoring the biochemical and molecular details of gene activation during somatic embryogenesis from single-celled beginnings, this goal has scarcely been achieved. The current status of research in this area has been reviewed by Raghavan (1983).

A comparative analysis of nucleic acid and protein synthesis in somatic cells growing in the presence or absence of auxin in the medium offers a valuable means of exploring the independence of macromolecule synthesis involved in embryogenesis from that concerned with undifferentiated growth. Fujimura, Komamine and Matsumoto (1980) showed that embryogenic cells of carrot growing in the basal medium synthesize RNA and proteins at a higher rate than do nonembryogenic cells growing in an auxin-enriched medium. The augmentation of macromolecule synthesis is noted beginning about 2 days after transfer of cells to the respective media, although the amounts synthesized are too small to have any impact on the RNA and protein contents of cells. The increased RNA synthesis in embryogenic cells is also associated with accelerated synthesis of enzymes of the pyrimidine nucleotide pathway (Ashihara, Fujimura and Komamine, 1981). These results are in agreement with another work in which the focus was on RNA and protein synthesis in the embryogenic and nonembryogenic cells at earlier periods of culture (Sengupta and Raghavan, 1980a). This work showed that in short-term labeling experiments, RNA and protein synthesis in the embryogenic cells of carrot increases appreciably over that of nonembryogenic cells as early as 2 to 4 hours after transfer to new media. The increased RNA synthetic activity in the embryogenic cells continues up to 12 hours after transfer to a medium lacking auxin. As no new cells are formed in the suspension culture during the first 12 hours of growth in a fresh medium, the increased incorporation of label into RNA is probably attributable to the synthesis of cellular RNA. Further evidence derives from an analysis, by acrylamide gel electrophoresis and affinity chromatography on an oligo-(dT) cellulose column, of the types of RNA synthesized by carrot cells during the early hours of embryogenic induction (Sengupta and Raghavan, 1980b). The results showed that as early as 6 hours after transfer of cells to a medium lacking auxin, a decrease in the synthesis of rRNA occurs concomitant with an increased synthesis of minor species of RNA in 12S and 18S regions. More importantly, embryogenic cells synthesize more poly(A)RNA (i.e., mRNA) than do nonembryogenic cells, and the periods of observed increase correlate well with the early periods of embryogenic induction when RNA synthesis appears to pick up. This evidence favors the view that removal of auxin from the medium controls the biochemical events of embryogenesis by eliciting the synthesis of mRNA. That no significant replication of rRNA genes is involved in carrot cell embryogenesis is also supported by the results of experiments in which the amounts of ribosomal DNA (rDNA) in embryogenic and nonembryogenic cells were determined by hybridization of DNA with rRNA (Masuda, Kikuta and Okazawa, 1984).

Yet, it has not appeared unequivocally clear that gene activity for embryogenic induction in carrot cells is initiated upon their transfer to a medium lacking auxin. The issue is whether proteins synthesized in cells nurtured in the auxin-free medium are coded on newly formed mRNA or on mRNA transcribed when cells are in contact with auxin. Results of experiments using cordycepin, an inhibitor of polyadenylation, have pointed to the tentative conclusion that embryogenic induction in carrot cells growing in an auxin-free medium is controlled both at the transcriptional and translational levels (Raghavan, 1983). For example, when cells are transferred to an auxin-free medium supplemented with cordycepin, which inhibits polyadenylation by 90 percent, embryoid formation is observed to proceed to an arrested globular stage. Because inhibition of polyadenylation does not inhibit protein synthesis in the embryogenic cells, it appears that proteins synthesized by these cells during the early hours of their growth in an auxin-free medium are not translated on newly transcribed mRNA. Presumably, mRNA made by cells while they are still in contact with auxin is able to sustain protein synthesis during the early period of their growth in a medium devoid of auxin.

It seems probable that rather than being solely involved in promoting undifferentiated growth of callus, 2,4-D might have a subtle effect at the level of the gene in directing a repertoire of molecular changes leading to somatic embryogenesis. This is not surprising in view of the fact that in a carrot cell suspension cell clumps with the potential to form embryoids are visible as soon as the callus is launched from the explant, although formation of organized structures is delayed until auxin is removed from the medium. The view that 2,4-D modulates gene activity for embryogenic induction in carrot cells has been reinforced in a study of the translational profile of cells growing in the presence or absence of auxin in the medium (Sung and Okimoto, 1981). In this work, cells at different times after transfer to fresh medium were incubated in ^{35}S-methionine and labeled proteins analyzed by two-dimensional electrophoresis and autoradiography. Comparison of the spectrum of proteins synthesized in cells growing for 12 days in the presence or absence of auxin shows no pronounced differences in the nearly 200 or so polypeptides spotted on the gels, except that the embryogenic cells have two additional proteins, E1 and E2 (Fig. 7.2). The surprising finding is that regardless of the presence or absence of auxin in the medium, these two proteins are synthesized in cells during the early days of their growth in fresh medium but, in the presence of 2,4-D, they gradually diminish and finally disappear. The synthesis of embryogenic proteins appears to be an early event of embryogenic induction triggered by auxin but, by its very presence in the medium, auxin also bars the continued synthesis of these proteins necessary for the formation of organized structures or embryoids.

Tangible evidence of gene expression is reflected by other biochemical changes that occur during somatic embryogenesis. One concerns polyamines. This group of compounds, of which putrescine, spermidine and spermine are the best known, is characteristically basic in nature and is distinguished from amino acids by the absence of a carboxyl group. Their ability to bind with acidic groups, especially with those in nucleic acids, makes them prime candidates for gene regulation. Montague, Koppenbrink and Jaworski (1978) demonstrated a significant difference in polyamine metabolism between embryogenic and nonembryogenic cells of carrot. This difference is particularly apparent in the concentration of putrescine in the embryogenic cells, which increases nearly twofold over that in the control within 24 hours of transfer to a medium lacking 2,4-D. As recently shown by Fienberg, Choi, Lubich and Sung (1984), the increases in spermidine and spermine levels are even more striking, the former registering a threefold

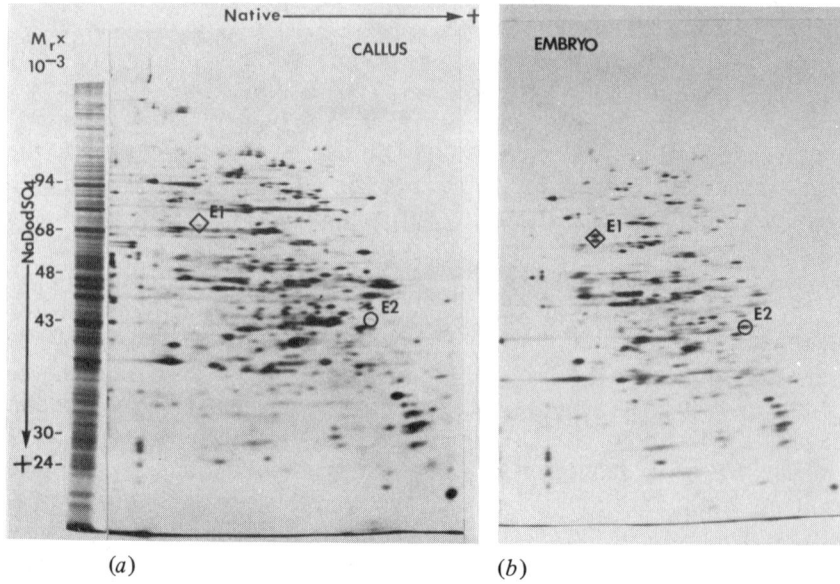

FIGURE 7.2. Autoradiographs of soluble proteins of carrot cells grown in media with or without 2,4-D for 12 days. Proteins were labeled with ^{35}S-methionine and separated by SDS-acrylamide gel electrophoresis. The one-dimensional gel of proteins of cells grown in a medium containing 2,4-D is shown on the left. (a) Protein profile of nonembryogenic cells grown in the presence of 2,4-D. (b) Protein profile of embryogenic cells grown in the absence of 2,4-D. E1 and E2 are the embryogenic proteins. (From Sung and Okimoto, 1981; photograph supplied by Dr. Z. R. Sung.)

increase and the latter a sixfold increase in the embryogenic cells over that in nondifferentiating controls. An embryoid-less mutant cell line of carrot exhibiting a growth rate similar to that of the wild-type cells does not show increased polyamine levels. In line with these observations, activities of arginine decarboxylase, an enzyme involved in the synthesis of putrescine from arginine and of S-adenosylmethionine decarboxylase, which catalyzes the decarboxylation of S-adenosylmethionine in the putrescine to spermidine pathway, are found to be higher in the embryogenic cells than in their nonembryogenic counterparts (Montague, Armstrong and Jaworski, 1979; Fienberg et al., 1984)). A role for arginine decarboxylase in somatic embryogenesis of carrot suggested by these results has been substantiated by Feirer, Mignon and Litvay (1984), who found a significant reduction in the embryogenic potential, with a concomitant decrease in the levels of putrescine and spermidine in carrot cells treated with α-difluoromethylarginine, an inhibitor of arginine decarboxylase, as shown in Figure 7.3.

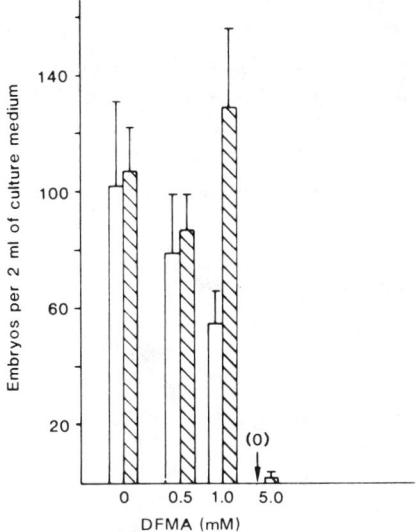

FIGURE 7.3. Effect of α-difluoromethylarginine (DFMA) on somatic embryogenesis in carrot and its reversal by putrescine. ☐ = inhibitor only, ▨ = 0.1 mM putrescine added. (From Feirer et al., 1984. *Science* 223: 1433–1435. © 1984 by the American Association for the Advancement of Science.)

Electrophoretic variations have been reported for several isozymes of carrot cells grown in embryogenic and nonembryogenic media (Lee and Dougall, 1973). The greatest differences are noted in the pattern of glutamate dehydrogenase, which is represented by only slowly migrating bands in the embryogenic cells, whereas nonembryogenic cells have in addition faster migrating bands. According to Kochba, Lavee and Spiegel-Roy (1977), an embryogenic line of Shamouti orange callus showed an upsurge of peroxidase activity concomitant with the appearance of embryoids. A particularly interesting finding is that at the time when the rise in enzyme activity is under way, a new band typical of the embryogenic cell line appears in the isozyme profile. Significant changes in protein composition as indicated by increasing or decreasing staining intensity of bands on acrylamide gels have also been noted during embryogenesis in petiole explants of celery (Zee, Wu and Yue, 1979).

Since genes for embryogenesis are apparently transcribed under special cultural conditions that inhibit undifferentiated growth of cells, is it conceivable that embryogenesis is associated with changes in the quality or quantity of chromosomal proteins? One way to answer this question is to isolate chromatin (a complex mixture of DNA, histones, acid proteins and possibly some RNA) and monitor the properties of histones and nonhistone nuclear proteins as a function of embryogenic development. Of the two classes of nuclear proteins, histones have long been favored as regulators of gene activity, particularly as a chemical mechanism to silence gene expression by inhibiting transcription. However, other studies have tended to minimize the role of histones in gene expression because they seem to lack specificity. Consequently, considerable attention has been directed toward nonhistone chromosomal proteins as regulators of gene expression.

Although no qualitative differences are noted in the histone composition between embryogenic and nonembryogenic cells of carrot, the percentage of histone H1 to the total histone is found to be lower in the embryogenic than in the nonembryogenic cells (Gregor, Reinert and Matsumoto, 1974; Fujimura et al., 1981). This is not surprising because some minor changes in the conventional histone pattern could be required for structural alterations in the chromatin of embryogenic cells to facilitate gene expression. However, of greater interest from our point of view are the changes observed in the activity spectrum of the nonhistone proteins. Although only minor differences are seen in the electrophoretic profile of nonhistone proteins between embryogenic and nonembryogenic cells of carrot, preparations of nonhistone proteins from 14-day-old embryogenic cells are more effective in restoring histone-inhibited DNA-directed RNA synthesis than are similar preparations from nonembryogenic cells (Matsumoto, Gregor and Reinert, 1975). Of course, conditions may be different *in vivo* to determine from these results the extent to which nonhistone nuclear proteins regulate gene expression during embryogenesis by making chromatin accessible to transcription.

The limited extent to which these studies have been carried forward only indicates the possibility that changes in gene expression occur at different stages of somatic embryogenesis. It is hoped that this information will serve as a background for a molecular definition of the gene-to-protein sequence of events during the embryogenic commitment of somatic cells and the expression of this commitment.

Pollen embryogenesis

During its normal ontogeny, the angiosperm pollen grain acquires the necessary developmental potential soon after it is released from the tetrad. The fundamental premise of gene regulation during pollen embryogenesis is that a pollen grain programmed for terminal differentiation into gametes is diverted to a pathway of continued divisions and increasingly complex morphology of an embryoid. From what we know about the molecular biology of cell differentiation, we would predict that these changes depend on the input of newly synthesized mRNA and proteins. From this point of view, there are a number of questions to be considered. What is the evidence for new mRNA synthesis connected with the subversion of the gametophytic program of a uninucleate pollen grain in the embryogenic pathway? What types of RNA are made by the generative cell and vegetative cell of a bicellular pollen grain during their normal development and as they become part of an embryoid? Do histones and nonhistone nuclear proteins play a role in the embryogenic development of a pollen grain? Much of the work addressed to these questions has been done in the author's laboratory on *Hyoscyamus niger* using autoradiography of incorporation of labeled precursors to localize RNA and protein synthesis and *in situ* hybridization to localize sites of mRNA accumulation. As background, reference is made to the discussion on the cytology of pollen embryogenesis in *H. niger* and some aspects of RNA and protein synthesis presented in Chapter 6. A review of this work has also been published (Raghavan, 1981a).

In *Hyoscyamus niger*, embryogenic development is initiated in a small number of pollen grains of anthers and anther segments cultured at the uninucleate, nonvacuolate stage of pollen development. Analysis of RNA synthesis by ^{3}H-uridine incorporation indicates that transcription in the pollen grain during its normal development occurs maximally at the uninucleate, nonvacuolate stage and in the vegetative cell of the bicellular pollen (Reynolds and Raghavan, 1982). The RNA synthesized by uninucleate pollen grains is likely to be rRNA, an interpretation supported by the absence of *in situ* ^{3}H-poly(U) binding sites in pollen grains. As most mRNAs of plant, animal and viral origins are covalently linked to a poly(A) segment, absence of ^{3}H-poly(U) binding in pollen grains under saturating conditions can be equated with the absence of accessible poly(A) tracts, hence of putative mRNA molecules. In cultured anther segments, uninucleate embryogenic pollen grains incorporate ^{3}H-uridine beginning at very early periods after culture. The apparent similarity in the timing of RNA synthesis in the pollen grains of intact anthers and cultured anther segments suggests that gene activation for rRNA synthesis is not causal for embryogenesis. Two important arguments have been made to support the view that embryogenic induction in pollen grains of cultured anther segments is associated with the synthesis of new mRNA. The first concerns the observation that embryogenic pollen grains begin to bind ^{3}H-poly(U) within the first hour of culture of anther segments and that this capacity increases with time (Fig. 7.4). By contrast, the large majority of pollen grains that disintegrate in culture and become nonembryogenic do not bind the label. The second argument derives from experiments using actinomycin D. In anther segments grown continuously in a medium containing the drug, pollen embryoid formation is inhibited during an experimental period when segments of the same anther grown in the basal medium produce globular or heart-shaped embryos. When anther segments

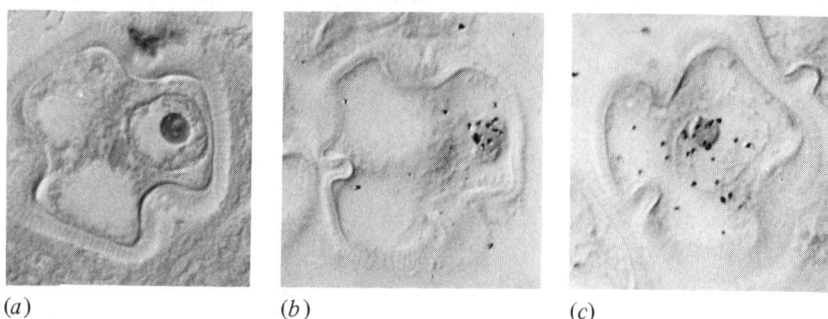

FIGURE 7.4. Autoradiographs of uninucleate embryogenic pollen grains of *Hyoscyamus niger* following *in situ* hybridization with ³H-polyuridylic acid of sections of anther segments at different times after culture. (*a*) Absence of binding in the pollen grain at the time of culture. (*b*) Binding in the nucleus 1 hour after culture. (*c*) Binding in the nucleus and cytoplasm 2 hours after culture. (From Raghavan, 1981b; reproduced from the *Journal of Cell Biology* 1981, 89: 593–606 by copyright permission of the Rockefeller University Press.)

are treated with the drug after various periods in the basal medium, there is a progressive increase in the number of embryoids formed. To the extent that actinomycin D acts here as it does in other systems, namely, by inhibiting mRNA synthesis, these findings are consistent with the view that initiation of embryogenic divisions in pollen grains of cultured anthers relies on newly formed mRNA (Raghavan 1979a, 1981b). Apparently, in a small number of pollen grains a genetic program for embryogenic development is superimposed upon the program for gametophytic development, which becomes redundant in the process.

Evidence for some change in gene expression in the generative cell and vegetative cell associated with embryogenesis comes from a comparison of ³H-uridine incorporation and ³H-poly(U) binding into these cells during the normal ontogeny of the pollen and during its embryogenic development. While the generative cell of the pollen grain in the intact anther does not incorporate any significant amount of ³H-uridine, a substantial increase in incorporation occurs after the first haploid mitosis in culture. Moreover, the generative cell of the normal bicellular pollen is enriched for ³H-poly(U)-binding sites only for a brief period during its life, but in the embryogenic pollen grain there is continued binding of ³H-poly(U) in the generative cell and its division products. These observations, and more importantly, the fact that embryoids are derived from bicellular pollen grains in which the generative cell or both generative and vegetative cells synthesize or accumulate rRNA and mRNA, make a good case that intensive transcription in the generative cell is an important event in the embryogenic development of pollen grains. It is interesting to note that although the vegetative cell of a normal pollen grain does not synthesize rRNA or or accumulate mRNA beyond the stage of starch accumulation, in pollen grains of cultured anthers continued transcription for gametophytic program occurs in this cell leading to abnormally enlarged, starch-filled nonembryogenic pollen grains (Raghavan, 1979b, 1981b).

We are now in a position to pull together these observations and attempt to construct a picture of regulation of gene activity during pollen embryogenesis in *Hyoscyamus niger*. It appears that as a result of the trauma of excision and culture of the anther, a small number of the enclosed pollen grains synthesize new mRNA which probably codes for proteins necessary for triggering the first haploid mitosis in culture. As shown by ^{3}H-amino acid incorporation experiments, this includes the synthesis of certain basic and general proteins (Raghavan, 1984). Subsequent embryogenic divisions of the pollen grains are mediated by the synthesis of additional rRNA and mRNA in the generative cell or in both generative and vegetative cells. On the other hand, synthesis of RNA species in the vegetative cell alone codes for proteins that perpetuate part of the gametophytic program. Whether proteins encoded by new mRNA in the embryogenic pollen grains include histones or nonhistone nuclear proteins involved in the regulation of gene activity requires further study.

Since pollen embryogenesis in *Hyoscyamus niger* occurs by the division of the generative cell, investigations on other systems, especially those in which the vegetative cell functions as the embryo mother cell will be required before the significance of these results could be placed in their proper perspective.

Regulation of gene activity in embryo mutants

Having reviewed the regulation of gene activity during embryogenesis from the biochemist's vantage point, we shall briefly look at it the geneticist's way. This is by disturbing the process by mutagenesis so that embryos arrested in developmental or nutritional functions will result. Arrest of development at a particular stage by a genetic mutation implies that the embryo is unable to complete a metabolic reaction or synthesize a specific nutrient substance required for completion of the next stage. This in turn assumes that (1) the defective function is controlled by a specific gene that identifies this function and (2) the phenotype caused by the mutation will provide clues to the regulation of gene action. Theoretically it is possible to induce mutations in embryos blocked at landmark stages of embryogenesis, characterize the mutation physiologically and biochemically, and assign genes for the defective function, but this has hardly been achieved.

The search for auxotrophs – mutants that are unable to synthesize specific metabolites for their growth – among angiosperm embryos has proved futile; thus far the only successful characterization of an auxotroph has been a proline-requiring mutant from maize. A class of mutants that promises the opportunity to study regulation of gene action during embryogenesis by identifying specific genes and following their expression is the developmental mutants. Although several embryo developmental mutants have been isolated from maize and *Arabidopsis*, these have not been analyzed in terms of the nature of the defects at the cellular or biochemical levels. For additional discussion, see Chapter 4.

Some mutation-selection experiments using carrot cell lines have yielded new insights into the mechanism underlying somatic embryogenesis as a function of gene regulation that are not easily observed in studies of somatic embryogenesis under normal conditions. The variant cell lines studied are those that make embryo-specific and callus-specific proteins, that exhibit resistance to amino acid analogues and drugs and that grow at restricted temperatures only. Aspects of the metabolic changes affecting variant cell lines and their impact on somatic embryogenesis have been considered in Chapter 5.

Concluding comments

The underlying assumption in the work reviewed in this chapter is that specific genes are activated at predetermined times to ensure progressive development of the embryo, be it from a fertilized egg, a somatic cell or a pollen grain. Although in many cases this assumption may well hold true, certain questions remain to be solved before we have a complete understanding of the control of gene expression during embryogenesis in angiosperms. Foremost among these concerns is the likely persistence of maternal mRNA in the fertilized egg and its role in early embryogenesis. Especially important is the need for a study, currently feasible only at the cytological level, of the pattern of gene expression during the first few division cycles of the zygote that carve out cells destined for different functions in the adult embryo. The discrete molecular differences seen in the embryogenic induction in an egg, a somatic cell and a pollen grain are also in need of elucidation.

Much of the currently available information on gene expression during zygotic, somatic and pollen embryogenesis is based on the work done on the so-called model systems. Since the validity of these results to other systems cannot be taken for granted, we need to look at many more examples and to integrate the information on the molecular and synthetic activities of embryos in each of the embryogenic classes into a general concept of gene regulation.

8
Applied aspects of embryogenesis

Production of the fruits and seeds on which the survival of our civilization depends is the direct outcome of processes related to pollination, fertilization and embryogenesis. This is a truism, the importance of which is not fully appreciated by the layman when the dependence of crop production on the knowledge of pre- and postfertilization events is considered. As the theme of this book is embryogenesis, we begin by introducing the principal product of embryogenic development in an angiosperm – the seed. The seed is a mature ovule that encloses the embryo and stored food materials of the endosperm within a protective coat or coats. Completion of embryogenesis inside the ovule marks the beginning of formation of the seed. Between the termination of embryogenic development and the formation of a mature seed, the embryo desiccates rapidly to a low moisture content, and the embryo alone or together with the endosperm comes to occupy the largest volume of the mature seed. The extent to which the embryo or the endosperm dominates the volume of a seed varies considerably. In legumes, for example, the endosperm is totally consumed during embryogenesis and the embryo with its massive cotyledons occupies the major part of the seed. On the other hand, in castor bean seed, the embryo is reduced in size and the endosperm with its rich endowment of storage lipids contributes to the largest volume of the seed. The caryopses of cereals and grasses also enclose a small embryo surrounded by a large mass of endosperm. When we consider that seeds are used in a variety of ways in our daily lives, for example, as major dietary components of food, for production of beverages like coffee, cocoa and beer, for extraction of edible oils from soybean, cotton, rape, sunflower, coconut and numerous other seeds, as condiments and nuts and as a source of fiber as in cotton, the practical impact of embryogenesis in our agriculture and economy hardly needs emphasis. For this reason, it is not difficult to account for the prodigious amount of published information about these and other uses of seeds. For a concise account of this information, the book by Duffus and Slaughter (1980) is recommended.

The purpose of this chapter is to review the possible applications of the knowledge derived from experimental embryogenesis – as used in a broad sense – to the solution of practical problems in agriculture and plant breed-

208 *Applied aspects of embryogenesis*

ing. As we have seen in Chapters 4, 5 and 6, advances in the field of tissue culture, particularly those relating to the culture of embryos, somatic embryogenesis and pollen embryogenesis have introduced new dimensions into our concepts of developmental potency of embryos, somatic cells and pollen grains of angiosperms. Such developments have paid off handsomely in devising methods to recover embryos from inviable crosses, in producing genetically identical plants, in breeding and selecting haploid plants for desirable characteristics and in preserving valuable germ plasm. Here we will attempt to establish the current state of these arts in a general sense and for specific crops. In the references to and descriptions of crosses given in this chapter, the female parent is named first.

Embryo rescue in inviable crosses

It has been known for a long time that most of our crop species have large reservoirs of unexplored or inadequately explored genetic variability and for this reason, breeding plants for newer and better characteristics has been a preoccupation with plant breeders. Today, on a worldwide basis, interspecific and intervarietal crosses followed by selection have accounted for improvement in the yield and quality of practically all our major crop species. However, the situation changes noticeably as breeders attempt to integrate into existing genotypes characteristics from other species and genera. One of the key problems in present-day breeding practices is that crosses between distantly related species and genera of plants have not yielded many agriculturally beneficial hybrids. From a genetic point of view, this is primarily because in attempts to introduce beneficial alien genes across interspecific and intergeneric barriers, many deleterious genes that are difficult to eliminate also find their way into the hybrid genome. Over the years many workers have expressed the conviction that this might cause a disturbance in the equilibrium between the growth of the maternal tissues, embryo and endosperm, leading to embryo lethality and collapse of seeds. In a general way we can say that union of the egg and sperm from members of the same species with identical chromosome numbers is a basic necessity for successful hybridization and that even a slight perturbation at the species level or chromosomal level can cause loss of embryos.

Why do hybrid embryos fail?

From a developmental point of view, embryogenesis and endosperm formation are profoundly affected in seeds of unsuccessful crosses. There are some close parallels in the pattern of embryogenesis seen in several unsuccessful crosses, and in virtually all investigations an early stage in embryogenesis when developmental anomalies begin to surface has been identified. A recent investigation of seed development in a cross between *Trifolium ambiguum* and *T. repens* is a representative example of the numerous studies that have been undertaken on inviable hybrids (Williams and White, 1976). In selfed *T. ambiguum,* the zygote enters the embryogenic pathway soon after fertilization and proceeds through the globular, heart-shaped, torpedo-shaped and mature stages of embryogenesis with precise regularity. Early growth of the hybrid embryo almost betrays its developmental fate, since it outgrows the control during the first 2 days after pollination in the number of cells formed. Growth of the embryo subsequently slows down, and in most of the hybrid seeds the embryo produces at best a tightly knit mass of about 1,000 cells and does not differentiate beyond the heart-shaped stage. In still others, the embryo begins to disintegrate soon after attaining this stage. In a *T. repens* × *T. medium* cross, where development does not proceed beyond the linear proembryo or globular embryo stage, compared with the selfed *T. repens* there is a delay in the development of both the embryo and endosperm suggesting that the effects of the deleterious genes are manifest at the time of fertilization or shortly thereafter (Kazimierska, 1978). In crosses involving different species of *Phaseolus* a gradation is found in the stages at which embryos suffer developmental blocks. In *P. vulgaris* × *P. lunatus* (Mok, Mok and Rabakoarihanta, 1978) and the reciprocal cross (Rabakoarihanta, Mok and Mok, 1979), embryos cease to grow at the pre-heart-shaped stage and four-celled stage, respectively, whereas in *P. vulgaris* × *P. acutifolius* (Mok et al., 1978) and *P. coccineus* × *P. vulgaris* (Shii, Rabakoarihanta, Mok and Mok, 1982), they reach adult form but fail to attain maturity. Mature embryos not susceptible to any block in development are formed in *P. vulgaris* × *P. coccineus* hybrids (Shii et al., 1982). In wide hybridization experiments using barley, rye and wheat as female parents and different closely or distantly related genera of Gramineae as pollinators, embryos have been shown to succumb to the action of the deleterious genes mostly at the globular stage (Zenkteler and Nitzsche, 1984).

The apparently normal early divisions of the hybrid embryo bring to the forefront some questions of general importance concerning the causes for its subsequent failure. Prominent among these are the cellular changes in the hybrid embryo during its life-span. In a comparative ultrastructural study of embryo development in selfed *Hibiscus costatus* and *H. aculeatus* and *H. costatus* × *H. furcellatus* hybrids, Ashley (1972) found that the embryogenic architecture of the selfed parent begins with a pronounced shrinkage of the zygote and a marked segregation of organelles towards the chalazal end. In contrast, zygote of the hybrid hardly changes physically or cytologically. Moreover, in contrast to the cells of the selfed embryo endowed with a generous supply of organelles, there is a paucity of organelles and extensive vacuolation in the cells of the hybrid embryo that they do not bear even the most tenuous resemblance to each other. Early degenerative changes observed in embryos arising from *Medicago sativa* × *M. scutellata* crosses are confined to the suspensor, which is much more reduced than suspensors of intraspecific embryos in the production of cells and in the distribution of organelles in each cell. The interspecific cross also affects the functional capacity of the suspensor, since the characteristic wall invaginations mediating metabolic interchange at the cell surface are absent (Sangduen et al., 1983b). These results imply that cellular and physiological malfunctions in the embryo probably account for its disturbed development, leading to hybrid failure.

Since early division-phase embryos are heterotrophic in nature and depend on the nutritional resources of the endosperm, considerable attention has been focused on the development of this latter tissue in seeds of inviable crosses. It is now well established that abortion of the embryo in interspecific and intergeneric crosses is preceded by the disintegration of the endosperm, thereby depriving the embryo of the source of nutrients. This is evident during seed development in *Trifolium ambiguum* × *T. repens* hybrid, referred to earlier (Williams and White, 1976). Although the division of the endosperm nucleus ceases at different stages of development of the hybrid seed, in no case does the endosperm proceed beyond the 128-nucleate stage; in many hybrids endosperm development ceases at even earlier stages. In contrast, selfed *T. ambiguum* seed has a complete endosperm including a well-developed chalazal haustorium. The extent to which disturbances in the endosperm in inviable crosses may cause starvation of the embryo remains unknown. However, the frequent correlation between embryo failure and the onset of endosperm disintegration observed in many plants suggests that embryo starvation is probably due to the inability of the physiologically disturbed endosperm to manufacture the exacting nutrients for embryo growth or absorb them from the surrounding maternal tissues and transmit them to the embryo. The idea that a reduced cytokinin biosynthesis in the endosperm is responsible for the abortion of the embryo is supported by a comparative study of the levels of this hormone in ovules of selfed *Phaseolus vulgaris* and of *P. vulgaris* × *P. acutifolius* cross. Whereas the concentration of cytokinin in the endosperm of selfed ovules is high and is closely correlated with periods cell division activity in the embryo, the level of the hormone in the hybrid ovule is lower by about an order of magnitude (Nesling and Morris, 1979).

Other points of view have also been evoked to account for embryo abortion in unsuccessful crosses. Brink and Cooper (1941) found that in *Nicotiana rustica* × *N. tabacum* and *N. rustica* × *N. glutinosa* hybrids, embryo failure is associated with retarded growth of the endosperm and a deficiency in the development of the conducting elements within the seed. The consequent abnormal distribution of nutrients in the seed leads to secondary effects such as hyperplastic growth of the nucellus. The term somatoplastic sterility is used to describe this type of growth of the maternal tissues of the ovule, and failure of embryos to complete full development following interspecific crosses in *Nicotiana* and other plants has been ascribed to this phenomenon. According to Esen and Soost (1973) the presence of a growing embryo might weaken the endosperm and induce abnormalities in its development. This viewpoint is based on the frequent observation that in $2x \times 4x$ crosses in *Citrus* many embryoless seeds have normal endosperm, while in those containing an embryo the endosperm invariably degenerates. In *Medicago sativa* ovules, cells of the nucellus and integumentary tapetum appear to serve as a nurse tissue from which lipid and starch are metabolized and transmitted to the endosperm and the embryo. In *M. sativa* × *M. scutellata* hybrid ovules, the storage reserves accumulate in the maternal tissues due to a block in their hydrolysis, thus starving the embryo of essential nutrients (Sangduen et al., 1983b).

The presence of mitotic abnormalities in hybrid embryos involving members of Gramineae has given rise to the possibility that hybrid failure is caused by the disparity in the mean cell doubling times of embryos of the parents. Forster and Dale (1983b) made intervarietal crosses between different barley and rye genotypes having different or similar cell doubling times and found that more and better developed hybrids are produced from crosses between parents with close cell doubling times than from parents whose mean cell doubling times are far apart. This is related to the breakdown of the endosperm that occurs earlier in the incompatible than in the compatible crosses.

These observations are obviously in line with the view that hybrid embryos are somehow deprived of a continuing supply of nutrients to complete their development in the ovule. Let us now consider the techniques employed to rescue hybrid embryos and surmount the postfertilization barriers to crossability in plants.

How are hybrid embryos rescued?

The problem of rescuing hybrid embryos has many facets and requires different kinds of approaches. The unique nutritional relationship between the embryo and endosperm tends to suggest that the cryptic potentiality of the hybrid embryo to complete its development may be realized if it is allowed to grow in an artificial medium supplied with nutrient substances that are normally identified with the endosperm. Early in the development of embryo culture as a research tool, Laibach (1925) demonstrated that progenies from nonviable seeds of *Linum perenne* × *L. austriacum* hybrid can be recovered by excision and culture of embryos before the latter begin to disintegrate. Since this pioneering work, embryo culture methods have been widely used to obtain transplantable seedlings from seeds of interspecific and intergeneric crosses which are traditionally condemned as being incapable of germination. In certain cases continued growth of the hybrid embryo is obtained by implanting it on a normal endosperm, which is then cultured, thereby essentially initiating a nurse culture. In a limited number of cases, efforts directed toward rescuing hybrid embryos by culturing ovules and ovaries have proved rewarding. Finally, a potentially useful, but largely untapped method of obtaining viable seedlings from distant crosses is by organogenesis or by somatic embryogenesis from an embryogenic callus.

Rescue by embryo culture. The numerous examples in which hybrid embryo rescue operations have been successfully mounted by embryo culture methods have been reviewed in earlier publications (Raghavan, 1976b, 1977b) and, for the sake of brevity, they will not be considered here. Rather, several recent attempts to retrieve interspecific and intergeneric hybrids from crop plants by the embryo culture method are discussed. Although nutrient media employed to culture embryos are as varied as the number of species investigated, it can be stated as a secure generalization that small hybrid embryos require combinations of vitamins, growth hormones, amino acids and natural endosperm extracts such as coconut milk, perhaps supplemented with an osmoticum to foster continued embryogenic development in culture. Larger embryos grow on a limited diet consisting of mineral salts and sucrose. After the embryos have grown into plantlets *in vitro,* they are removed from the original medium and nurtured in sterilized soil or vermiculite and grown to maturity in the greenhouse.

In legumes, prospects of rescuing embryos from interspecific crosses, raising fertile hybrid plants and eventual release of hybrid cultivars have improved dramatically by the use of embryo culture. Table 8.1 lists the more important legume hybrids reared by embryo culture. Although culture of embryos excised from ovules following incompatible pollination of flowers is often successful, in certain cases treatment of incompatibly pollinated flowers with hormones has been shown to prevent premature abscission of flowers, thereby allowing the hybrid embryo to grow further *in vivo* (Sastri and Moss, 1982; Gosal and Bajaj, 1983). Cereals are another group of plants in which embryo culture has been used to advantage to rescue hybrids. Barley and rice breeding programs have long been concerned with crosses between cultivated and wild species with a view to transfer genes for disease and pest resistance, capacity to withstand unfavorable environmental conditions and for high yields, and invariably the final step to obtain transplantable seedlings has been excision and culture of embryos. In more recent years, hopes of facilitating improvements in the breeding programs of barley and rye have been boosted by success obtained in rearing intergeneric hybrid seedlings by embryo culture methods from *Hordeum vulgare* × *Triticum aestivum* (Kruse, 1973; Fedak, 1977b; Mujeeb et al., 1978), *H. vulgare* × *T. dicoccum, H. vulgare* × *T. monococcum* (Kruse, 1973), *H. vulgare* × *T. turgidum* (Mujeeb et al., 1978), *H. vulgare* × *Secale cereale* (Pickering and Thomas, 1979; Bajaj, Verma and Dhanju, 1980), (*Hordeum vulgare* × *Triticum aestivum*) × *S. cereale* (Fedak and Armstrong, 1980) and *Elymus canadensis* × *S. cereale* crosses (Hang and Franckowiak, 1984).

In barley and certain other cereals, an important utilization of the embryo culture method is in the production of monoploids. This method, known as the bulbosum method, consists of crossing cultivated *Hordeum vulgare* with the wild *H. bulbosum*. After fertilization, which occurs without any hindrance, the zygote begins to divide in a normal way simultaneously as the *H. bulbosum* genome is selectively eliminated. Since the haploid embryos begin to abort in the caryopses, they are cultured *in vitro* and raised as seedlings (Kasha and Kao, 1970; Jensen, 1977). Monoploids have also been raised from crosses between several other species of *Hordeum* and *H. bulbosum* (Subrahmanyam, 1977, 1979, 1980) and from the intergeneric crosses *Triticum aestivum* × *H. bulbosum* (Barclay, 1975), *T. ventricosum* × *H. bulbosum* (Fedak, 1983), *Aegilops crassa* × *H. bulbosum* (Shigenobu and Sakamoto, 1977) and *H. vulgare* × *Secale cereale* (Fedak, 1977a) by culturing embryos from which the chromatin of the male parent is lost.

The successful hybridization of wheat and rye to produce the synthetic hybrid *Triticale* is partly attributable to the utilization of methods to culture hybrid embryos in quantity. The crossability between wheat and rye is known to be controlled by two dominant genes present in wheat. These genes apparently interfere with the growth of pollen tubes and thus thwart fertilization. Another mitigating factor in wheat–rye cross is the ploidy level of wheat, with hexaploid wheat holding a slight edge over tetraploid wheat in crossing with a given rye parent. Although embryos in hexaploid wheat–rye crosses develop to a respectable size *in vivo*, the percentage of yield of viable embryos is considerably enhanced by culturing them *in vitro*. On the other hand, since embryos from tetraploid wheat–rye crosses abort within 2 weeks after pollination, embryo culture is a prerequisite to raise plants (Oettler, 1983). Among attempts made to enhance the production of viable plants from tetraploid wheat–rye crosses, culture of hybrid embryos in a high nitrogen-containing medium fortified with growth hormones and an extract of macerated wheat endosperm has been found to be beneficial (Bajaj, Gill and Sandha, 1978). The injection of ε-amino-*n*-caproic acid or lysine into wheat caryopses pollinated with rye pollen, which improves the normal *in vivo* development of embryos, has been suggested as a means to improve their growth *in vitro* (Taira and Larter,

Table 8.1. *Embryo culture of interspecific legume hybrids*

Hybrids	Medium composition[a]	Seedling characteristics	References
Arachis hypogaea × *A. villosa*	Murashige-Skoog medium with IAA (4 mg/liter) and kinetin (2.0 mg/liter)	Normal plants with chromosome number intermediate between parents	Bajaj, Kumar, Singh and Labana (1982)
A. hypogaea × *A. monticola*, *A. glabrata*	Probably Murashige-Skoog or White's medium	Normal-looking seedlings formed in culture tubes	Sastri and Moss (1982)
Lathyrus clymenum × *L. articulatus*	Inorganic medium of Bonner with thiamin and nicotinic acid; initially grown in a medium containing 8–12% sucrose and subsequently transferred to a medium containing 4% sucrose; 10% coconut milk necessary for growth of very small embryos	Seedlings transferred to soil	Pecket and Selim (1965)
Lotus japonicus × *L. filicaulis*; *L. japonicus* × *L. alpinus*; *L. japonicus* × *L. frondosus*; *L. japonicus* × *L. schoelleri*	Randolph and Cox's medium substituting chelated iron for ferrous sulfate	Plants grown in the greenhouse; hybrids with characters intermediate between parents	Grant, Bullen and de Nettancourt (1962)
Medicago sativa and unnamed interspecific hybrids	Crone's medium	Normal seedlings	Fridriksson and Bolton (1963)
Melilotus officinalis × *M. alba*	Crone's or Randolph and Cox's medium	Plants reared to maturity; characters intermediate between parents	Webster (1955); Schlosser-Szigat (1962)
Phaseolus vulgaris × *P. acutifolius*	White's, Randolph and Cox's or modified Crone's medium	Mature plants producing flowers, fruits and viable seeds	Honma (1955)

Cross	Medium	Result	Reference
P. vulgaris × *P. ritensis*	White's medium with 0.1% casein hydrolyzate; transferred to a medium lacking casein hydrolyzate after 4 weeks	Transplantable seedlings	Braak and Kooistra (1975)
P. vulgaris × *P. lunatus*; *P. vulgaris* × *P. acutifolius*; *P. acutifolius* × *P. vulgaris*	Murashige-Skoog medium; survival of embryos enhanced by addition of glutamine	Normal seedlings showing limited growth	Mok et al. (1978)
P. vulgaris × *P. coccineus*	Medium of Emsweller, Asen and Uhring with 4% sucrose	Hybrid plants produce pods with normal seeds after backcross, self-pollination or interhybrid pollination	Alvarez, Ascher and Davis (1981)
(*P. vulgaris* × *P. coccineus*) × *P. acutifolius*	Medium as above	Plants with low fertility	
P. coccineus × *P. acutifolius*	Medium as above	Plants with flowers resembling those of the female parent	
P. coccineus × *P. vulgaris*	Murashige-Skoog medium	Plants from small underdeveloped embryos have low survival rate; plants from large shrunken embryos have high survival rate and reach flowering	Shii et al. (1982)
Trifolium ambiguum × *T. hybridum*; *T. repens* × *T. nigrescens*; *T. nigrescens* × *T. repens*; *T. repens* × *T. uniflorum*; *T. uniflorum* × *T. repens*; *T. hybridum* × *T. ambiguum*	Randolph and Cox's or Tukey's medium	Healthy plants grown in pots in soil; some hybrid seedlings flower	Keim (1953); Evans (1962)

Table 8.1 (cont.)

Hybrids	Medium composition[a]	Seedling characteristics	References
T. sarosiense × *T. pratense*	Medium with a high concentration of sucrose, a moderate level of auxin and low cytokinin; viable embryos subsequently transferred to a medium with reduced level of sucrose	Normal hybrid plants with distinct traits of both parents	Phillips, Collins and Taylor (1982)
Vigna radiata × *V. angularis*	Vacin and Went's medium	Plants intermediate between parents in regard to germination habit, petiole length and shape of primary leaves	Ahn and Hartmann (1978a)
V. umbellata × *V. angularis*	Medium as above	Healthy plants with characteristics intermediate between parents in regard to color and shape of leaves, pods and seeds	Ahn and Hartmann (1978b)
V. mungo × *V. radiata*	Murashige-Skoog medium, with IAA (1 mg/liter), kinetin (0.2 mg/liter), coconut milk (70 ml/liter) or casein hydrolyzate (500 mg/liter)	Plants established in soil; leaf color and other morphological characters intermediate between parents	Gosal and Bajaj (1983)

[a] All media contained mineral salts as well as a carbon energy source such as sucrose or glucose.

1977). Excision and culture of embryos appears to be the only feasible method to obtain hybrid seedlings from other wide crosses involving different species of *Triticum,* such as *Triticum sativum* × *Aegilops speltoides* (Chueca, Cauderon and Tempé, 1977) and *T. crassum* × *H. vulgare* (Fedak and Nakamura, 1981) and in *Aegilops squarrosa* × *T. boeoticum* (Gill, Waines and Sharma, 1981) and *Agropyron spicatum* × *A. intermedium* crosses (Napier and Walton, 1984).

Finally we may consider the possible use of embryo culture method to raise seedlings from hybrids of other crop plants. Investigations to retrieve embryos from several inviable interspecific and intervarietal crosses in cotton, tomato, okra, jute, cabbage and fruit trees have proceeded on a modest level along parallel lines with studies on legumes and cereals (Raghavan, 1976b); among recent examples are hybrids from *Solanum melongena* × *S. khasianum* (Sharma, Chowdhury, Ahuja and Dhankhar, 1980), *Allium cepa* × *A. fistulosum* and the reciprocal combination (Doležel, Novák and Lužný, 1980) and cultivated × wild species of *Gossypium* (Gill and Bajaj, 1984). In the cultivation and breeding of fruit trees, embryo culture finds a variety of applications. A classic example is the "Makapuno" coconut in which the endosperm generally degenerates early during the development of the fruit, thus depriving the embryo of the essential source of nutrients for germination. However, these embryos can be rescued by culture in a nutrient medium to a size large enough for field planting (de Guzman and del Rosario, 1964).

In certain fruit trees in which dormancy of seeds and slow growth of seedlings necessitate long breeding seasons, embryo culture technique is of special value in reducing the breeding cycle of new varieties. Another use of embryo culture is in inducing germination of embryos of seeds of the early-ripening varieties of fruit trees. These seeds, which have low viability and which do not germinate even after the appropriate afterripening treatments, can be propagated by embryo culture. The singular advantage of embryo culture method of germination is that it is possible to preserve varieties for development of still earlier ripening characteristics; these studies have been reviewed by Ramming (1983). Recently, Skene and Barlass (1983) showed that *Persea americana* (avocado) embryos excised from seeds that abscise before maturity can be grown in a medium supplemented with BA to yield multiple shoots suitable for grafting onto seedling root stocks.

Production of new genetic combinations of economically important crop plants is continually being attempted by regular breeding programs followed by embryo culture to rescue aborted embryos in various agricultural research establishments throughout the world. Although some of the results are published in refereed journals, a good number of successful attempts also appear in annual reports, experimental station bulletins, crop newsletters and similar types of publications. It should therefore be recognized that the account given here is based on a small cross section of an active field of research with practical overtones. In general, the task of isolating and culturing embryos from hundreds of ovules might appear a formidable undertaking but, for a breeder who endures the repeated sight of failed crosses, the advantages of embryo culture are compelling.

Embryo implantation. The effectiveness of a medium to support the growth of hybrid embryos generated in a cross is based on its ability to compensate for the nutrient substances present in the endosperm of the aborted seed. Although in most cases this is empirically met by growth hormones, amino acids or coconut milk, alone or in combination, due to the inherent differences in the chemical composition of endosperms of different plants, the medium formulated to foster growth of embryos of one hybrid combination may not be suitable for another. Moreover, in critical cases the concoction of a medium which corresponds closely with the composition of the endosperm is limited by a lack of knowledge of the chemical composition of the endosperm. From these perspectives, transplantation or implantation of the hybrid embryo onto a normal endosperm might overcome the constraints of the artificial medium in inducing growth of embryos.

Pissarev and Vinogradova (1944) first showed that the crossability between a low crossable wheat and rye could be improved if the wheat plants were raised from embryos transplanted on rye endosperm during the early phase of germination. In recent years the role of the endosperm in inducing the growth of hybrid embryo has been highlighted by culturing the latter directly on the endosperm of the female parent. In a method described by Kruse (1974) to rescue hybrid embryos from *Hordeum × Triticale, Hordeum × Agropyron* and *Hordeum × Secale* crosses, the embryo is removed from a dehulled hybrid caryopsis. Next the endosperm is extruded out of the pericarp of *Hordeum* caryopsis and transferred either singly or in groups of five to ten to the surface of a solidified mineral salt medium. The hybrid embryo is then placed in the correct position in the endosperm at the site of the original embryo. Best results in terms of the yield of hybrid plants and ability to induce growth in relatively small embryos are obtained when the donor endosperm is cultured in a medium containing 25 percent coconut milk. An analogous technique has been used to rescue hybrid embryos from interspecific crosses in legume genera *Trifolium, Lotus, Ornithopus* (Williams and de Lautour, 1980; de Lautour, Jones and Ross, 1978) and *Glycine* (Broué, Douglass, Grace and Marshall, 1982). In the work on *Trifolium* and related genera it was found that hybrid embryos excised at early to late heart-shaped stages do not grow when cultured directly on the surface of a nutrient medium. However, when they are enveloped in a nurse endosperm dissected from normally developing ovules and cultured, growth, differentiation and recovery rates of embryos are high. Generally, a medium containing casein hydrolyzate and vitamins is superior for growth of early heart-shaped embryos whereas advanced-stage embryos require less complex additives to the medium.

The application of embryo implantation technique on a wider level to rescue hybrid embryos is hampered by the delicate nature of the operations, making field trials particularly exasperating. Despite this limitation, the technique offers a promising means of obtaining progenies from crosses involving unrelated parents where direct culture of embryos may not be successful.

Ovule and ovary culture. When embryo abortion occurs relatively early in embryogenesis of the hybrid plant, removal of the embryo without injury may be a deterrent to their culture. Granted that small hybrid embryos are isolated without injury, lack of a suitable medium may prevent their further growth. From this point of view, ovule culture has proved to be useful as an alternate method to rear to maturity embryos from certain inviable crosses. One crop plant in which ovule culture has been successfully employed to rescue hybrid embryos is cotton. Among cultivated cotton, *Gossypium arboreum* and *G. herbaceum* are diploids, while *G. barbadense* and *G. hirsutum* are allotetraploids. While hybrid embryos from tetraploid × diploid crosses have been grown to maturity by culturing them in a defined medium, only sporadic success has been reported in rescuing embryos from reciprocal crosses. However, Stewart and Hsu (1978) obtained triploid hybrid plants by ovule culture from all possible crosses between diploid *G. arboreum* and *G. herbaceum* and tetraploid *G. barbadense* and *G. hirsutum* except from the cross *G. herbaceum* and *G. barbadense*. The procedure followed generally consists of culturing ovules 2 to 4 days after anthesis in a high salt medium containing 4 percent sucrose. Germinated embryos are subsequently removed and allowed to root and form seedlings in a low salt medium before transplantation to soil. Although cultured *G. hirsutum* ovules yield well-developed hybrid embryos in the basal medium without the addition of IAA, those of *G. barbadense*, *G. arboreum* and *G. herbaceum* require the presence of the hormone for favorable embryo yields.

In another study, Reed and Collins (1978) found that whereas seeds obtained from crosses between certain wild species of *Nicotiana* (*N. stocktonii*, *N. nesophila* and *N. repanda*) and the cultivated *N. tabacum* do not germinate, it is possible to obtain seedlings from all the three desired hybrids by culture of ovules and recovering germinated embryos. Hybrids between cultivated soybean and a wild perennial relative (*Glycine tomentella*) have been generated for the first time by ovule culture (Newell and

Hymowitz, 1982). To enhance the hybridization of seedless grapes in which embryo development stops soon after fertilization, *in ovulo* embryo culture has a great potential (Emershad and Ramming, 1984).

Another technique in this category is to culture ovaries, instead of ovules, following wide crosses. In an extensive investigation of crosses between *Brassica campestris* and *B. oleracea,* Inomata (1978) obtained seeds with well-developed embryos protruding out when ovaries are explanted four days after pollination onto White's medium containing casein hydrolyzate. Subsequent subculture of embryos in the same medium leads to the development of plantlets with roots and leaves. In the reciprocal cross, ovule culture is found superior to ovary culture for the production of hybrid plants (Takeshita, Kato and Tokumasu, 1980). These results have led to the expectation that both ovule culture and ovary culture might be of use in raising plants from wide crosses in crops where embryo culture and embryo implantation techniques are less satisfactory.

Plant regeneration from callus. Efficiency of embryo rescue operations thus far described is handicapped by the need to culture a large number of embryos or ovules to obtain a sizable population of seedlings for further analysis. Toward our goal of producing a large number of plants of the same genetic constitution from crosses between unrelated species, regeneration by organogenesis or embryogenesis from a callus of hybrid embryo origin has a decided advantage over other methods of embryo rescue.

In general, utilization of an embryo-derived callus to obtain hybrid plants has been beset by the slow growth of the callus and failure of seedlings to survive. Despite this, a few attempts have been made to utilize the regeneration capabilities of hybrid embryo calluses to produce seedling plants. In one case, isolated parts of seedlings generated from cultured ovules of *Nicotiana stocktonii* × *N. tabacum* and *N. nesophila* × *N. tabacum* hybrids are induced to form a callus and plants regenerated from the callus by organogenesis are grown to maturity (Reed and Collins, 1978). In other studies, cultured embryos from *Hordeum distichum* × *Secale cereale* (Cooper et al., 1978) and *H. vulgare* × *S. cereale* (Forster and Dale, 1983b) hybrids are found to give rise to friable calluses which regenerate roots and green shoots, when cultured in the same medium or in a different medium. Clones of the hybrid plant have been established from green shoots formed from *H. distichum* × *S. cereale* hybrid embryo callus. According to Thomas and Pratt (1981), a callus is formed when underdeveloped hybrid seeds obtained from a cross between *Lycopersicon esculentum* and *L. peruvianum* are cultured in a hormone-enriched medium; by manipulation of the medium composition, plantlet regeneration from the callus is subsequently obtained. Despite the fact that the callus originates from cultured seeds, the regenerated plants display morphological traits, ploidy values, isozyme banding patterns and fertility relationships expected of embryo-derived hybrids. Finally, as shown in hybridization of *Trifolium sarosiense* with *T. pratense*, transformation of the cultured hybrid embryo into a callus occasionally facilitates shoot formation through somatic embryogenesis (Phillips et al., 1982). Overall, these results are encouraging and suggest that current techniques to regenerate normal plants from cultured cells and tissues may permit production of a large number of hybrid plants from rescued embryos. Unfortunately, the number of species to which the technique has been extended is too limited to provide much backlog of information.

Clonal multiplication of plants

Clonal multiplication is an age-old horticultural practice in which conventional propagation by seeds is bypassed to produce, rather rapidly and in large numbers, plants of the same genotype. Leaf, stem and root cuttings and axillary buds are traditional starting materials that facilitate multiplication of the same genotype under greenhouse conditions. With the demonstration that cultured organs and tissues of certain plants produce a randomly proliferating callus that can be induced to regenerate leafy buds and roots by subtle changes of the hormonal constituents of the medium, plant propagation *in vitro* (micropropagation) became a serious laboratory exercise. This method is being used increasingly to enhance the rate of multiplication of plants that have thus far been asexually propagated in the greenhouse. But it is with reference to the ability of somatic cells of carrot to go through definitive stages of embryogenesis to the formation of embryoids and plantlets that the dramatic impact of mass cloning of plants in the laboratory and the role of embryogenesis in the process came to be recognized. This is because the production of adventitious buds is often slow, requires a precise balanced mixture of hormones, is unpredictable and the yield of buds is abysmally low. On the other hand, somatic embryogenesis is a single-step process involving transfer of callus cells from a hormone-enriched medium to one lacking the hormone; in terms of numbers, production of somatic embryos in tissue culture is staggering. According to Steward (1963), an aliquot of cell suspension from a cultured carrot embryo spread on a nutrient medium produces more than 100,000 embryoids. Recently, Drew (1980) constructed an apparatus for large-scale cultivation of embryogenic cell suspension of carrot and has reported that as many as 1.35×10^6 embryoids are produced in 1 liter of the suspension.

These remarkable results with carrot are not unique to that system, as similar faculties have been shown to reside in a number of plants including a host of economically important species (see Chapter 5). An up-to-date list of these plants is provided in the review by Ammirato (1983a).

Applied aspects of embryogenesis

Is micropropagation by somatic embryogenesis currently being used for large-scale plantings in commercial operations? According to my information this is not being presently done and there is no immediate promise of applying this method to any crop plant in the field. To quote Krikorian (1982), "plantlet production via somatic embryos derived from cells is still, and will long continue to be, without question the most difficult of the various methods of aseptic culture for clonal propagation to be implemented. This is especially so since there is as yet no theoretical foundation on which new investigations involving untested species can be based." In retrospect, a number of reasons can be assigned for the lack of enthusiasm on the part of the farmer to embrace the new technology. Many published reports on somatic embryogenesis have not emphasized the rapid propagation or mass cloning advantages of the method. Where seeds are readily available and where methods have been perfected for raising crops year after year from seeds, it is doubtful whether any particular advantage can be found in the propagation of progeny by somatic embryogenesis. Since an enormous number of embryoids are involved, there is the question of storing them for short periods of time until conditions are ripe for planting, but no methods have been devised for embryoids to mimic dormancy of the seed embryo. Finally, there is a need for information on the ability of cells to undergo embryogenic development in long-term cultures, especially with reference to species considered for commercial use. In carrot cell suspension cultures in which somatic embryogenesis is expressed relatively predictably, the regenerative ability continually diminishes over time and subculturing. The decrease in embryogenic potential correlates with a rise in the percentage of cytologically unstable cells in culture (Smith and Street, 1974). This problem of genetic instability of embryogenic cultures, which might lead to the production of phenotypically altered plants, has not been addressed by investigators.

However, the use of somatic embryos for clonal multiplication of plants provides a number of advantages. Foremost is the fact that these vegetative propagules have developed through a simulated process of embryogenesis and, unlike a bud regenerated on a callus, the embryoid comes packaged with a well-developed shoot and root meristems. If cultures are grown under conditions that permit separation of cells from the callus and subsequent organization of cells into embryoids, the necessity of physical separation of individual clones is obviated. Lastly, the advantage of having a large number of embryoids of the same genotype for planting purposes cannot be overlooked. Perhaps highly heterozygous species that exhibit extreme levels of incompatibility or that seed poorly or whose seeds germinate with difficulty are prime candidates for initiating a pilot program of clonal multiplication by somatic embryogenesis.

Haploids for breeding and selection

Much has been written about the practical significance, for genetics and plant breeding, of the demonstrated potential of pollen grains to form embryoids and plantlets with the haploid set of chromosomes. Since the technology in this area has been covered in Chapter 6, we shall focus here on the possible uses of haploids. It should be noted that because of spontaneous chromosome doubling and endopolyploidy, not all species whose pollen grains divide in the embryogenic pathway yield haploid plants. Moreover, in many cases where haploid plants are readily formed in anther or pollen cultures, their prospects for appreciable longevity with the same chromosome number are bleak.

The use of haploids generated by anther and pollen culture techniques has received complete legitimacy and recognized value in the production of homozygous diploid plants by either spontaneous chromosome doubling or by colchicine treatment. In this respect, some comments are in order regarding both the number of haploids produced by anther culture as opposed to other means, and the time frame involved in the production of isogenic lines by the doubled-haploid method versus by conventional means. Until the discovery of anther culture method to generate haploids in plants, occurrence of haploid plants has been reported only sporadically in angiosperms, but *in vitro* culture of anthers has provided a method *par excellence* for generating haploids in large numbers in a matter of weeks. Although it is possible to develop pure lines by classic genetic methods of inbreeding and backcrossing, the procedures are time consuming and may take up to several years to complete, while doubled haploids can be obtained in a single generation by anther culture. With these advantages it is no wonder that anther culture technique is being used extensively throughout the world for crop improvement. In the People's Republic of China, several new varieties of rice, incorporating genes for high yield, resistance to bacterial blight and adaptability to extreme environmental conditions, wheat, incorporating genes for large spike and many grains, and tobacco for disease resistance and leaf qualities, obtained by anther culture are already under cultivation; investigations in developing new varieties of rubber, maize and sugarcane are at an advanced stage (Zeng, 1983). In the United States, investigations on the anther culture of rice have led to the recovery of plants with both deleterious and useful genes, such as for dwarfness, reduced seed and leaf size and seed weight (Schaeffer, 1982) and for increased seed protein content (Schaeffer, Sharpe, and Cregan, 1984). *Brassica* is another plant in which anther culture methodology has been advantageously employed to select superior strains of plants (Nitzsche and Wenzel, 1977).

Selection of plants with desirable traits by mutation is a major objective in plant breeding. Haploid systems are especially useful for mutational studies, as the presence of a single set of chromosomes causes the immediate expression of the mutated gene without any allelism effect. However, the theoretical promise for the production of useful mutants by the use of physical or chemical mutagens has not materialized in the limited work done with haploid plants. In one of the early experiments it was found that when haploid tobacco plants are subjected to γ-irradiation, a number of phenotypic abnormalities occur in the transplanted seedlings (Nitsch, Nitsch and Péreau-Leroy, 1969). Several abnormal phenotypes of tobacco have also been obtained when X-irradiated anthers are cultured (Devreux and Saccardo, 1971). Another work showed the occurrence of white-flowered mutants when anthers of tobacco are grown in a medium supplemented with N-3-nitrophenyl-N-phenylurea (Nitsch, 1972). In retrospect, certain factors appear to mitigate against selecting whole plant mutants using pollen grains or pollen-derived haploids. One is the multicellular nature of the target, which makes chimera formation very likely. Second, in using pollen grains as starting material, high toxicity of chemical mutagens may kill them, making the probability of occurrence of useful mutation very low. Third, the deleterious nature of most mutations limits their usefulness.

An offshoot of the production of haploid plants by anther and pollen culture methods is the use of haploid cells for mutation screening experiments. Carlson (1970) first successfully utilized a technique modeled after a system designed for use with mammalian cell cultures for isolating auxotrophic mutants from haploid tobacco cells mutagenized by treatment with EMS. The selection method depends upon the incorporation of 5-bromo-2'-deoxyuridine (BUDR) into DNA of wild-type cells and their subsequent lethality as well as the lack of incorporation of BUDR into DNA of mutants and their survival. Although auxotrophs for amino acids, vitamins and nucleic acids were isolated by this method, the mutants appeared to be leaky and eventually continued to grow even in the unsupplemented medium. The selection of streptomycin-resistant cell lines was attempted by Binding, Binding and Straub (1970) from haploid cells of *Petunia hybrida*. These experiments were extended by Maliga, Sz-Breznovits and Márton (1973), who regenerated streptomycin-resistant diploid plants from a haploid tobacco. The plants gave rise to seeds and seedlings that proved resistant to the antibiotic. In other studies, haploid tobacco cell lines resistant to cycloheximide and BUDR have also been isolated (Maliga, Márton and Sz-Breznovits, 1973; Maliga, Lázár, Sváb and Nagy, 1976). Since these pioneering experiments, haploid cell lines from diverse plants resistant to amino acids and their analogues, salts, herbicides, drugs and temperature conditions have been selected (Dix and Street, 1975, 1976; Dix, Joó and

Maliga, 1977; Bourgin, 1978; Tyagi et al., 1981a; He, Xu, Xu and Loo, 1984).

Methods used for protoplast isolation provide yet another level of use for haploids in mutational studies. Carlson (1973) treated haploid cells and protoplasts of tobacco with a mutagen and selected calluses from regenerated cells resistant to methionine sulfoximine, presumed to be an analogue of both methionine and the wildfire toxin produced by *Pseudomonas tabaci*. The regenerates showed a considerably low level of infection when inoculated with the wildfire pathogen and increased level of free methionine. From this it was concluded that methionine sulfoximine-resistant plants are resistant to the wildfire pathogen because of the structural similarity of the compound to the toxin. Although there is some question as to whether methionine sulfoximine and the wildfire toxin are analogues and as to whether the former acts as an analogue of methionine, the results indicate the possibility of producing mutants from haploid cells with increased capacity for biosynthesis of specific metabolites.

The ease of induction of haploid plants from *Hyoscyamus muticus* by pollen embryogenesis has stimulated attempts to use this system to isolate biochemical mutants from haploid protoplasts. Strauss, Bucher and King (1981) treated mesophyll protoplasts with nitrosoguanidine sufficient to cause 90 percent lethality and incubated them on a complete medium supporting wall regeneration and organized growth. Out of more than 3,000 individual clones of calluses isolated, one is found to grow well on the complete medium but only slightly on the minimal medium. The clone is, however, rescued by supplementing the basal medium with casein hydrolyzate or ammonium as a source of N, but not by nitrate. This, as well as the fact that under inductive conditions the clone exhibits negligible nitrate reductase activity seem to suggest that the phenotypic alteration of the callus is caused by a derangement in nitrate assimilation. By modification of the mutation-selection procedure, variant cell lines from mesophyll protoplasts of *H. muticus* showing an absolute requirement for specific amino acids like histidine and tryptophan, or a vitamin, nicotinamide or sensitivity to high temperature have been isolated (Gebhardt, Schnebli and King, 1981).

Preservation of the germ plasm

Because the germ plasm of angiosperms is traditionally maintained in the form of seeds, as the gene bank of the seed, the embryo occupies a preeminent role in the preservation of germ plasm. Just as money in a bank is slowly exhausted by periodic withdrawls, so too does the gene bank become depleted of its reserves with the passage of time. A tangible expression of this is the decrease in seedling viability resulting in the germination of fewer and fewer individuals, the longer a stock of seeds is stored.

While farmers have been continually striving to prevent the deterioration of their seed stocks and thus improve the viability of embryos for short term benefits, in recent years there has been an increasing awareness of the need to preserve valuable germ plasm of the original wild-type varieties for long-term benefits, especially to increase the genetic diversity in future breeding programs. One important reason for this is the realization that the selective crossing and dissemination of improved varieties portend a virtual extinction of the original genotype. There is also an urgent need to preserve many wild species of plants that face extinction chiefly through the destruction of their habitats.

For most of our crop plants, the long-term storage of germ plasm as seeds presents no special problems; the method also appears to be ideal for international exchange of germ plasm. However, this method is not suitable for seeds that do not withstand long periods of storage under conditions of low humidity and temperature; the method is also too risky for seeds of rare and endangered species. Plants that are propagated vegetatively as well as those whose seeds have low viability present other kinds of problems regarding long-term preservation.

In recent years, practical applications for long-term preservation of germ plasm as seeds, embryos, embryogenic cells, embryoids, anthers, calluses, protoplasts and shoot tips have developed from cryopreservation, the science of preserving living tissues at a very low temperature, often as low as $-196°C$, the temperature of liquid nitrogen. It is not the intention here to survey the many reported successes in the storage of germ plasm in these various forms; rather this section considers specific examples of embryogenic systems that have been stored by cryopreservation methods as well as their survival characteristics. Basic techniques used for cryopreservation of all materials are more or less the same and involve freezing by slow cooling in the presence of a cryoprotectant, storage at the appropriate low temperature, rapid thawing, washing off the cryoprotectant and culture of the preserved material in a suitable medium. Slow cooling permits flow of water from the cells, thereby promoting the formation of ice crystals in the extracellular fluid rather than in the intracellular environment. Since this creates a high solute concentration in the cell, a cryoprotectant such as dimethylsulfoxide (DMSO) has been used to alleviate the effect of the electrolytes. A number of other compounds, such as glucose, glycerol and proline, to name a few, may also be used in place of DMSO to confer cryoprotection during the freezing process. Rapid thawing of cells is advocated to prevent ice crystal formation, while washing off the cryoprotectant eliminates any toxic effect of the compound on the subsequent growth of the tissue in culture. Details of the development of the methodology used in general for cryopreservation of plant materials are presented by Bajaj and Reinert (1977).

Zygotic embryos enclosed within the protective covering of the seed coat or coats have been retrieved in a viable state after freezing to liquid N temperatures (Sakai and Noshiro, 1975; Stanwood and Bass, 1978). However, only a few studies have been done on the effects of long-term cryopreservation of seeds and isolated embryos on their subsequent survival. According to Bajaj (1981b), rice grains preserved for 3 weeks in liquid N germinate in as high percentages as do the unfrozen control samples, while excised embryos exhibit some loss of viability upon freezing. The survival of immature embryos of wheat and rice in liquid N is much less than that of mature embryos, although embryos that survive show morphogenetic responses typical of normal unfrozen embryos (Bajaj, 1984). From the point of view of long-term genetic conservation of germ plasm, a recent attempt of preservation of embryos of oil palm is of interest (Grout, Shelton and Pritchard, 1983). Seeds of oil palm have a high water content and suffer loss of viability when the water content is reduced below a certain level. The longest period of viability reported for seed stocks is about 15 months when the water content is reduced to 15 percent of fresh weight. While the preservation of whole seeds dessicated to different water contents at $-18°C$ or $-196°C$ results in a total loss of viability, excised embryos at 10.4 percent water content cooled to $-196°C$ remain viable for as long as 8 months.

In the context of conservation of germ plasm of plants that have recalcitrant seeds or that produce seeds only rarely, storage of embryogenic cells or embryoids is an important alternative to soil maintenance of the species. Like studies on many other aspects of somatic embryogenesis, embryogenic cells of carrot have been extensively used for basic investigations in germ plasm conservation of embryogenic cells. According to one report, deliberate intervention in the growth of embryoids by withholding sucrose from the medium is an effective means of storage of embryoids at ordinary laboratory conditions for periods of up to 2 years. The removal of sucrose from the medium presents a condition probably analogous to the normal seed maturation processes when sugars are converted to storage products (Jones, 1974a). Nag and Street (1973) showed that cryopreservation of carrot cells growing in a medium containing 2,4-D at $-196°C$ for 45 days does not affect their embryogenic competence when they are subsequently challenged in a medium lacking the auxin. Dougall and Wetherell (1974) extended the storage period of carrot cells to 12 months without impairment of their embryogenic capacity. It has become apparent that this property is not unique to carrot cells, since embryogenic competence of the callus of *Phoenix dactylifera* (date palm) has also been restored after cold storage for several weeks (Ulrich, Finkle and Tisserat, 1982). Although attempts to cryopreserve somatic embryos of carrot by the method that was so successfully used with embryogenic cells were not effective, Withers

(1979) has developed a modification of this procedure termed dry freezing to obtain very high yields of viable embryoids. An important change in the protocol that bestowed success is the removal of superficial moisture from cryoprotectant-treated embryoids and enclosing them in a foil envelope before freezing.

Long-term storage techniques are particularly useful for maintaining gene banks of pollen grains at the sensitive stage of development for embryogenic induction, as well as of haploid embryoids and cell lines. Obvious reasons are the loss of embryogenic potency of pollen grains at earlier or later stages of development and the loss of haploidy in embryoids and cell lines by endopolyploidy. Although it has not been possible to preserve the embryogenic potency of pollen grains of frozen anthers, pollen embryoids of *Nicotiana tabacum* and *Atropa belladonna* (Bajaj, 1977) and cultured anthers of *N. tabacum, A. belladonna, Petunia hybrida* (Bajaj, 1978) and *Primula obconica* (Bajaj, 1981a) containing embryoids of various stages of development and of wheat and rice enclosing embryogenic calluses (Bajaj, 1984) have been successfully revived into plantlets after cryopreservation for periods ranging from one minute to two months. Survival rates of embryoids vary according to their stage of development at the time of freezing, young globular embryoids generally showing a better survivability than older heart-shaped embryoids.

In summary, it appears that cryopreservation techniques have scarcely been exploited for germ plasm maintenance of excised embryos, embryoids, embryogenic cells and embryogenic pollen grains. The examples considered here clearly indicate that techniques that have been perfected for protoplasts, stem tips and other cultured materials can be used with minor modifications for embryogenic systems. Especially for species that can be propagated only vegetatively, cryopreservation of embryogenic cells holds great promise as an alternative method of preserving the germ plasm.

Concluding comments

During the last few years major developments have occurred in the study of angiosperm embryogenesis which point toward some significant applications in our agriculture. Thus far embryo culture methods are outstanding in their applications to rescue progenies from crosses between unrelated plants. Nevertheless, the methods need to be improved and modified to obtain hybrid seedlings in large numbers for field operations and for genetic analysis. It is also necessary to integrate embryo culture protocol into hybridization programs between distant species in order to have applied value in the foreseeable future.

Advances made in the study of somatic embryogenesis extend considerably beyond the production of facsimiles of embryos. Nevertheless, no serious attempts have been made to utilize this method, which produces uniform genotypes with relative ease, for practical purposes. The most potential for micropropagation of agronomic crops by somatic embryogenesis might be with species that are propagated solely by vegetative means or for the production of clones from embryos of distant crosses.

As haploidy is the only practical source of homozygous lines for most crops, our understanding of the importance of haploid plants has come into clearer focus since the discovery of pollen embryogenesis. Within the coming decade, major achievements can be anticipated in the production of whole plant mutants for desirable characteristics such as salt tolerance, herbicide resistance and disease resistance by anther and pollen culture techniques. This will not only broaden the genetic base for these characters in plants but will also alter the strategy employed in breeding crop plants for favorable traits.

As far as it applies to embryogenesis, the new science of cryopreservation has emphasized the importance of preserving the germ plasm as seeds, embryos, embryoids or embryogenic cells. Not only should our efforts be directed toward conserving the ancient genes in our crop plants, but we should also make the existing genes function more effectively to our advantage. It is from this point of view that the future of angiosperm embryogenesis is to be viewed. Although the potential applications of genetic engineering techniques in embryogenesis have not been emphasized in this chapter, this new technology is expected to play a significant role in the manipulation of plant embryos in attempts to increase the quality and quantity of food stored in embryos and endosperms of some of our important crops.

References

Abo El-Nil, M.M., and Hildebrandt, A.C. (1973). Origin of androgenetic callus and haploid Geranium plants. *Can. J. Bot.* 51: 2107-2109.

Ackerson, R.C. (1984a). Regulation of soybean embryogenesis by abscisic acid. *J. Expt. Bot.* 35: 403-413.

Ackerson, R.C. (1984b). Abscisic acid and precocious germination in soybeans. *J. Expt. Bot.* 35: 414-421.

Ahée, J., Arthuis, P., Cas, G., Duval, Y., Guénin, G., Hanower, J., Hanower, P., Lievoux, D., Lioret, C., Malaurie, B., Pannetier, C., Raillot, D., Varechon, C., and Zuckerman, L. (1981). La multiplication végétative *in vitro* du palmier à huile par embryogenèse somatique. *Oleagineux* 36: 113-116.

Ahloowalia, B.S., and Maretzki, A. (1983). Plant regeneration via somatic embryogenesis in sugarcane. *Plant Cell Rep.* 2: 21-25.

Ahn, C.S., and Hartmann, R.W. (1978a). Interspecific hybridization between mung bean [*Vigna radiata* (L.) Wilczek] and adzuki bean [*V. angularis* (Willd.) Ohwi & Ohashi]. *J. Am. Soc. Hort. Sci.* 103: 3-6.

Ahn, C.S., and Hartmann, R.W. (1978b). Interspecific hybridization between rice bean [*Vigna umbellata* (Thunb.) Ohwi & Ohashi] and adzuki bean [*Vigna angularis* (Willd.) Ohwi & Ohashi]. *J. Am. Soc. Hort. Sci.* 103: 435-438.

Al-Abta, S., and Collin, H.A. (1978). Control of embryoid development in tissue cultures of celery. *Ann. Bot.* 42: 773-782.

Al-Abta, S., and Collin, H.A. (1979). Endogenous auxin and cytokinin changes during embryoid development in celery tissue. *New Phytol.* 82: 29-35.

Alpi, A., Tognoni, F., and D'Amato, F. (1975). Growth regulator levels in embryo and suspensor of *Phaseolus coccineus* at two stages of development. *Planta* 127: 153-162.

Alvarez, M.N., Ascher, P.D., and Davis, D.W. (1981). Interspecific hybridization in *Euphaseolus* through embryo rescue. *HortSci.* 16: 541-543.

Alvarez, M.R., and Sagawa, Y. (1965). A histochemical study of embryo development in *Vanda* (Orchidaceae). *Caryologia* 18: 251-261.

Ammirato, P.V. (1974). The effects of abscisic acid on the development of somatic embryos from cells of caraway (*Carum carvi* L.). *Bot. Gaz.* 135: 328-337.

Ammirato, P.V. (1977). Hormonal control of somatic embryo development from cultured cells of caraway. *Plant Physiol.* 59: 579-586.

Ammirato, P.V. (1983a). Embryogenesis. In: *Handbook of Plant Cell Culture*, Vol. 1, ed. D. A. Evans, W. R. Sharp, P. V. Ammirato and Y. Yamada, pp. 82-123. New York: Macmillan Publishing Co.

Ammirato, P.V. (1983b). The regulation of somatic embryo development in plant cell cultures: Suspension culture techniques and hormone requirements. *Bio/Technology* 1: 68-74.

Amos, J.A., and Scholl, R.L. (1978). Induction of haploid callus from anthers of four species of *Arabidopsis*. *Z. Pflanzenphysiol.* 90: 33-43.
Anand, V.V., and Arekal, G.D. (1979). *In vitro* culture of excised anthers of *Solanum mammosum* L. *Indian J. Expt. Biol.* 17: 444-446.
Anand, V.V., Arekal, G.D., and Swamy, B.G.L. (1980). Chimeral embryoids of pollen origin in tobacco. *Curr. Sci.* 49: 603-604.
Andrews, C.J., and Simpson, G.M. (1969). Dormancy studies in seed of *Avena fatua*. 6. Germinability of the immature caryopsis. *Can. J. Bot.* 47: 1841-1849.
Andronescu, D.I. (1919). Germination and further development of the embryo of *Zea mays* separated from the endosperm. *Am. J. Bot.* 6: 443-452.
Arditti, J. (1979). Aspects of the physiology of orchids. *Adv. Bot. Res.* 7: 421-655.
Ashihara, H., Fujimura, T., and Komamine, A. (1981). Pyrimidine nucleotide biosynthesis during somatic embryogenesis in a carrot cell suspension culture. *Z. Pflanzenphysiol.* 104: 129-137.
Ashley, T. (1972). Zygote shrinkage and subsequent development in some *Hibiscus* hybrids. *Planta* 108: 303-317.
Augsten, H. (1956). Wachstumsversuche mit isolierten Weizen-Embryonen. *Planta* 48: 24-46.
Avanzi, S., Cionini, P.G., and D'Amato, F. (1970). Cytochemical and autoradiographic analyses on the embryo suspensor cells of *Phaseolus coccineus*. *Caryologia* 23: 605-638.
Avanzi, S., Durante, M., Cionini, P.G., and D'Amato, F. (1972). Cytological localization of ribosomal cistrons in polytene chromosomes of *Phaseolus coccineus*. *Chromosoma* 39: 191-203.
Ba, L.T., Cavé, G., Henry, M., and Guignard, J. (1978). Embryogénie des Potomogetonacées. Etude en microscopie électronique à balayage de l'origine du cotylédon chez *Potomogeton lucens* L. *Compt. Rend. Acad. Sci. Paris* 286D: 1351-1353.
Babbar, S.B., and Gupta, S.C. (1980). Chilling induced androgenesis in anthers of *Petunia hybrida* without any culture medium. *Z. Pflanzenphysiol.* 100: 279-283.
Babbar, S.B., Mittal, A. (née Vishnoi), and Gupta, S.C. (1980). *In vitro* induction of androgenesis, callus formation and organogenesis in *Iberis amara* Linn. anthers. *Z. Pflanzenphysiol.* 100: 409-414.
Backs-Hüsemann, D., and Reinert, J. (1970). Embryobildung durch isolierte Einzelzellen aus Gewebekulturen von *Daucus carota*. *Protoplasma* 70: 49-60.
Bajaj, Y.P.S. (1977). Survival of *Atropa* and *Nicotiana* pollen-embryos frozen at $-196°C$. *Curr. Sci.* 46: 305-307.
Bajaj, Y.P.S. (1978). Effect of super-low temperature on excised anthers and pollen-embryos of *Atropa*, *Nicotiana* and *Petunia*. *Phytomorphology* 28: 171-176.
Bajaj, Y.P.S. (1981a). Regeneration of plants from ultra-low frozen anthers of *Primula obconica*. *Scientia Hort.* 14: 93-95.
Bajaj, Y.P.S. (1981b). Growth and morphogenesis in frozen ($-196°C$) endosperm and embryos of rice. *Curr. Sci.* 50: 947-948.
Bajaj, Y.P.S. (1983). *In vitro* production of haploids. In: *Handbook of Plant Cell Culture*, Vol. 1, ed. D.A. Evans, W.R. Sharp, P.V. Ammirato and Y. Yamada, pp. 228-287. New York: Macmillan Publishing Co.
Bajaj, Y.P.S. (1984). The regeneration of plants from frozen pollen embryos and zygotic embryos of wheat and rice. *Theor. Appl. Genet.* 67: 525-528.
Bajaj, Y.P.S., Gill, K.S., and Sandha, G.S. (1978). Some factors enhancing the *in vitro* production of hexaploid Triticale (*Triticum durum* × *Secale cereale*). *Crop Improv.* 5: 62-72.

Bajaj, Y.P.S., Gosch, G., Ottma, M., Weber, A., and Gröbler, A. (1978). Production of polyploid and aneuploid plants from anthers and mesophyll protoplasts of *Atropa belladonna* and *Nicotiana tabacum. Indian J. Expt. Biol.* 16: 947-953.
Bajaj, Y.P.S., Kumar, P., Singh, M.M., and Labana, K.S. (1982). Interspecific hybridization in the genus *Arachis* through embryo culture. *Euphytica* 31: 365-370.
Bajaj, Y.P.S., Labana, K.S., and Dhanju, M.S. (1980). Induction of pollen-embryos and pollen-callus in anther cultures of *Arachis hypogaea* and *A. glabrata. Protoplasma* 103: 397-399.
Bajaj, Y.P.S., Ram, A.K., Labana, K.S., Singh, H. (1981). Regeneration of genetically variable plants from the anther-derived callus of *Arachis hypogaea* and *Arachis villosa. Plant Sci. Lett.* 23: 35-39.
Bajaj, Y.P.S., and Reinert, J. (1977). Cryobiology of plant cell cultures and establishment of gene-banks. In: *Applied and Fundamental Aspects of Plant Cell, Tissue, and Organ Culture,* ed. J. Reinert and Y.P.S. Bajaj, pp. 757-777. Berlin: Springer-Verlag.
Bajaj, Y.P.S., and Singh, H. (1980). *In vitro* induction of androgenesis in mung bean *Phaseolus aureus* L. *Indian J. Expt. Biol.* 18: 1316-1318.
Bajaj, Y.P.S., Singh, H., and Gosal, S.S. (1980). Haploid embryogenesis in anther cultures of pigeon-pea (*Cajanus cajan*). *Theor. Appl. Genet.* 58: 157-159.
Bajaj, Y.P.S., Verma, M.M., and Dhanju, M.S. (1980). Barley × rye hybrids (*Hordecale*) through embryo culture. *Curr. Sci.* 49: 362-363.
Balfour, E. (1957). The development of the vascular systems in *Macropiper excelsum* Forst. I. The embryo and the seedling. *Phytomorphology* 7: 354-364.
Banerjee, S., and Gupta, S. (1975). Embryoid and plantlet formation from stock cultures of *Nigella* tissues. *Physiol. Plant.* 34: 243-245.
Bapat, V.A., and Rao, P.S. (1979). Somatic embryogenesis and plantlet formation in tissue cultures of sandalwood (*Santalum album* L.). *Ann. Bot.* 44: 629-630.
Bapat, V.A., and Wenzel, G. (1982). *In vitro* haploid plantlet induction in *Physalis ixocarpa* Brot. through microspore embryogenesis. *Plant Cell Rep.* 1: 154-156.
Barclay, I.R. (1975). High frequencies of haploid production in wheat (*T. aestivum*) by chromosome elimination. *Nature* (*London*) 256: 410-411.
Barrow, J., Katterman, F., and Williams, D. (1978). Haploid and diploid callus from cotton anthers. *Crop Sci.* 18: 619-622.
Barthe, P., and Bulard, C. (1978). Bound and free abscisic acid levels in dormant and after-ripened embryos of *Pyrus malus* L. cv. Golden Delicious. *Z. Pflanzenphysiol.* 90: 201-208.
Battaglia, E. (1963). Apomixis. In: *Recent Advances in the Embryology of Angiosperms,* ed. P. Maheshwari, pp. 221-264. Delhi: International Society of Plant Morphologists.
Batygina, T.B. (1969). On the possibility of separation of a new type of embryogenesis in angiospermae. *Rev. Cytol. Biol. Veg.* 32: 335-341.
Batygina, T.B., and Vasilyeva, V.E. (1981). Experimental study of embryo differentiation in angiosperms. *Acta Soc. Bot. Poloniae* 50: 257-263.
Bayliss, M.W., and Dunn, S.D.M. (1979). Factors affecting callus formation from embryos of barley (*Hordeum vulgare*). *Plant Sci. Lett.* 14: 311-316.
Beachy, R.N., Barton, K.A., Thompson, J.F., and Madison, J.T. (1980). *In vitro* synthesis of the α and α' subunits of the 7S storage proteins (conglycinin) of soybean seeds. *Plant Physiol.* 65: 990-994.
Beachy, R.N., Thompson, J.F., and Madison, J.T. (1978). Isolation of polyribosomes and messenger RNA active in *in vitro* synthesis of soybean seed proteins. *Plant Physiol.* 61: 139-144.

Beevers, L., and Poulson, R. (1972). Protein synthesis in cotyledons of *Pisum sativum* L. I. Changes in cell-free amino acid incorporation capacity during seed development and maturation. *Plant Physiol.* 49: 476-481.
Ben-Hayyim, G., and Neumann, H. (1983). Stimulatory effect of glycerol on growth and somatic embryogenesis in *Citrus* callus cultures. *Z. Pflanzenphysiol.* 110: 331-337.
Bennett, M.D., and Hughes, W.G. (1972). Additional mitosis in wheat pollen induced by ethrel. *Nature (London)* 240: 566-568.
Bennett, M.D., Rao, M.K., Smith, J.B., and Bayliss, M.W. (1973). Cell development in the anther, the ovule, and the young seed of *Triticum aestivum* L. var. Chinese Spring. *Phil. Trans. Roy. Soc. London* 266B: 39-81.
Bennett, M.D., and Smith, J.B. (1976). The nuclear DNA content of the egg, the zygote and young proembryo cells in *Hordeum*. *Caryologia* 29: 435-446.
Bennett, M.D., Smith, J.B., and Barclay, I. (1975). Early seed development in the Triticale. *Phil. Trans. Roy. Soc. London* 272B: 199-227.
Bennici, A., and Cionini, P.G. (1979). Cytokinins and *in vitro* development of *Phaseolus coccineus* embryos. *Planta* 147: 27-29.
Bennici, A., Cionini, P.G., and D'Amato, F. (1976). Callus formation from the suspensor of *Phaseolus coccineus* in hormone-free medium: A cytological and DNA cytophotometric study. *Protoplasma* 89: 251-261.
Beranger-Novat, N., and Monin, J. (1971). A propos de la levée de dormance des embryons d'*Evonymus europaeus* L. par l'acide gibbérellique. *Compt. Rend. Acad. Sci. Paris* 272D: 1368-1371.
Berjak, P., and Villiers, T.A. (1972). Ageing in plant embryos. IV. Loss of regulatory control in aged embryos. *New Phytol.* 71: 1069-1074.
Bernard, S. (1971). Développement d'embryons haploides à partir d'anthères cultivées *in vitro*. Etude cytologique comparée chez le tabac et le petunia. *Rev. Cytol. Biol. Veg.* 34: 165-188.
Beversdorf, W.D., and Bingham, E.T. (1977). Degrees of differentiation obtained in tissue cultures of *Glycine* species. *Crop Sci.* 17: 307-311.
Beyl, C.A., and Sharma, G.C. (1983). Picloram induced somatic embryogenesis in *Gasteria* and *Haworthia*. *Plant Cell Tissue Organ Culture* 2: 123-132.
Bhalla, P.L., Singh, M.B., and Malik, C.P. (1979). Physiology of sexual reproduction. IV. Embryogenesis in *Tropaeolum majus* L. Enzyme changes. *Acta Bot. Indica* 7: 72-86.
Bhalla, P.L., Singh, M.B., and Malik, C.P. (1980a). Dark fixation of CO_2 by embryo-suspensors of *Nasturtium* (*Tropaeolum majus*). *Biochem. Physiol. Pflanzen* 175: 263-267.
Bhalla, P.L., Singh, M.B., and Malik, C.P. (1980b). Hydrolase and oxido-reductase activities during embryogeny of okra, *Abelmoschus esculentus* (L.) Moench. *Proc. Indian Acad. Sci.* (*Plant Sci.*) 89: 315-321.
Bhalla, P.L., Singh, M.B., and Malik, C.P. (1980c). Changes in level of major biochemical constituents and some enzymes of carbohydrate metabolism during embryogeny in okra (*Abelmoschus esculentus* (L.) Moench. *Indian J. Expt. Biol.* 18: 686-689.
Bhalla, P.L., Singh, M.B., and Malik, C.P. (1981). Studies on the comparative biosynthetic activities of embryo and suspensor in *Tropaeolum majus* L. *Z. Pflanzenphysiol.* 103: 115-119.
Bhatnagar, S.P., and Johri, B.M. (1972). Development of angiosperm seeds. In: *Seed Biology*, Vol. 1, ed. T.T. Kozlowski, pp. 77-149. New York: Academic Press.

Bhatnagar, S.P., and Kallarackal, J. (1980). Cytochemical studies on the endosperm of *Linaria bipartita* (Vent.) Willd. with a note on the role of endosperm haustoria. *Cytologia* 45: 247-256.

Bhojwani, S.S., and Bhatnagar, S.P. (1978). *The Embryology of Angiosperms*, IIIrd Revised ed. New Delhi: Vikas Publishing House.

Bhojwani, S.S., Dunwell, J.M., and Sunderland, N. (1973). Nucleic-acid and protein contents of embryogenic tobacco pollen. *J. Expt. Bot.* 24: 863-871.

Bianco, J., and Bulard, C. (1977). Etude de la dormance embryonnaire chez *Sorbus aucuparia* L. *Trav. Scient. Parc Natl. Vanoise* 8: 147-155.

Binding, H., Binding, K., and Straub, J. (1970). Selektion in Gewebekulturen mit haploiden Zellen. *Naturwissenschaften* 57: 138-139.

Bohdanowicz, J. (1973). Karyological anatomy of the suspensor in *Alisma* L. I. *Alisma plantago-aquatica* L. *Acta Biol. Cracov. Ser. Bot.* 16: 235-246.

Bollini, R., and Chrispeels, M.J. (1979). The rough endoplasmic reticulum is the site of reserve-protein synthesis in developing *Phaseolus vulgaris* cotyledons. *Planta* 146: 487-501.

Borthwick, H.A. (1931). Development of the macrogametophyte and embryo of *Daucus carota*. *Bot. Gaz.* 92: 23-44.

Botti, C., and Vasil, I.K. (1983). Plant regeneration by somatic embryogenesis from parts of cultured mature embryos of *Pennisetum americanum* (L.) K. Schum. *Z. Pflanzenphysiol.* 111: 319-325.

Botti, C., and Vasil, I.K. (1984). Ontogeny of somatic embryos of *Pennisetum americanum*. II. In cultured immature inflorescences. *Can. J. Bot.* 62: 1629-1635.

Bouharmont, J. (1977). Cytology of microspores and calli after anther culture in *Hordeum vulgare*. *Caryologia* 30: 351-360.

Bourgin, J. (1978). Valine-resistant plants from *in vitro* selected tobacco cells. *Mol. Gen. Genet.* 161: 225-230.

Bowes, B.G. (1976). *In vitro* morphogenesis of *Crambe maritima* L. *Protoplasma* 89: 185-188.

Boyes, C.J., and Vasil, I.K. (1984). Plant regeneration by somatic embryogenesis from cultured young inflorescences of *Sorghum arundinaceum* (Desv.) Stapf. var. *sudanense* (Sudan grass). *Plant Sci. Lett.* 35: 153-157.

Boyle, S.A., and Yeung, E.C. (1983). Embryogeny of *Phaseolus:* Developmental pattern of lactate and alcohol dehydrogenases. *Phytochemistry* 22: 2413-2416.

Braak, J.P., and Kooistra, E. (1975). A successful cross between *Phaseolus vulgaris* L. and *P. ritensis* Jones with the aid of embryo culture. *Euphytica* 24: 669-679.

Bradley, P.M., El-Fiki, F., and Giles, K.L. (1984). Polyamines and arginine affect somatic embryogenesis of *Daucus carota*. *Plant Sci. Lett..* 34: 397-401.

Brady, T. (1973). Feulgen cytophotometric determination of the DNA content of the embryo proper and suspensor cells of *Phaseolus coccineus*. *Cell Diff.* 2: 65-75.

Brady, T., and Clutter, M.E. (1972). Cytolocalization of ribosomal cistrons in plant polytene chromosomes. *J. Cell Biol.* 53: 827-832.

Brady, T., and Clutter, M.E. (1974). Structure and replication of *Phaseolus* polytene chromosomes. *Chromosoma* 45: 63-79.

Brady, T., and Walthall, E.D. (1985). The effect of the suspensor and gibberellic acid on *Phaseolus vulgaris* embryo protein content. *Develop. Biol.* 107: 531-536.

Bray, C.M., and Chow, T.-Y. (1976a). Lesions in post-ribosomal supernatant fractions associated with loss of viability in pea (*Pisum arvense*) seed. *Biochim. Biophys. Acta* 442: 1-13.

Bray, C.M., and Chow, T.-Y. (1976b). Lesions in the ribosomes of non-viable pea (*Pisum arvense*) embryonic axis tissue. *Biochim. Biophys. Acta* 442: 14-23.

Bray, C.M., and Dasgupta, J. (1976). Ribonucleic acid synthesis and loss of viability in pea seed. *Planta* 132: 103-108.

Breton, A.M., and Sung, Z.R. (1982). Temperature-sensitive carrot variants impaired in somatic embryogenesis. *Develop. Biol.* 90: 58-66.

Brettell, R.I.S., Thomas, E., and Wernicke, W. (1981). Production of haploid maize plants by anther culture. *Maydica* 26: 101-111.

Brettell, R.I.S., Wernicke, W., and Thomas, E. (1980). Embryogenesis from cultured immature inflorescences of *Sorghum bicolor*. *Protoplasma* 104: 141-148.

Bright, S.W.J., Wood, E.A., and Miflin, B.J. (1978). The effect of aspartate-derived amino acids (lysine, threonine, methionine) on the growth of excised embryos of wheat and barley. *Planta* 139: 113-117.

Brink, R.A., and Cooper, D.C. (1941). Incomplete seed failure as a result of somatoplastic sterility. *Genetics* 26: 487-505.

Brocklehurst, P.A., and Fraser, R.S.S. (1980). Ribosomal RNA integrity and rate of seed germination. *Planta* 148: 417-421.

Broekaert, D., and van Parijs, R. (1978). The relationship between the endomitotic cell cycle and the enhanced capacity for protein synthesis in Leguminosae embryogeny. *Z. Pflanzenphysiol.* 86: 165-175.

Broué, P., Douglass, J., Grace, J.P., and Marshall, D.R. (1982). Interspecific hybridisation of soybeans and perennial *Glycine* species indigenous to Australia via embryo culture. *Euphytica* 31: 715-724.

Brown, J.W.S., Ersland, D.R., and Hall, T.C. (1982). Molecular aspects of storage protein synthesis during seed development. In: *The Physiology and Biochemistry of Seed Development, Dormancy and Germination*, IInd ed., ed. A.A. Khan, pp. 3-42. Amsterdam: Elsevier Biomedical Press.

Brown, S., Wetherell, D.F., and Dougall, D.K. (1976). The potassium requirement for growth and embryogenesis in wild carrot suspension cultures. *Physiol. Plant.* 37: 73-79.

Buell, K.M. (1952). Developmental morphology in *Dianthus*. I. Structure of the pistil and seed development. *Am. J. Bot.* 39: 194-210.

Buffard-Morel, J. (1968). Effets du glucose, du lévulose, du maltose et du saccharose sur le développement des embryons de palmier à huile (*Elaeis guineensis* Jacq. var. *Dura* Bec.) en culture *in vitro*. *Compt. Rend. Acad. Sci. Paris* 267D: 185-188.

Bui Dang Ha, D., Norreel, B., and Masset, A. (1975). Regeneration of *Asparagus officinalis* L. through callus cultures derived from protoplasts. *J. Expt. Bot.* 26: 263-270.

Bui Dang Ha, D., and Pernes, J. (1982). Androgenesis in pearl millet I. Analysis of plants obtained from microspore culture. *Z. Pflanzenphysiol.* 108: 317-327.

Bulard, C., and Monin, J. (1963). Etude du comportement d'embryons de *Fraxinus excelsior* L. prélevés dans des graines dormantes et cultivés *in vitro*. *Phyton* 20: 115-125.

Burghardtová, K., and Tupý, J. (1980). Utilization of exogenous sugars by excised maize embryos in culture. *Biol. Plant.* 22: 57-64.

Button, J., and Botha, C.E.J. (1975). Enzymic maceration of *Citrus* callus and the regeneration of plants from single cells. *J. Expt. Bot.* 26: 723-729.

Button, J., Kochba, J., and Bornman, C.H. (1974). Fine structure of and embryoid development from embryogenic ovular callus of 'Shamouti' orange (*Citrus sinensis* Osb.). *J. Expt. Bot.* 25: 446-457.

Cadic, A., and Sangwan-Norreel, B.S. (1983). Modifications ultrastructurales provoqueés par des traitements promoteurs de l'androgenèse chez le *Datura innoxia* Mill. *Ann. Sci. Naturl. XIII Bot.* 5: 97-114.

Camefort, H., and Sangwan, R.S. (1979). Effets d'un choc thermique sur certaines ultrastructures des grains de pollen embryogènes du *Datura metel* L. *Compt. Rend. Acad. Sci. Paris* 288D: 1383-1386.

Cameron-Mills, V., and Duffus, C.M. (1977). The *in vitro* culture of immature barley embryos on different culture media. *Ann. Bot.* 41: 1117-1127.

Cameron-Mills, V., and Duffus, C.M. (1979). Sucrose transport in isolated immature barley embryos. *Ann. Bot.* 43: 559-569.

Carasco, J.F., Croy, R., Derbyshire, E., and Boulter, D. (1978). The isolation and characterization of the major polypeptides of the seed globulin of cowpea (*Vigna unguiculata* L. Walp) and their sequential synthesis in developing seeds. *J. Expt. Bot.* 29: 309-323.

Carlson, P.S. (1970). Induction and isolation of auxotrophic mutants in somatic cell cultures of *Nicotiana tabacum*. *Science* 168: 487-489.

Carlson, P.S. (1973). Methionine sulfoximine-resistant mutants of tobacco. *Science* 180: 1366-1368.

Carniel, K. (1967). Uber die Embryobildung in der Gattung *Paeonia*. *Osterr. Bot. Z.* 114: 4-19.

Cass, D.D., and Karas, I. (1974). Ultrastructural organization of the egg of *Plumbago zeylanica*. *Protoplasma* 81: 49-62.

Cave, M.S., Arnott, H.J., and Cook, S.A. (1961). Embryogeny in the California peonies with reference to their taxonomic position. *Am. J. Bot.* 48: 397-404.

Ceccarelli, N., Lorenzi, R., and Alpi, A. (1979). Kaurene and kaurenol biosynthesis in cell-free system of *Phaseolus coccineus* suspensors. *Phytochemistry* 18: 1657-1658.

Ceccarelli, N., Lorenzi, R., and Alpi, A. (1981a). Gibberellin biosynthesis in *Phaseolus coccineus* suspensor. *Z. Pflanzenphysiol.* 102: 37-44.

Ceccarelli, N., Lorenzi, R., and Alpi, A. (1981b). Kaurene metabolism in cell-free extracts of *Phaseolus coccineus* suspensors. *Plant Sci. Lett.* 21: 325-332.

Chaleff, R.S., and Stolarz, A. (1981). Factors influencing the frequency of callus formation among cultured rice (*Oryza sativa*) anthers. *Physiol. Plant.* 51: 201-206.

Chang, W., and Hsing, Y. (1980a). *In vitro* flowering of embryoids derived from mature root callus of ginseng (*Panax ginseng*). *Nature (London)* 284: 341-342.

Chang, W.C., and Hsing, Y.I. (1980b). Plant regeneration through somatic embryogenesis in root-derived callus of ginseng (*Panax ginseng* C.A. Meyer). *Theor. Appl. Genet.* 57: 133-135.

Chatterjee, A., Saha, P.K., Das Gupta, P., Ganguly, S.N., and Sircar, S.M. (1976). Chemical examination of viable and non-viable rice seeds. *Physiol. Plant.* 38: 307-308.

Cheah, K.S.E., and Osborne, D.J. (1978). DNA lesions occur with loss of viability in embryos of ageing rye seed. *Nature (London)* 272: 593-599.

Chen, C. (1977). *In vitro* development of plants from microspores of rice. *In Vitro* 13: 484-489.

Chen, C. (1978). Effects of sucrose concentration on plant production in anther culture of rice. *Crop Sci.* 18: 905-906.

Chen, C., and Lin, M. (1976). Induction of rice plantlets from anther culture. *Bot. Bull. Acad. Sinica* 17: 18-24.
Chen, Y., Wang, R., Tian, W., Zuo, Q., Zheng, S., Lu, D., and Zhang, G. (1980). Studies on pollen culture *in vitro* and induction of plantlets in *Oryza sativa* subsp. Keng. *Acta Genet. Sinica* 7: 46-47.
Chen, Z., Qian, C., Qin, M., Xu, X., and Xiao, Y. (1982). Recent advances in anther culture of *Hevea brasiliensis* (Muell.-Arg.). *Theor. Appl. Genet.* 62: 103-108.
Cheng, J., and Raghavan, V. (1985). Somatic embryogenesis and plant regeneration in *Hyoscyamus niger. Am. J. Bot.* 72: 580-587.
Chlan, C.A., and Dure, L. III. (1983). Plant seed embryogenesis as a tool for molecular biology. *Mol. Cell. Biochem.* 55: 5-15.
Choinski, J.S., Jr., and Trelease, R.N. (1978). Control of enzyme actvities in cotton cotyledons during maturation and germination. II. Glyoxysomal enzyme development in embryos. *Plant Physiol.* 62: 141-145.
Christianson, M.L., Warnick, D.A., and Carlson, P.S. (1983). A morphogenetically competent soybean suspension culture. *Science* 222: 632-634.
Chu, C. (1982). Haploids in plant improvement. In: *Plant Improvement and Somatic Cell Genetics,* ed. I. K. Vasil, W. R. Scowcroft and K. J. Frey, pp. 129-158. New York: Academic Press.
Chu, C., Liu, H., and Du, R. (1982). Microphotometric determination of DNA contents of early developmental pollen grains in tobacco anther culture. *Acta Bot. Sinica* 24: 1-9.
Chu, C., Wang, C., Sun, C., Hsü, C., Yin, K., Chu, C., and Bi, P. (1975). Establishment of an efficient medium for anther culture of rice through comparative experiments on the nitrogen sources. *Scientia Sinica* 18: 659-668.
Chu, C.C., Sun, C.S., Chen, X., Zhang, W.X., and Du, Z.H. (1984). Somatic embryogenesis and plant regeneration in callus from inflorescences of *Hordeum vulgare* × *Triticum aestivum* hybrids. *Theor. Appl. Genet.* 68: 375-379.
Chu, Z., Wang, C., Sun, C., Chien, N., Yin, K., and Hsü, C. (1973). Investigations on the induction and morphogenesis of wheat (*Triticum vulgare*) pollen plants. *Acta Bot. Sinica* 15: 1-11.
Chueca, M., Cauderon, Y., and Tempé, J. (1977). Technique d'obtention d'hybrides blé tendre × *Aegilops* par culture *in vitro* d'embryons immatures. *Ann. Amelior. Plant.* 27: 539-547.
Cionini, P.G., and Avanzi, S. (1972). Pattern of binding of tritiated actinomycin D to *Phaseolus coccineus* polytene chromosomes. *Expt. Cell Res.* 75: 154-158.
Cionini, P.G., Bennici, A., Alpi, A., and D'Amato, F. (1976). Suspensor, gibberellin and *in vitro* development of *Phaseolus coccineus* embryos. *Planta* 131: 115-117.
Clapham, D. (1971). *In vitro* development of callus from the pollen of *Lolium* and *Hordeum. Z. Pflanzenzuchtg.* 65: 285-292.
Clapham, D.H. (1977). Haploid induction in cereals. In: *Applied and Fundamental Aspects of Plant Cell, Tissue, and Organ Culture,* ed. J. Reinert and Y.P.S. Bajaj, pp. 279-298. Berlin: Springer-Verlag.
Clowes, F.A.L. (1978a). Origin of the quiescent centre in *Zea mays. New Phytol.* 80: 409-419.
Clowes, F.A.L. (1978b). Origin of quiescence at the root pole of pea embryos. *Ann. Bot.* 42: 1237-1239.
Clutter, M., Brady, T., Walbot, V., and Sussex, I. (1974). Macromolecular synthesis during plant embryogeny. Cellular rates of RNA synthesis in diploid and polytene cells in bean embryos. *J. Cell Biol.* 63: 1097-1102.

Cocucci, A., and Jensen, W.A. (1969a). Orchid embryology: The mature megagametophyte of *Epidendrum scutella*. *Kurtziana* 5: 23-38.
Cocucci, A., and Jensen, W.A. (1969b). Orchid embryology: Megagametophyte of *Epidendrum scutella* following fertilization. *Am. J. Bot.* 56: 629-640.
Collins, G.B., and Sunderland, N. (1974). Pollen-derived haploids of *Nicotiana knightiana*, *N. raimondii*, and *N. attenuata*. *J. Expt. Bot.* 25: 1030-1035.
Côme, D., and Durand, M. (1971). Influence de l'acide gibbérellique sur la levée de dormance des embryons de pommier (*Pirus malus* L.) par le froid. *Compt. Rend. Acad. Sci. Paris* 273D: 1937-1940.
Côme, D., and Thévenot, C. (1982). Environmental control of embryo dormancy and germination. In: *The Physiology and Biochemistry of Seed Development, Dormancy and Germination*, IInd ed., ed. A.A. Khan, pp. 271-298. Amsterdam: Elsevier Biomedical Press.
Conger, B.V., Hanning, G.E., Gray, D.J., and McDaniel, J.K. (1983). Direct embryogenesis from mesophyll cells of orchardgrass. *Science* 221: 850-851.
Cooper, K.V., Dale, J.E., Dyer, A.F., Lyne, R.L., and Walker, J.T. (1978). Hybrid plants from the barley × rye cross. *Plant Sci. Lett.* 12: 293-298.
Corduan, G. (1975). Regeneration of anther-derived plants of *Hyoscyamus niger* L. *Planta* 127: 27-36.
Corduan, G., and Spix, C. (1975). Haploid callus and regeneration of plants from anthers of *Digitalis purpurea* L. *Planta* 124: 1-11.
Cornejo-Martin, M.J., and Primo-Millo, E. (1981). Anther and pollen grain culture of rice (*Oryza sativa* L.). *Euphytica* 30: 541-546.
Corsi, G. (1972). The suspensor of *Eruca sativa* Miller (Cruciferae) during embryogenesis *in vitro*. *Giorn. Bot. Ital.* 106: 41-54.
Corsi, G., Renzoni, G.C., and Viegi, L. (1973). A DNA cytophotometric investigation on the suspensor of *Eruca sativa* Miller. *Caryologia* 26: 531-540.
Cremonini, R., and Cionini, P.G. (1977). Extra DNA synthesis in embryo suspensor cells of *Phaseolus coccineus*. *Protoplasma* 91: 303-313.
Crété, P. (1963). Embryo. In: *Recent Advances in the Embryology of Angiosperms*, ed. P. Maheshwari, pp. 171-220. Delhi: International Society of Plant Morphologists.
Cronauer, S., and Krikorian, A.D. (1983). Somatic embryos from cultured tissues of triploid plantains (*Musa* 'ABB'). *Plant Cell Rep.* 2: 289-291.
Cross, J.W., and Adams, W.R., Jr. (1983). Embryo-specific globulins from *Zea mays* L. and their subunit composition. *J. Agric. Food Chem.* 31: 534-538.
Crouch, M.L. (1982). Non-zygotic embryos of *Brassica napus* L. contain embryo-specific storage proteins. *Planta* 156: 520-524.
Crouch, M.L., and Sussex, I.M. (1981). Development and storage-protein synthesis in *Brassica napus* L. embryos *in vivo* and *in vitro*. *Planta* 153: 64-74.
Crouch, M.L., Tenbarge, K.L., Simon, A.E., and Ferl, R. (1983). cDNA clones for *Brassica napus* seed storage proteins: Evidence from nucleotide sequence analysis that both subunits of napin are cleaved from a precursor polypeptide. *J. Mol. Appl. Genet.* 2: 273-283.
Cullis, C.A. (1976). Chromatin-bound DNA-dependent RNA polymerase in developing pea cotyledons. *Planta* 131: 293-298.
Cullis, C.A. (1978). Chromatin-bound DNA-dependent RNA polymerase in developing pea cotyledons II. Polymerase activity and template availability under different growth conditions. *Planta* 144: 57-62.
Cummings, D.P., Green, C.E., and Stuthman, D.D. (1976). Callus induction and plant regeneration in oats. *Crop Sci.* 16: 465-470.

D'Alascio-Deschamps, R. (1973). Organisation du sac embryonnaire du *Linum catharticum* L., espèce récoltée en station naturelle; étude ultrastructurale. *Bull. Soc. Bot. France* 120: 189-200.

D'Alascio-Deschamps, R. (1981). Embryologie du *Linum catharticum* L. Le zygote: étude ultrastructurale. *Bull. Soc. Bot. France Lett. Bot.* 128: 269-278.

Dale, P.J. (1975). Pollen dimorphism and anther culture in barley. *Planta* 127: 213-220.

Dale, P.J. (1980). Embryoids from cultured immature embryos of *Lolium multiflorum*. *Z. Pflanzenphysiol.* 100: 73-77.

Dale, P.J., and Deambrogio, E. (1979). A comparison of callus induction and plant regeneration from different explants of *Hordeum vulgare*. *Z. Pflanzenphysiol.* 94: 65-77.

Dale, P.J., Thomas, E., Brettell, R.I.S., and Wernicke, W. (1981). Embryogenesis from cultured immature inflorescences and nodes of *Lolium multiflorum*. *Plant Cell Tissue Organ Culture* 1: 47-55.

Dauphiné, A., and Rivière, S. (1940). Sur la présence de tubes criblés dans des embryons de graines non germées. *Compt. Rend. Acad. Sci. Paris* 211: 359-361.

Davidson, E.H. (1976). *Gene Activity in Early Development*, IInd ed. New York: Academic Press.

Davies, D.R., and Bedford, I.D. (1982). Abscisic acid and storage protein accumulation in *Pisum sativum* embryos grown *in vitro*. *Plant Sci. Lett.* 27: 337-343.

Davies, D.R., and Brewster, V. (1975). Studies of seed development in *Pisum sativum*. II. Ribosomal RNA contents in reciprocal crosses. *Planta* 124: 303-309.

Davis, B.D. (1983). Growth of excised pea embryonic axes on different sugars. *Am. J. Bot.* 70: 816-820.

Davis, G.L. (1966). *Systematic Embryology of the Angiosperms*. New York: John Wiley & Sons.

Debata, B.K., and Patnaik, S.N. (1983). *In vitro* culture of anther of *Crotalaria pallida* Ait. for induction of haploid. *Indian J. Expt. Biol.* 21: 44-46.

Debergh, P., and Nitsch, C. (1973). Premiers résultats sur la culture *in vitro* de grains de pollen isolés chez la tomate. *Compt. Rend. Acad. Sci. Paris* 276D: 1281-1284.

de Guzman, E.V., del Rosario, A.G., and Eusebio, E.C. (1971). The growth and development of coconut "makapuno" embryo *in vitro*. III. Resumption of root growth in high sugar media. *Philipp. Agric.* 53: 566-579.

de Guzman, E.V., and del Rosario, D.A. (1964). The growth and development of *Cocos nucifera* L. (Makapuno) embryos *in vitro*. *Philipp. Agric.* 48: 82-94.

de la Roche, A.I., and Keller, W.A. (1977). The morphogenetic control of erucic acid synthesis in *Brassica campestris*. *Z. Pflanzenzuchtg.* 78: 319-326.

de Lautour, G., Jones, W.T., and Ross, M.D. (1978). Production of interspecific hybrids in *Lotus* aided by endosperm transplants. *New Zealand J. Bot.* 16: 61-68.

Dell'Aquila, A., de Leo, P., Caldiroli, E., and Zocchi, G. (1978). Damages at translational level in aged wheat embryos. *Plant Sci. Lett.* 12: 217-226.

del Rosario, A.G., and de Guzman, E.V. (1978). The growth of coconut "makapuno"embryos *in vitro* as affected by mineral composition and sugar level of the medium during the liquid and solid cultures. *Philipp. J. Sci.* 105: 215-222.

Deschamps, R. (1969). Premiers stades du développement de l'embryon et de l'albumen du lin: Etude au microscope électronique. *Rev. Cytol. Biol. Veg.* 32: 379-390.

Deumling, B., and Nagl, W. (1978). DNA characterization, satellite DNA localization, and nuclear organization in *Tropaeolum majus*. *Cytobiologie* 16: 412-420.

Devi, H.M., and Pullaiah, T. (1976). Embryological investigations in the Melampodinae. I. *Melampodium divaricatum. Phytomorphology* 26: 77-86.
Devreux, M., and Saccardo, F. (1971). Mutazioni sperimental osservate su piante aploidi di tabacco ottenute per colture *in vitro* di antere irradiate. *Atti. Ass. Genet. Ital.* 16: 69-71.
Dhillon, S.S., and Miksche, J.P. (1983). DNA, RNA, protein and heterochromatin changes during embryo development and germination of soybean (*Glycine max* L.). *Histochem. J.* 15: 21-37.
Diboll, A.G. (1968). Fine structural development of the megagametophyte of *Zea mays* following fertilization. *Am. J. Bot.* 55: 797-806.
Diboll, A.G., and Larson, D.A. (1966). An electron microscopic study of the mature megagametophyte in *Zea mays*. *Am. J. Bot.* 53: 391-402.
Diez, J.L., and Cionini, P.G. (1971). DNA/histone ratio in different regions of polytene chromosomes in the embryo suspensor cells of *Phaseolus coccineus*. *Caryologia* 24: 463-470.
Dix, P.J., Joó, F., and Maliga, P. (1977). A cell line of *Nicotiana sylvestris* with resistance to kanamycin and streptomycin. *Mol. Gen. Genet.* 157: 285-290.
Dix, P.J., and Street, H.E. (1975). Sodium chloride-resistant cultured cell lines from *Nicotiana sylvestris* and *Capsicum annuum*. *Plant Sci. Lett.* 5: 231-237.
Dix, P.J., and Street, H.E. (1976). Selection of plant cell lines with enhanced chilling resistance. *Ann. Bot.* 40: 903-910.
Doležel, J., Novák, F.J., and Lužný, J. (1980). Embryo development and *in vitro* culture of *Allium cepa* and its interspecific hybrids. *Z. Pflanzenzuchtg.* 85: 177-184.
Dollmantel, H.-J., and Reinert, J. (1980). Auxin levels, antiauxin(s) and androgenic plantlet formation in isolated pollen cultures of *Nicotiana tabacum*. *Protoplasma* 103: 155-162.
Domoney, C., Davies, D.R., and Casey, R. (1980). The initiation of legumin synthesis in immature embryos of *Pisum sativum* L. grown *in vivo* and *in vitro*. *Planta* 149: 454-460.
Dorion, N., Chupeau, Y., and Bourgin, J.P. (1975). Isolation, culture and regeneration into plants of *Ranunculus sceleratus* L. leaf protoplasts. *Plant Sci. Lett.* 5: 325-331.
Dorion, N., Godin, B., and Bigot, C. (1984). Embryogenèse somatique à partir de cultures issues de protoplastes foliaires de *Ranunculus sceleratus*. *Can. J. Bot.* 62: 2345-2355.
Dougall, D.K., and Verma, D.C. (1978). Growth and embryo formation in wild-carrot suspension cultures with ammonium ion as a sole nitrogen source. *In Vitro* 14: 180-182.
Dougall, D.K., and Wetherell, D.F. (1974). Storage of wild carrot cultures in the frozen state. *Cryobiology* 11: 410-415.
Drew, R.L.K. (1979). Effect of activated charcoal on embryogenesis and regeneration of plantlets from suspension cultures of carrot (*Daucus carota* L.). *Ann. Bot.* 44: 387-389.
Drew, R.L.K. (1980). A cheap, simple apparatus for growing large batches of plant tissue in submerged liquid culture. *Plant Sci. Lett.* 17: 227-236.
Drira, N., and Benbadis, A. (1975). Analyse, par culture d'anthères *in vitro*, des potentialités androgénétiques de deux espéces de *Citrus* (*Citrus medica* L. et *Citrus limon* L. Burm.). *Compt. Rend. Acad. Sci. Paris* 281D: 1321-1324.
Dublin, P. (1981). Embryogenèse somatique directe sur fragments de feuilles de caféier arabusta. *Cafe Cacao The* 25: 237-242.

Duffus, C., and Slaughter, C. (1980). *Seeds and Their Uses.* Chichester: John Wiley & Sons.
Duffus, C.M., and Rosie, R. (1975). Biochemical changes during embryogeny in *Hordeum distichum. Phytochemistry* 14: 319-323.
Duncan, E.J., and Heberle, E. (1976). Effect of temperature shock on nuclear phenomena in microspores of *Nicotiana tabacum* and consequently on plantlet production. *Protoplasma* 90: 173-177.
Dunwell, J.M. (1979). Anther culture of *Nicotiana tabacum:* The role of the culture vessel atmosphere in pollen embryo induction and growth. *J. Expt. Bot.* 30: 419-428.
Dunwell, J.M. (1981a). Stimulation of pollen embryo induction in tobacco by pretreatment of excised anthers in a water-saturated atmosphere. *Plant Sci. Lett.* 21: 9-13.
Dunwell, J.M. (1981b). Influence of genotype and environment on growth of barley embryos *in vitro. Ann. Bot.* 48: 535-542.
Dunwell, J.M., Cornish, M., and de Courcel, A.G.L. (1985). Influence of genotype, plant growth temperature and anther incubation temperature on microspore embryo production in *Brassica napus* spp. *oleifera. J. Expt. Bot.* 36: 679-689.
Dunwell, J.M., and Sunderland, N. (1974a). Pollen ultrastructure in anther cultures of *Nicotiana tabacum.* I. Early stages of culture. *J. Expt. Bot.* 25: 352-361.
Dunwell, J.M., and Sunderland, N. (1974b). Pollen ultrastructure in anther cultures of *Nicotiana tabacum.* II. Changes associated with embryogenesis. *J. Expt. Bot.* 25: 363-373.
Dunwell, J.M., and Sunderland, N. (1975). Pollen ultrastructure in anther cultures of *Nicotiana tabacum.* III. The first sporophytic division. *J. Expt. Bot.* 26: 240-252.
Dunwell, J.M., and Sunderland, N. (1976). Pollen ultrastructure in anther cultures of *Datura innoxia.* I. Division of the presumptive vegetative cell. *J. Cell Sci.* 22: 469-480.
Durand, M., Thévenot, C., and Côme, D. (1973). Influences de l'acide abscissique sur la germination et la levée de dormance des embryons de pommier (*Pirus malus* L.). *Compt. Rend. Acad. Sci. Paris* 277D: 53-55.
Durante, M., Cionini, P.G., Avanzi, S., Cremonini, R., and D'Amato, F. (1977). Cytological localization of the genes for the four classes of ribosomal RNA (25S, 18S, 5.8S and 5S) in polytene chromosomes of *Phaseolus coccineus. Chromosoma* 60: 269-282.
Dure, L. III, and Chlan, C. (1981). Developmental biochemistry of cottonseed embryogenesis and germination. XII. Purification and properties of principal storage proteins. *Plant Physiol.* 68: 180-186.
Dure, L. III, and Galau, G.A. (1981). Developmental biochemistry of cottonseed embryogenesis and germination. XIII. Regulation of biosynthesis of principal storage proteins. *Plant Physiol.* 68: 187-194.
Dure, L. III, Galau, G., Chlan, C., and Pyle, J. (1983). Developmentally regulated gene sets in cotton embryogenesis. In: *Plant Molecular Biology,* ed. R. B. Goldberg, pp. 331-342. New York: Alan R. Liss.
Dure, L. III, Greenway, S.C., and Galau, G.A. (1981). Developmental biochemistry of cottonseed embryogenesis and germination: Changing messenger ribonucleic acid populations as shown by *in vitro* and *in vivo* protein synthesis. *Biochemistry* 20: 4162-4168.
Dure, L. III, Pyle, J.B., Chlan, C.A., Baker, J.C., and Galau, G.A. (1983).

Developmental biochemistry of cottonseed embryogenesis and germination. XVII. Developmental expression of genes for the principal storage proteins. *Plant Mol. Biol.* 2: 199-206.

Dure, L.S., and Jensen, W.A. (1957). The influence of gibberellic acid and indoleacetic acid on cotton embryos cultured *in vitro*. *Bot. Gaz.* 118: 254-261.

Eeuwens, C.J., and Schwabe, W.W. (1975). Seed and pod wall development in *Pisum sativum* L. in relation to extracted and applied hormones. *J. Expt. Bot.* 26: 1-14.

Egorova, N.A., and Reznikova, S.A. (1982). Investigation of isolated anther culture in connection with induction of androgenesis in coriander. *Soviet Plant Physiol.* 29: 125-131.

Eichholtz, D.A., Robitaille, H. A., and Hasegawa, P.M. (1979). Adventive embryony in apple. *HortSci.* 14: 699-700.

Eid, A.A.H., de Langhe, E., and Waterkeyn, L. (1973). *In vitro* culture of fertilized cotton ovules. I. The growth of cotton embryos. *La Cellule* 69: 361-371.

Emershad, R.L., and Ramming, D.W. (1984). *In-ovulo* embryo culture of *Vitis vinifera* L. cv. 'Thompson Seedless'. *Am. J. Bot.* 71: 873-877.

Emerson, C.P., and Humphreys, T. (1971). A simple and sensitive method for quantitative measurement of cellular RNA synthesis. *Anal. Biochem.* 40: 254-266.

Engvild, K.C., Linde-Laursen, I., and Lundqvist, A. (1972). Anther cultures of *Datura innoxia:* Flower bud stage and embryoid level of ploidy. *Hereditas* 72: 331-332.

Epstein, E., Kochba, J., and Neumann, H. (1977). Metabolism of indoleacetic acid by embryogenic and non-embryogenic callus lines of "Shamouti" orange (*Citrus sinensis* Osb.). *Z. Pflanzenphysiol.* 85: 263-268.

Erdelská, O. (1980). Some structural aspects of flax embryo nutrition. *Biologia* 35: 243-249.

Ernst, D., and Oesterhelt, D. (1984). Effect of exogenous cytokinins on growth and somatic embryogenesis in anise cells (*Pimpinella anisum* L.). *Planta* 161: 246-248.

Esau, K. (1965). *Plant Anatomy,* IInd ed. New York: John Wiley & Sons.

Esen, A., and Soost, R.K. (1973). Seed development in *Citrus* with special reference to 2x × 4x crosses. *Am. J. Bot.* 60: 448-462.

Esen, A., and Soost, R.K. (1977). Adventive embryogenesis in *Citrus* and its relation to pollination and fertilization. *Am. J. Bot.* 64: 607-614.

Evans, A.M. (1962). Species hybridization in *Trifolium* I. Methods of overcoming species incompatibility. *Euphytica* 11: 164-176.

Evans, I.M., Gatehouse, J.A., Croy, R.R.D., and Boulter, D. (1984). Regulation of the transcription of storage-protein mRNA in nuclei isolated from developing pea (*Pisum sativum* L.) cotyledons. *Planta* 160: 559-568.

Facciotti, D., and Pilet, P.-E. (1979). Plants and embryoids from haploid *Nicotiana sylvestris* protoplasts. *Plant Sci. Lett.* 15: 1-6.

Favre-Duchartre, M. (1978). Oogenèse chez les angiospermes et autres plantes ovulées. *Rev. Cytol. Biol. Veg.-Bot.* 1: 79-95.

Fedak, G. (1977a). Haploids from barley = rye crosses. *Can. J. Genet. Cytol.* 19: 15-19.

Fedak, G. (1977b). Increased homoeologeous chromosome pairing in *Hordeum vulgare = Triticum aestivum* hybrids. *Nature (London)* 266: 529-530.

Fedak, G. (1983). Haploids in *Triticum ventricosum* via intergeneric hybridization with *Hordeum bulbosum*. *Can. J. Genet. Cytol.* 25: 104-106.

Fedak, G., and Armstrong, K.C. (1980). Production of trigeneric (barley × wheat) × rye hybrids. *Theor. Appl. Genet.* 56: 221-224.

Fedak, G., and Nakamura, C. (1981). Intergeneric hybrids between *Triticum crassum* and *Hordeum vulgare*. *Theor. Appl. Genet.* 60: 349-352.

Feirer, R.P., Mignon, G., and Litvay, J.D. (1984). Arginine decarboxylase and polyamines required for embryogenesis in the wild carrot. *Science* 223: 1433-1435.

Fienberg, A.A., Choi, J.H., Lubich, W.P., and Sung, Z.R. (1984). Developmental regulation of polyamine metabolism in growth and differentiation of carrot culture. *Planta* 162: 532-539.

Finkelstein, R.R., and Crouch, M.L. (1984). Precociously germinating rapeseed embryos retain characteristics of embryogeny. *Planta* 162: 125-131.

Fischer, R.L., and Goldberg, R.B. (1982). Structure and flanking regions of soybean seed protein genes. *Cell* 29: 651-660.

Fisher, D.B., and Jensen, W.A. (1972). Nuclear and cytoplasmic DNA synthesis in cotton embryos: A correlated light and electron microscope autoradiographic study. *Histochemie* 32: 1-22.

Fitch, M.M., and Moore, P.H. (1983). Haploid production from anther culture of *Saccharum spontaneum* L. *Z. Pflanzenphysiol.* 109: 197-206.

Floris, C. (1970). Ageing in *Triticum durum* seeds: Behaviour of embryos and endosperms from aged seeds as revealed by the embryo-transplantation technique. *J. Expt. Bot.* 21: 462-468.

Floris, C., and Anguillesi, M.C. (1974). Ageing of isolated embryos and endosperms of durum wheat: An analysis of chromosome damage. *Mutation Res.* 22: 133-138.

Floris, C., Giovannozzi-Sermanni, G., and Meletti, P. (1972). Seed germination and growth in *Triticum*. I. Biological activity of extracts from *T. durum* endosperms. *Plant Cell Physiol.* 13: 331-336.

Foard, D.E., and Haber, A.H. (1962). Use of growth characteristics in studies of morphologic relations. I. Similarities between epiblast and coleorhiza. *Am. J. Bot.* 49: 520-523.

Folsom, M.W., and Peterson, C.M. (1984). Ultrastructural aspects of the mature embryo sac of soybean, *Glycine max* (L.) Merr. *Bot. Gaz.* 145: 1-10.

Fong, F., Smith, J.D., and Koehler, D.E. (1983). Early events in maize seed development. 1-Methyl-3-phenyl-5-(3-[trifluoromethyl]phenyl)-4-(1H)-pyridinone induction of vivipary. *Plant Physiol.* 73: 899-901.

Forino, L.M.C., Tagliasacchi, A.M., and Avanzi, S. (1979). Different structure of polytene chromosomes of *Phaseolus coccineus* suspensors during early embryogenesis. 1. Nucleolus organizing chromosome pairs S_1 and S_2. *Protoplasma* 101: 231-246.

Forman, M., and Jensen, W.A. (1965). Respiration and embryogenesis in cotton. *Plant Physiol.* 40: 765-769.

Foroughi-Wehr, B., Friedt, W., and Wenzel, G. (1982). On the genetic improvement of androgenetic haploid formation in *Hordeum vulgare* L. *Theor. Appl. Genet.* 62: 233-239.

Foroughi-Wehr, B., Mix, G., Gaul, H., and Wilson, H.M. (1976). Plant production from cultured anthers of *Hordeum vulgare* L. *Z. Pflanzenzuchtg.* 77: 198-204.

Forster, B.P., and Dale, J.E. (1983a). A comparative study of early seed development in genotypes of barley and rye. *Ann. Bot.* 52: 603-612.

Forster, B.P., and Dale, J.E. (1983b). Effects of parental embryo and endosperm mitotic cycle times on development of hybrids between barley and rye. *Ann. Bot.* 52: 613-620.

Freed, H.J., and Grant, W.F. (1976). Polytene chromosomes in the suspensor cells of *Lotus* (Fabaceae). *Caryologia* 29: 387-390.

Fridborg, G., Pedersén, M., Landström, L., and Eriksson, T. (1978). The effect of activated charcoal on tissue cultures: Adsorption of metabolites inhibiting morphogenesis. *Physiol. Plant.* 43: 104-106.

Fridriksson, S., and Bolton, J.L. (1963). Preliminary report on the culture of alfalfa embryos. *Can. J. Bot.* 41: 439-440.

Fujimura, T., and Komamine, A. (1975). Effects of various growth regulators on the embryogenesis in a carrot cell suspension culture. *Plant Sci. Lett.* 5: 359-364.

Fujimura, T., and Komamine, A. (1979a). Involvement of endogenous auxin in somatic embryogenesis in a carrot cell suspension culture. *Z. Pflanzenphysiol.* 95: 13-19.

Fujimura, T., and Komamine, A. (1979b). Synchronization of somatic embryogenesis in a carrot cell suspension culture. *Plant Physiol.* 64: 162-164.

Fujimura, T., and Komamine, A. (1980a). Mode of action of 2,4-D and zeatin on somatic embryogenesis in a carrot cell suspension culture. *Z. Pflanzenphysiol.* 99: 1-8.

Fujimura, T., and Komamine, A. (1980b). The serial observation of embryogenesis in a carrot cell suspension culture. *New Phytol.* 86: 213-218.

Fujimura, T., Komamine, A., and Matsumoto, H. (1980). Aspects of DNA, RNA and protein synthesis during somatic embryogenesis in a carrot cell suspension culture. *Physiol. Plant.* 49: 255-260.

Fujimura, T., Komamine, A., and Matsumoto, H. (1981). Changes in chromosomal proteins during early stages of synchronized embryogenesis in a carrot cell suspension culture. *Z. Pflanzenphysiol.* 102: 293-298.

Furuya, M., and Soma, K. (1957). The effects of auxins on the development of bean embryos cultivated *in vitro*. *J. Fac. Sci. Univ. Tokyo Sect. III Bot.* 7: 163-198.

Galau, G.A., Chlan, C.A., and Dure, L. III (1983). Developmental biochemistry of cottonseed embryogenesis and germination. XVI. Analysis of the principal storage protein gene family with cloned cDNA probes. *Plant Mol. Biol.* 2: 189-198.

Galau, G.A., and Dure, L. III. (1981). Developmental biochemistry of cottonseed embryogenesis and germination: Changing messenger ribonucleic acid populations as shown by reciprocal heterologous complementary deoxyribonucleic acid-messenger ribonucleic acid hybridization. *Biochemistry* 20: 4169-4178.

Galitz, D.S., and Howell, R.W. (1965). Measurement of ribonucleic acids and total free nucleotides of developing soybean seeds. *Physiol. Plant.* 18: 1018-1021.

Gamborg, O.L., Constabel, F., and Miller, R.A. (1970). Embryogenesis and production of albino plants from cell cultures of *Bromus inermis*. *Planta* 95: 355-358.

Gärtner, P.-J., and Nagl, W. (1980). Acid phosphatase activity in plastids (plastolysomes) of senescing embryo-suspensor cells. *Planta* 149: 341-349.

Gatehouse, J.A., Evans, I.M., Bown, D., Croy, R.R.D., and Boulter, D. (1982). Control of storage-protein synthesis during seed development in pea (*Pisum sativum* L.). *Biochem. J.* 208: 119-127.

Gayler, K.R., and Sykes, G.E. (1981). β-conglycinins in developing soybean seeds. *Plant Physiol.* 67: 958-961.

Gebhardt, C., Schnebli, V., and King, P.J. (1981). Isolation of biochemical mutants using haploid mesophyll protoplasts of *Hyoscyamus muticus*. II. Auxotrophic and temperature-sensitive clones. *Planta* 153: 81-89.

Genovesi, A.D., and Collins, G.B. (1982). *In vitro* production of haploid plants of corn via anther culture. *Crop Sci.* 22: 1137-1144.

Genovesi, A.D., and Magill, C.W. (1979). Improved rate of callus and green plant production from rice anther culture following cold shock. *Crop Sci.* 19: 662-664.
George, L., and Rao, P.S. (1982). In vitro induction of pollen embryos and plantlets in *Brassica juncea* through anther culture. *Plant Sci. Lett.* 26: 111-116.
Gharyal, P.K., and Maheshwari, S.C. (1981). In vitro differentiation of somatic embryoids in a leguminous tree – *Albizzia lebbeck* L. *Naturwissenschaften* 67: 379.
Gharyal, P.K., Rashid, A., and Maheshwari, S.C. (1983a). Androgenic response from cultured anthers of a leguminous tree, *Cassia siamea* Lam. *Protoplasma* 118: 91-93.
Gharyal, P.K., Rashid, A., and Maheshwari, S.C. (1983b). Production of haploid plantlets in anther cultures of *Albizzia lebbeck* L. *Plant Cell Rep.* 2: 308-309.
Gill, B.S., Waines, J.G., and Sharma, H.C. (1981). Endosperm abortion and the production of viable *Aegilops squarrosa* × *Triticum boeoticum* hybrids by embryo culture. *Plant Sci. Lett.* 23: 181-187.
Gill, M.S., and Bajaj, Y.P.S. (1984). Interspecific hybridization in the genus *Gossypium* through embryo culture. *Euphytica* 33: 305-311.
Giuliano, G., Rosellini, D., and Terzi, M. (1983). A new method for the purification of the different stages of carrot embryoids. *Plant Cell Rep.* 2: 216-218.
Gleddie, S., Keller, W., and Setterfield, G. (1983). Somatic embryogenesis and plant regeneration from leaf explants and cell suspensions of *Solanum melongena* (eggplant). *Can. J. Bot.* 61: 656-666.
Godineau, J.-C. (1966). Ultrastructure du sac embryonnaire du *Crepis tectorum* L.: Les cellules du pôle micropylaire. *Compt. Rend. Acad. Sci. Paris* 263D: 852-855.
Goel, S., Mudgal, A.K., and Gupta, S.C. (1981). In vitro induction of divisions in pollen, callus formation and plantlet regeneration in anthers of *Lobularia maritima*. *Z. Pflanzenphysiol.* 104: 187-191.
Goldberg, R.B., Hoschek, G., Ditta, G.S., and Breidenbach, R.W. (1981a). Developmental regulation of cloned superabundant embryo mRNAs in soybean. *Develop. Biol.* 83: 218-231.
Goldberg, R.B., Hoschek, G., Tam, S.H., Ditta, G.S., and Breidenbach, R.W. (1981b). Abundance, diversity, and regulation of mRNA sequence sets in soybean embryogenesis. *Develop. Biol.* 83: 201-217.
Goldberg, R.B., Hoschek, G., and Vodkin, L.O. (1983). An insertion sequence blocks the expression of a soybean lectin gene. *Cell* 33: 465-475.
Gonzalez-Medina, M., and Bouharmont, J. (1978). Experiments on anther culture in barley. Influence of culture methods on cell proliferation and organ differentiation. *Euphytica* 27: 553-559.
Gosal, S.S., and Bajaj, Y.P.S. (1983). Interspecific hybridization between *Vigna mungo* and *Vigna radiata* through embryo culture. *Euphytica* 32: 129-137.
Gosch, G., Bajaj, Y.P.S., and Reinert, J. (1975). Isolation, culture, and induction of embryogenesis in protoplasts from cell-suspensions of *Atropa belladonna*. *Protoplasma* 86: 405-410.
Graham, T.A., and Gunning, B.E.S. (1970). Localization of legumin and vicilin in bean cotyledon cells using fluorescent antibodies. *Nature (London)* 228: 81-82.
Grambow, H.J., Kao, K.N., Miller, R.A., and Gamborg, O.L. (1972). Cell division and plant development from protoplasts of carrot cell suspension cultures. *Planta* 103: 348-355.
Granatek, C.H., and Cockerline, A.W. (1978). Callus formation versus differentiation of cultured barley embryos: Hormonal and osmotic interactions. *In Vitro* 14: 212-217.

Grant, W.F., Bullen, M.R., and de Nettancourt, D. (1962). The cytogenetics of *Lotus*. I. Embryo-cultured interspecific diploid hybrids closely related to *L. corniculatus*. L. *Can. J. Genet. Cytol.* 4: 105-128.

Gray, D., Ward, J.A., and Steckel, J.R.A. (1984). Endosperm and embryo development in *Daucus carota* L. *J. Expt. Bot.* 35: 459-465.

Gray, D.J., Conger, B.V., and Hanning, G.E. (1984). Somatic embryogenesis in suspension and suspension-derived callus cultures of *Dactylis glomerata*. *Protoplasma* 122: 196-202.

Green, C.E., and Donovan, C.M. (1980). Effect of aspartate-derived amino acids and aminoethyl cysteine on growth of excised mature embryos of maize. *Crop Sci.* 20: 358-362.

Green, C.E., and Phillips, R.L. (1974). Potential selection system for mutants with increased lysine, threonine, and methionine in cereal crops. *Crop Sci.* 14: 827-830.

Green, C.E., and Phillips, R.L. (1975). Plant regeneration from tissue cultures of maize. *Crop Sci.* 15: 417-421.

Gregor, D., Reinert, J., and Matsumoto, H. (1974). Changes in chromosomal proteins from embryo induced carrot cells. *Plant Cell Physiol.* 15: 875-881.

Grewal, S., Sachdeva, U., and Atal, C.K. (1976). Regeneration of plants by embryogenesis from hypocotyl cultures of *Ammi majus* L. *Indian J. Expt. Biol.* 14: 716-717.

Grout, B.W.W., Shelton, K., and Pritchard, H.W. (1983). Orthodox behaviour of oil palm seed and cryopreservation of the excised embryo for genetic conservation. *Ann. Bot.* 52: 381-384.

Grunewaldt, J., and Malepszy, S. (1975). Beobachtungen an Antherenkallus von *Hordeum vulgare* L. *Z. Pflanzenzuchtg.* 75: 55-61.

Gu, S., Gui, Y., and Xu, T. (1984). Effects of physical and chemical factors on induction frequency of the pollen plantlet and changes in starch formation in anthers. *Acta Bot. Sinica* 26: 156-162.

Gu, Z., and Cheng, K. (1982). Studies on induction of pollen plantlets from the anther cultures of lily. *Acta Bot. Sinica* 24: 28-32.

Guha, S., Iyer, R.D., Gupta, N., and Swaminathan, M.S. (1970). Totipotency of gametic cells and the production of haploids in rice. *Curr. Sci.* 39: 174-176.

Guha, S., and Maheshwari, S.C. (1964). *In vitro* production of embryos from anthers of *Datura*. *Nature (London)* 204: 497.

Guha, S., and Maheshwari, S.C. (1966). Cell division and differentiation of embryos in the pollen grains of *Datura in vitro*. *Nature (London)* 212: 97-98.

Guha-Mukherjee, S. (1973). Genotypic differences in the *in vitro* formation of embryoids from rice pollen. *J. Expt. Bot.* 24: 139-144.

Gui, Y., Mu, X., and Xu, T. (1982). Studies on morphological differentiation of endosperm plantlets of Chinese gooseberry *in vitro*. *Acta Bot. Sinica* 24: 216-221.

Guignard, J.L. (1975). Du cotylédon des monocotylédones. *Phytomorphology* 25: 193-200.

Guignard, J.-L. (1984). The development of cotyledon and shoot apex in monocotyledons. *Can. J. Bot.* 62: 1316-1318.

Gupta, S. (1976). Morphogenetic response of haploid callus tissue of *Pisum sativum* (var. B 22). *Indian Agric.* 20: 11-21.

Gupta, S.C., and Babbar, S.B. (1980). Enhancement of plantlet formation in anther cultures of *Datura metel* L. by pre-chilling of buds. *Z. Pflanzenphysiol.* 96: 465-470.

Haccius, B. (1963). Restitution in acidity damaged plant embryos – Regeneration or regulation? *Phytomorphology* 13: 107-115.
Haccius, B. (1978). Question of unicellular origin of non-zygotic embryos in callus cultures. *Phytomorphology* 28: 74-81.
Haccius, B., and Hausner, G. (1976). Ein transplantierbarer embryogener Callus aus Nucellusgewebe von *Cynanchum vincetoxicum* und die Rolle globulärer Vorstadien in der Entwicklungsgeschichte nicht-zygotischer Embryonen. *Protoplasma* 90: 265-282.
Haccius, B., and Reichert, H. (1964). Restitutionserscheinungen an pflanzlichen Meristemen nach Röntgenbestrahlung. II. Adventiv-Embryonie nach Samenbestrahlung von *Eranthis hiemalis*. *Planta* 62: 355-372.
Hall, T.C., Ma, Y., Buchbinder, B.U., Pyne, J.W., Sun, S.M., and Bliss, F.A. (1978). Messenger RNA for G1 protein of French bean seeds: Cell-free translation and product characterization. *Proc. Natl. Acad. Sci. U.S.A.* 75: 3196-3200.
Halperin, W. (1966). Alternative morphogenetic events in cell suspensions. *Am. J. Bot.* 53: 443-453.
Halperin, W., and Jensen, W.A. (1967). Ultrastructural changes during growth and embryogenesis in carrot cell cultures. *J. Ultrastr. Res.* 18: 428-443.
Halperin, W., and Wetherell, D.F. (1964). Adventive embryony in tissue cultures of wild carrot, *Daucus carota*. *Am. J. Bot.* 51: 274-283.
Halperin, W., and Wetherell, D.F. (1965). Ammonium requirement for embryogenesis *in vitro*. *Nature (London)* 205: 519-520.
Hammerschlag, F.A. (1983). Factors influencing the frequency of callus formation among cultured peach anthers. *HortSci.* 18: 210-211.
Hanawa, J. (1960). Late embryogeny and histogenesis in *Sesasum indicum* L. *Bot. Mag. Tokyo* 73: 369-376.
Hang, A., and Franckowiak, J.D. (1984). Cytological analysis of an *Elymus canadensis* × *Secale cereale* hybrid. *J. Hered.* 75: 235-236.
Hannig, E. (1904). Zur Physiologie pflanzlicher Embryonen. *Bot. Ztg.* 62: 45-80.
Hanning, G.E., and Conger, B.V. (1982). Embryoid and plantlet formation from leaf segments of *Dactylis glomerata* L. *Theor. Appl. Genet.* 63: 155-159.
Hardham, A.R. (1976). Structural aspects of the pathways of nutrient flow to the developing embryo and cotyledons of *Pisum sativum* L. *Austr. J. Bot.* 24: 711-721.
Harris, G.P. (1956). Amino acids as sources of nitrogen for the growth of isolated oat embryos. *New Phytol.* 55: 253-268.
Harry, E., Mestre, J., and Guignard, J. (1977). Etude du rôle des divers types cellulaires d'une souche embryogène de cal de carotte dans la différenciation des embryoides. *Compt. Rend. Acad. Sci. Paris* 284D: 2099-2101.
Hasitschka-Jenschke, G. (1959). Bemerkenswerte Kernstrukturen im Endosperm und im Suspensor zweier Helobiae. *Osterr. Bot. Z.* 106: 301-314.
Hasitschka-Jenschke, G. (1962). Notizen über endopolyploide Kerne im Bereich der Samenanlage von Angiospermen. *Osterr. Bot. Z.* 109: 125-137.
Haydu, Z., and Vasil, I.K. (1981). Somatic embryogenesis and plant regeneration from leaf tissues and anthers of *Pennisetum purpureum* Schum. *Theor. Appl. Genet.* 59: 269-273.
He, Z. (Ho, C.), Xu, Z., Xu, S., and Loo, S. (1984). Norleucine-resistant haploid mutant cell line from *Nicotiana tabacum* L. *Acta Biol. Expt. Sinica* 17: 171-176.
Heberle-Bors, E. (1980). Interaction of activated charcoal and iron chelates in anther cultures of *Nicotiana* and *Atropa belladonna*. *Z. Pflanzenphysiol.* 99: 339-347.

Heberle-Bors, E. (1982a). In vitro pollen embryogenesis in Nicotiana tabacum L. and its relation to pollen sterility, sex balance, and floral induction of the pollen donor plants. Planta 156: 396-401.
Heberle-Bors, E. (1982b). On the time of embryogenic pollen grain induction during sexual development of Nicotiana tabacum L. plants. Planta 156: 402-406.
Heberle-Bors, E. (1984). Genotypic control of pollen plant formation in Nicotiana tabacum L. Theor. Appl. Genet. 68: 475-479.
Heberle-Bors, E., and Reinert, J. (1979). Androgenesis in isolated pollen cultures of Nicotiana tabacum: Dependence upon pollen development. Protoplasma 99: 237-245.
Heberle-Bors, E., and Reinert, J. (1980). Isolated pollen cultures and pollen dimorphism. Naturwissenschaften 67: 311.
Heberle-Bors, E., and Reinert, J. (1981). Environmental control and evidence for predetermination of pollen embryogenesis in Nicotiana tabacum pollen. Protoplasma 109: 249-255.
Heirwegh, K.M.G., Banerjee, N., van Nerum, K., and de Langhe, E. (1985). Somatic embryogenesis and plant regeneration in Cichorium intybus L. Plant Cell Rep. 4: 108-111.
Henry, Y., and de Buyser, J. (1981). Float culture of wheat anthers. Theor. Appl. Genet. 60: 77-79.
Hepler, P.K., and Palevitz, B.A. (1974). Microtubules and microfilaments. Annu. Rev. Plant Physiol. 25: 309-362.
Hesemann, C.U. (1980). Haploide Zellen in Kalli aus Antherenkulturen von Vicia faba. Z. Pflanzenzuchtg. 84: 18-22.
Heyser, J.W., and Nabors, M.W. (1982a). Long term plant regeneration, somatic embryogenesis and green spot formation in secondary oat (Avena sativa) callus. Z. Pflanzenphysiol. 107: 153-160.
Heyser, J.W., and Nabors, M.W. (1982b). Regeneration of proso millet from embryogenic calli derived from various plant parts. Crop Sci. 22: 1070-1074.
Hidaka, T., Yamada, Y., and Shichijo, T. (1979). In vitro differentiation of haploid plants by anther culture in Poncirus trifoliata (L.) Raf. Jap. J. Breed. 29: 248-254.
Hinchee, M.A.W., and Fitch, M.M.M. (1984). Culture of isolated microspores of Saccharum spontaneum. Z. Pflanzenphysiol. 113: 305-314.
Ho, W., and Vasil, I.K. (1983a). Somatic embryogenesis in sugarcane (Saccharum officinarum L.). I. The morphology and physiology of callus formation and the ontogeny of somatic embryos. Protoplasma 118: 169-180.
Ho, W., and Vasil, I.K. (1983b). Somatic embryogenesis in sugarcane (Saccharum officinarum L.): Growth and plant regeneration from embryogenic cell suspension cultures. Ann. Bot. 51: 719-726.
Hòmes, J.L.A. (1968). Influence de la concentration en glucose sur le développement et la différenciation d'embryons formés dans des tissus de carotte cultivés in vitro. In: Les cultures de tissus de plantes, pp. 49-60. Paris: Centre National de la Recherche Scientifique.
Honma, S. (1955). A technique for artificial culturing of bean embryos. Proc. Am. Soc. Hort. Sci. 65: 405-408.
Horner, M., McComb, J.A., McComb, A.J., and Street, H.E. (1977). Ethylene production and plantlet formation by Nicotiana anthers cultured in the presence and absence of charcoal. J. Expt. Bot. 28: 1365-1372.
Horner, M., and Mott, R.L. (1979). The frequency of embryogenic pollen grains is

not increased by *in vitro* anther culture in *Nicotiana tabacum* L. *Planta* 147: 156-158.
Horner, M., and Pratt, M.L. (1979). Amino acid analysis of *in vivo* and androgenic anthers of *Nicotiana tabacum*. *Protoplasma* 98: 279-282.
Horner, M., and Street, H.E. (1978a). Problems encountered in the culture of isolated pollen of a Burley cultivar of *Nicotiana tabacum*. *J. Expt. Bot.* 29: 217-226.
Horner, M., and Street, H.E. (1978b). Pollen dimorphism–Origin and significance in pollen plant formation by anther culture. *Ann. Bot.* 42: 763-771.
Hsu, F.C. (1979). Abscisic acid accumulation in developing seeds of *Phaseolus vulgaris* L. *Plant Physiol.* 63: 552-556.
Hu, C. (1976). Light-mediated inhibition of *in vitro* development of rudimentary embryos of *Ilex opaca*. *Am. J. Bot.* 63: 651-656.
Hu, C., Rogalski, F., and Ward, C. (1979). Factors maintaining *Ilex* rudimentary embryos in the quiescent state and the ultrastructural changes during *in vitro* activation. *Bot. Gaz.* 140: 272-279.
Hu, C.Y. (1975). *In vitro* culture of rudimentary embryos of eleven *Ilex* species. *J. Am. Soc. Hort. Sci.* 100: 221-225.
Hu, C.Y., and Sussex, I.M. (1971). *In vitro* development of embryoids on cotyledons of *Ilex aquifolium*. *Phytomorphology* 21: 103-107.
Hu, S., Zhu, C., and Zee, S.Y. (1983). Transfer cells in suspensor and endosperm during early embryogeny of *Vigna sinensis*. *Acta Bot. Sinica* 25: 1-7.
Huang, B., and Sunderland, N. (1982). Temperature–stress pretreatment in barley anther culture. *Ann. Bot.* 49: 77-88.
Huang, P.C., and Paddock, E.F. (1962). The time and site of the semidominant lethal action of "Wo" in *Lycopersicon esculentum*. *Am. J. Bot.* 49: 388-393.
Huang, Q., Yang, H., and Zhou, C. (1982). Embryological observations on ovary culture of unpollinated young flowers in *Hordeum vulgare* L. *Acta Bot. Sinica* 24: 295-300.
Hughes, K.W., Bell, S.L., and Caponetti, J.D. (1975). Anther-derived haploids of the African violet. *Can. J. Bot.* 53: 1442-1444.
Idzikowska, K. (1981). Mitochondria during androgenesis in *Hordeum vulgare*. *Acta Soc. Bot. Poloniae* 50: 359-366.
Idzikowska, K., Ponitka, A., and Młodzianowski, F. (1982). Pollen dimorphism and androgenesis in *Hordeum vulgare*. *Acta Soc. Bot. Poloniae* 51: 153-156.
Ihle, J.N., and Dure, L. III. (1970). Hormonal regulation of translation inhibition requiring RNA synthesis. *Biochem. Biophys. Res. Comm.* 38: 995-1001.
Ihle, J.N., and Dure, L.S. III. (1972). The developmental biochemistry of cottonseed embryogenesis and germination. III. Regulation of the biosynthesis of enzymes utilized in germination. *J. Biol. Chem.* 247: 5048-5055.
Imamura, J., and Harada, H. (1980a). Effects of abscisic acid and water stress on the embryo and plantlet formation in anther culture of *Nicotiana tabacum* cv. Samsun. *Z. Pflanzenphysiol.* 100: 285-289.
Imamura, J., and Harada, H. (1980b). Stimulatory effect of reduced atmospheric pressure on pollen embryogenesis. *Naturwissenschaften* 67: 357-358.
Imamura, J., and Harada, H. (1981). Stimulation of tobacco pollen embryogenesis by anaerobic treatments. *Z. Pflanzenphysiol.* 103: 259-263.
Imamura, J., Okabe, E., Kyo, M., and Harada, H. (1982). Embryogenesis and plantlet formation through direct culture of isolated pollen of *Nicotiana tabacum* cv. Samsun and *Nicotiana rustica* cv. Rustica. *Plant Cell Physiol.* 23: 713-716.
Inomata, N. (1978). Production of interspecific hybrids in *Brassica campestris* × *B.*

oleracea by culture *in vitro* of excised ovaries. I. Development of excised ovaries in the crosses of various cultivars. *Jap. J. Genet.* 53: 161-173.

Irikura, Y. (1975). Induction of haploid plants by anther culture in tuber-bearing species and interspecific hybrids of *Solanum*. *Potato Res.* 18: 133-140.

Isaia, A., and Bulard, C. (1978). Relative levels of some bound and free gibberellins in dormant and after-ripened embryos of *Pyrus malus* cv. Golden Delicious. *Z. Pflanzenphysiol.* 90: 409-414.

Iyer, R.D., and Raina, S.K. (1972). The early ontogeny of embryoids and callus from pollen and subsequent organogenesis in anther cultures of *Datura metel* and rice. *Planta* 104: 146-156.

Jacobsen, E., and Sopory, S.K. (1978). The influence and possible recombination of genotypes on the production of microspore embryoids in anther cultures of *Solanum tuberosum* and dihaploid hybrids. *Theor. Appl. Genet.* 52: 119-123.

Jalouzout, M.-F. (1975). Aspects ultrastructuraux du sac embryonnaire d'*Oenothera lamarckiana*. *Compt. Rend. Acad. Sci. Paris* 281D: 1305-1308.

Janick, J., Wright, D.C., and Hasegawa, P.M. (1982). *In vitro* production of cacao seed lipids. *J. Am. Soc. Hort. Sci.* 107: 919-922.

Jarvis, B.C. (1979). The influence of cotyledons on embryonic axes during induction of dormancy in *Corylus avellana*. *Physiol. Plant.* 45: 363-366.

Jarvis, B.C., and Shannon, P.R.M. (1981). Changes in poly(A) RNA metabolism in relation to storage and dormancy-breaking of hazel seed. *New Phytol.* 88: 31-40.

Jarvis, B.C., and Wilson, D. (1977). Gibberellin effects within hazel (*Corylus avellana* L.) seeds during the breaking of dormancy. I. A direct effect of gibberellin on the embryonic axis. *New Phytol.* 78: 397-401.

Jarvis, B.C., and Wilson, D.A. (1978). Factors influencing growth of embryonic axes from dormant seeds of hazel (*Corylus avellana* L.). *Planta* 138: 189-191.

Jarvis, B.C., Wilson, D.A., and Fowler, M.W. (1978). Growth of isolated embryonic axes from dormant seeds of hazel (*Corylus avellana* L.). *New Phytol.* 80: 117-123.

Jean, R., Linder, R., Hoffmann, F., Thomas, E., and Wenzel, G. (1976). Comportement *in vitro* de microspores d'*Oenothera coronifera* Renner. *Compt. Rend. Acad. Sci. Paris* 283D: 781-784.

Jensen, C.J. (1977). Monoploid production by chromosome elimination. In: *Applied and Fundamental Aspects of Plant Cell, Tissue, and Organ Culture*, ed. J. Reinert and Y.P.S. Bajaj, pp. 299-330. Berlin: Springer-Verlag.

Jensen, W.A. (1964). Cell development during plant embryogenesis. In: *Meristems and Differentiation*, Brookhaven Symp. Biol. 16: 179-202.

Jensen, W.A. (1965). The ultrastructure and composition of the egg and central cell of cotton. *Am. J. Bot.* 52: 781-797.

Jensen, W.A. (1968). Cotton embryogenesis: The zygote. *Planta* 79: 346-366.

Johansen, D.A. (1950). *Plant Embryology. Embryogeny of the Spermatophyta.* Waltham: Chronica Botanica Co.

Johansson, L. (1983). Effects of activated charcoal in anther cultures. *Physiol. Plant.* 59: 397-403.

Johansson, L., Andersson, B., and Eriksson, T. (1982). Improvement of anther culture technique: Activated charcoal bound in agar medium in combination with liquid medium and elevated CO_2 concentration. *Physiol. Plant.* 54: 24-30.

Johansson, L., and Eriksson, T. (1977). Induced embryo formation in anther cultures of several *Anemone* species. *Physiol. Plant.* 40: 172-174.

Johri, B.M. (ed.). (1982). *Experimental Embryology of Vascular Plants.* Berlin: Springer-Verlag.

Johri, B.M. (ed.). (1984). *Embryology of Angiosperms*. Berlin: Springer-Verlag.
Jones, L.H. (1974a). Long term survival of embryoids of carrot (*Daucus carota* L.). *Plant Sci. Lett.* 2: 221-224.
Jones, L.H. (1974b). Factors influencing embryogenesis in carrot cultures (*Daucus carota* L.). *Ann. Bot.* 38: 1077-1088.
Jones, P.A. (1977). Development of the quiescent center in maturing embryonic radicles of pea (*Pisum sativum* L. cv. Alaska). *Planta* 135: 233-240.
Jones, R.A., Larkins, B.A., and Tsai, C.Y. (1977a). Storage protein synthesis in maize. II. Reduced synthesis of a major zein component by the *opaque*-2 mutant of maize. *Plant Physiol.* 59: 525-529.
Jones, R.A., Larkins, B.A., and Tsai, C.Y. (1977b). Storage protein synthesis in maize. III. Developmental changes in membrane-bound polyribosome composition and *in vitro* protein synthesis of normal and *opaque*-2 maize. *Plant Physiol.* 59: 733-737.
Jordan, M. (1974). Multizelluläre Pollen bei *Prunus avium* nach *in-vitro*-Kultur. *Z. Pflanzenzuchtg.* 71: 358-363.
Joshi, P.C., and Johri, B.M. (1972). *In vitro* growth of ovules of *Gossypium hirsutum*. *Phytomorphology* 22: 195-209.
Juncosa, A.M. (1982). Developmental morphology of the embryo and seedling of *Rhizophora mangle* L. (Rhizophoracaea). *Am. J. Bot.* 69: 1599-1611.
Juncosa, A.M. (1984a). Embryogenesis and seedling development in *Cassipourea elliptica* (Sw.) Poir. (Rhizophoraceae). *Am. J. Bot.* 71: 170-179.
Juncosa, A.M. (1984b). Embryogenesis and developmental morphology of the seedling in *Bruguiera exaristata* Ding Hou (Rhizophoraceae). *Am. J. Bot.* 71: 180-191.
Kamada, H., and Harada, H. (1979a). Studies on the organogenesis in carrot tissue cultures. I. Effects of growth regulators on somatic embryogenesis and root formation. *Z. Pflanzenphysiol.* 91: 255-266.
Kamada, H., and Harada, H. (1979b). Studies on the organogenesis in carrot tissue cultures. II. Effects of amino acids and inorganic nitrogenous compounds on somatic embryogenesis. *Z. Pflanzenphysiol.* 91: 453-463.
Kamada, H., and Harada, H. (1981). Changes in the endogenous level and effects of abscisic acid during somatic embryogenesis of *Daucus carota* L. *Plant Cell Physiol* 22: 1423-1429.
Kameya, T., and Uchimiya, H. (1972). Embryoids derived from isolated protoplasts of carrot. *Planta* 103: 356-360.
Kao, K.N., and Michayluk, M.R. (1980). Plant regeneration from mesophyll protoplasts of alfalfa. *Z. Pflanzenphysiol.* 96: 135-141.
Kao, K.N., and Michayluk, M.R. (1981). Embryoid formation in alfalfa cell suspension cultures from different plants. *In Vitro* 17: 645-648.
Kaplan, D.R. (1969). Seed development in *Downingia*. *Phytomorphology* 19: 253-278.
Karssen, C.M., Brinkhorst-van der Swan, D.L.C., Breekland, A.E., and Koornneef, M. (1983). Induction of dormancy during seed development by endogenous abscisic acid: Studies on abscisic acid deficient genotypes of *Arabidopsis thaliana* (L.) Heynh. *Planta* 157: 158-165.
Kasha, K.J., and Kao, K.N. (1970). High frequency haploid production in barley (*Hordeum vulgare* L.). *Nature (London)* 225: 874-876.
Kato, H. (1968). The serial observations of the adventive embryogenesis in the microculture of carrot tissue. *Sci. Papers Coll. Gen. Edn. Univ. Tokyo* 18: 191-197.
Kato, H., and Takeuchi, M. (1963). Morphogenesis *in vitro* starting from single cells of carrot root. *Plant Cell Physiol.* 4: 243-245.

Kato, H., and Takeuchi, M. (1966). Embryogenesis from the epidermal cells of carrot hypocotyl. *Sci. Papers Coll. Gen. Edn. Univ. Tokyo* 16: 245-254.

Kavathekar, A.K., and Ganapathy, P.S. (1973). Embryoid differentiation in *Eschscholzia californica. Curr. Sci.* 42: 671-673.

Kavathekar, A.K., Ganapathy, P.S., and Johri, B.M. (1977). Chilling induces development of embryoids into plantlets in *Eschscholzia. Z. Pflanzenphysiol.* 81: 358-363.

Kazimierska, E.M. (1978). Embryological studies of cross compatibility in the genus *Trifolium* L. II. Fertilization, development of embryo and endosperm in crossing *T. repens* L. with *T. medium* L. *Genet. Polonica* 19: 15-24.

Keim, W.F. (1953). Interspecific hybridization in *Trifolium* using embryo culture techniques. *Agron. J.* 45: 601-606.

Keller, W.A., and Armstrong, K.C. (1978). High frequency production of microspore-derived plants from *Brassica napus* anther cultures. *Z. Pflanzenzuchtg.* 80: 100-108.

Keller, W.A., and Armstrong, K.C. (1979). Stimulation of embryogenesis and haploid production in *Brassica campestris* anther cultures by elevated temperature treatments. *Theor. Appl. Genet.* 55: 65-67.

Keller, W.A., Rajhathy, T., and Lacapra, J. (1975). In vitro production of plants from pollen in *Brassica campestris. Can. J. Genet. Cytol.* 17: 655-666.

Kent, N., and Brink, R.A. (1947). Growth *in vitro* of immature *Hordeum* embryos. *Science* 106: 547-548.

Keyes, G.J., Collins, G.B., and Taylor, N.L. (1980). Genetic variation in tissue cultures of red clover. *Theor. Appl. Genet.* 58: 265-271.

Khan, S.K., and Ghosh, P.D. (1983). *In vitro* induction of androgenesis and organogenesis in *Cicer arietinum* L. *Curr. Sci.* 52: 891-893.

Khavkin, E.E., Misharin, S.I., Ivanov, V.N., and Danovich, K.N. (1977). Embryonal antigens in maize caryopses: The temporal order of antigen accumulation during embryogenesis. *Planta* 135: 225-231.

King, R.W. (1976). Abscisic acid in developing wheat grains and its relationship to grain growth and maturation. *Planta* 132: 43-51.

Klimaszewska, K., and Keller, W.A. (1983). The production of haploids from *Brassica hirta* Moench (*Sinapis alba* L.) anther cultures. *Z. Pflanzenphysiol.* 109: 235-241.

Kochba, J., Ben-Hayyim, G., Spiegel-Roy, P., Saad, S., and Neumann, H. (1982). Selection of stable salt-tolerant callus cell lines and embryos in *Citrus sinensis* and *C. aurantium. Z. Pflanzenphysiol.* 106: 111-118.

Kochba, J., and Button, J. (1974). The stimulation of embryogenesis and embryoid development in habituated ovular callus from "Shamouti" orange (*Citrus sinensis*) as affected by tissue age and sucrose concentration. *Z. Pflanzenphysiol.* 73: 415-421.

Kochba, J., Lavee, S., and Spiegel-Roy, P. (1977). Differences in peroxidase activity and isoenzymes in embryogenic and non-embryogenic 'Shamouti' orange ovular callus lines. *Plant Cell Physiol.* 18: 463-467.

Kochba, J., and Spiegel-Roy, P. (1973). Effect of culture media on embryoid formation from ovular callus of 'Shamouti' orange (*Citrus sinensis*). *Z. Pflanzenzuchtg.* 69: 156-162.

Kochba, J., and Spiegel-Roy, P. (1977a). Embryogenesis in gamma-irradiated habituated ovular callus of the "Shamouti" orange as affected by auxin and by tissue age. *Environ. Expt. Bot.* 17: 151-159.

Kochba, J., and Spiegel-Roy, P. (1977b). The effects of auxins, cytokinins and inhibitors on embryogenesis in habituated ovular callus of the "Shamouti"orange (*Citrus sinensis*). *Z. Pflanzenphysiol.* 81: 283-288.
Kochba, J., Spiegel-Roy, P., Neumann, H., and Saad, S. (1978). Stimulation of embryogenesis in *Citrus* ovular callus by ABA, ethephon, CCC and Alar and its suppression by GA_3. *Z. Pflanzenphysiol.* 89: 427-432.
Kochba, J., Spiegel-Roy, P., Neumann, H., and Saad, S. (1982). Effect of carbohydrates on somatic embryogenesis in subcultured nucellar callus of *Citrus* cultivars. *Z. Pflanzenphysiol.* 105: 359-368.
Kochba, J., Spiegel-Roy, P., Saad, S., and Neumann, H. (1978). Stimulation of embryogenesis in *Citrus* tissue culture by galactose. *Naturwissenschaften* 65: 261.
Kohlenbach, H.W., Wenzel, G., and Hoffmann, F. (1982). Regeneration of *Brassica napus* plantlets in cultures from isolated protoplasts of haploid stem embryos as compared with leaf protoplasts. *Z. Pflanzenphysiol.* 105: 131-142.
Kohlenbach, H.W., and Wernicke, W. (1978). Investigations on the inhibitory effect of agar and the function of active carbon in anther culture. *Z. Pflanzenphysiol.* 86: 463-472.
Konar, R.N., and Nataraja, K. (1965). Experimental studies in *Ranunculus sceleratus* L. Development of embryos from the stem epidermis. *Phytomorphology* 15: 132-137.
Konar, R.N., Thomas, E., and Street, H.E. (1972a). Origin and structure of embryoids arising from epidermal cells of the stem of *Ranunculus sceleratus* L. *J. Cell Sci.* 11: 77-93.
Konar, R.N., Thomas, E., and Street, H.E. (1972b). The diversity of morphogenesis in suspension cultures of *Atropa belladonna* L. *Ann. Bot.* 36: 249-258.
Kononowicz, H., and Janick, J. (1984). Response of embryogenic callus of *Theobroma cacao* L. to gibberellic acid and inhibitors of gibberellic acid synthesis. *Z. Pflanzenphysiol.* 113: 359-366.
Kononowicz, H., Kononowicz, A.K., and Janick, J. (1984). Asexual embryogenesis via callus of *Theobroma cacao* L. *Z. Pflanzenphysiol.* 113: 347-358.
Kott, L.S., and Kasha, K.J. (1984). Initiation and morphological development of somatic embryoids from barley cell cultures. *Can. J. Bot.* 62: 1245-1249.
Koul, A.K., and Karihaloo, J.L. (1977). *In vivo* embryoids from anthers of *Narcissus biflorus*. Curt. *Euphytica* 26: 97-102.
Kowyama, Y. (1983). Cell-cycle dependency of radiosensitivity and mutagenesis in fertilized egg cells of rice, *Oryza sativa* L. 1. Autoradiographic determination of the first DNA synthetic phase. *Theor. Appl. Genet.* 65: 303-308.
Krikorian, A.D. (1982). Cloning higher plants from aseptically cultured tissues and cells. *Biol. Rev.* 57: 151-218.
Krul, W.R., and Worley, J.F. (1977). Formation of adventitious embryos in callus cultures of 'Seyval', a French hybrid grape. *J. Am. Soc. Hort. Sci.* 102: 360-363.
Krumbiegel-Schroeren, G., Finger, J., Schroeren, V., and Binding, H. (1984). Embryoid formation and plant regeneration from callus of *Secale cereale*. *Z. Pflanzenzuchtg.* 92: 89-94.
Kruse, A. (1973). *Hordeum* × *Triticum* hybrids. *Hereditas* 73: 157-161.
Kruse, A. (1974). An *in vivo/vitro* embryo culture technique. *Hereditas* 77: 219-224.
Ku, M., Cheng, W., Kuo, L., Kuan, Y., An, H., and Huang, C. (1978). Induction factors and morpho-cytological characteristics of pollen-derived plants in maize (*Zea mays*). In: *Proceedings of Symposium on Plant Tissue Culture*, pp. 35-42. Peking: Science Press.

Kubicki, B., Telezynska, J., and Milewska-Pawliczuk, E. (1975). Induction of embryoid development from apple pollen grains. *Acta Soc. Bot. Poloniae* 44: 631-635.
Lacadena, J. (1974). Spontaneous and induced parthenogenesis and androgenesis. In: *Haploids in Higher Plants. Advances and Potential*, ed. K. J. Kasha, pp. 13-32. Guelph: University of Guelph.
La Cour, L.F. (1949). Nuclear differentiation in the pollen grain. *Heredity* 3: 319-337.
Laibach, F. (1925). Das Taubwerden von Bastardsamen und die künstliche Aufzucht früh absterbender Bastardembryonen. *Z. Bot.* 17: 417-459.
Lanaud, C. (1981). Production de plantules de *C. canephora* par embryogenèse somatique réalisée à partir de culture *in vitro* d'ovules. *Cafe Cacao The* 25: 231-236.
Lang, H., and Kohlenbach, H.W. (1975). Morphogenese in Kulturen isolierter Mesophyllzellen von *Macleaya cordata*. In: *Form, Structure and Function in Plants*, ed. H.Y. Mohan Ram, J.J. Shah and C.K. Shah, pp. 125-133. Meerut: Sarita Prakashan.
Laroche-Raynal, M., Aspart, L., Delseny, M., and Penon, P. (1984). Characterization of radish mRNA at three developmental stages. *Plant Sci. Lett.* 35: 139-146.
Lazar, M.D., Baenziger, P.S., and Schaeffer, G.W. (1984). Combining abilities and heritability of callus formation and plantlet regeneration in wheat (*Triticum aestivum* L.) anther cultures. *Theor. Appl. Genet.* 68: 131-134.
Lee, D.W., and Dougall, D.K. (1973). Electrophoretic variation in glutamate dehydrogenase and other isozymes in wild carrot cells cultured in the presence and absence of 2,4-dichlorophenoxyacetic acid. *In Vitro* 8: 347-352.
Leelavathi, S., Reddy, V.S., and Sen, S.K. (1984). Somatic cell genetic studies in *Brassica* species. I. High frequency production of haploid plants in *Brassica alba* (L.) H. f. & T. *Plant Cell Rep.* 3: 102-105.
Lersten, N.R. (1983). Suspensors in Leguminosae. *Bot. Rev.* 49: 233-257.
Li, L., and Kohlenbach, H.W. (1982). Somatic embryogenesis in quite a direct way in cultures of mesophyll protoplasts of *Brassica napus* L. *Plant Cell Rep.* 1: 209-211.
Li, M. (1982). Pollen dimorphism and androgenesis of *Paeonia in vivo*. *Acta Bot. Sinica* 24: 17-20.
Liang, G.H., Sangduen, N., Heyne, E.G., and Sears, R.G. (1982). Polyhaploid production through anther culture in common wheat. *J. Hered.* 73: 360-364.
Lichter, R.(1982). Induction of haploid plants from isolated pollen of *Brassica napus*. *Z. Pflanzenphysiol.* 105: 427-434.
Lima-de-Faria, A., Pero, R., Avanzi, S., Durante, M., Ståhle, U., D'Amato, F., and Granström, H. (1975). Relation between ribosomal RNA genes and the DNA satellites of *Phaseolus coccineus*. *Hereditas* 79: 5-19.
Litz, R.E. (1984). *In vitro* somatic embryogenesis from callus of jaboticaba, *Myrciaria cauliflora*. *HortSci*. 19: 62-64.
Litz, R.E., and Conover, R.A. (1982). *In vitro* somatic embryogenesis and plant regeneration from *Carica papaya* L. ovular callus. *Plant Sci. Lett.* 26: 153-158.
Litz, R.E., and Conover, R.A. (1983). High-frequency somatic embryogenesis from *Carica* suspension cultures. *Ann. Bot.* 51: 683-686.
Litz, R.E., Knight, R.J., Jr., and Gazit, S. (1984). *In vitro* somatic embryogenesis from *Mangifera indica* L. callus. *Scientia Hort.* 22: 233-240.
Litz, R.E., Knight, R.L., and Gazit, S. (1982). Somatic embryos from cultured ovules of polyembryonic *Mangifera indica* L. *Plant Cell Rep.* 1: 264-266.
Liu, J.R., and Cantliffe, D.J. (1984). Somatic embryogenesis and plant regenera-

tion in tissue cultures of sweet potato (*Ipomea batatas* Poir.). *Plant Cell Rep.* 3: 112-115.
Liu, J.R., Sink, K.C., and Dennis, F.G., Jr. (1983). Adventive embryogenesis from leaf explants of apple seedlings. *HortSci.* 18: 871-873.
Loh, C.-S., and Ingram, D.S. (1983). The response of haploid secondary embryoids and secondary embryogenic tissues of winter oilseed rape to treatment with colchicine. *New Phytol.* 95: 359-366.
Long, S.R., Dale, R.M.K., and Sussex, I.M. (1981). Maturation and germination of *Phaseolus vulgaris* embryonic axes in culture. *Planta* 153: 405-415.
Lorenzi, R., Bennici, A., Cionini, P.G., Alpi, A., and D'Amato, F. (1978). Embryo-suspensor relations in *Phaseolus coccineus:* Cytokinins during seed development. *Planta* 143: 59-62.
Lörz, H., Potrykus, I., and Thomas, E. (1977). Somatic embryogenesis from tobacco protoplasts. *Naturwissenschaften* 64: 439.
Lu, C., Chandler, S.F., and Vasil, I.K. (1984). Somatic embryogenesis and plant regeneration from cultured immature embryos of rye (*Secale cereale* L.). *J. Plant Physiol.* 115: 237-244.
Lu, C., and Vasil, I.K. (1981a). Somatic embryogenesis and plant regeneration from leaf tissues of *Panicum maximum* Jacq. *Theor. Appl. Genet.* 59: 275-280.
Lu, C., and Vasil, I.K. (1981b). Somatic embryogenesis and plant regeneration from freely-suspended cells and cell groups of *Panicum maximum* Jacq. *Ann. Bot.* 48: 543-548.
Lu, C., and Vasil, I.K. (1982). Somatic embryogenesis and plant regeneration in tissue cultures of *Panicum maximum* Jacq. *Am. J. Bot.* 69: 77-81.
Lu, C., Vasil, V., and Vasil, I.K. (1981). Isolation and culture of protoplasts of *Panicum maximum* Jacq. (Guinea grass): Somatic embryogenesis and plantlet formation. *Z. Pflanzenphysiol.* 104: 311-318.
Lu, C., Vasil, V., and Vasil, I.K. (1983). Improved efficiency of somatic embryogenesis and plant regeneration in tissue cultures of maize (*Zea mays* L.). *Theor. Appl. Genet.* 66: 285-289.
Lu, D.Y., Davey, M.R., and Cocking, E.C. (1983). A comparison of the cultural behaviour of protoplasts from leaves, cotyledons and roots of *Medicago sativa*. *Plant Sci. Lett.* 31: 87-99.
Lu, W. (1978). Study on the induction of haploid plants derived from isolated pollen grains of *Nicotiana tabacum* L. cultured *in vitro*. *Scientia Sinica* 21: 669-675.
Lupi, C., Bennici, A., and Gennai, D. (1985). *In vitro* culture of *Bellevalia romana* (L.) Rchb. I. Plant regeneration through adventitious shoots and somatic embryos. *Protoplasma* 125: 185-189.
Maddock, S.E., Lancaster, V.A., Risiott, R., and Franklin, J. (1983). Plant regeneration from cultured immature embryos and inflorescences of 25 cultivars of wheat (*Triticum aestivum*). *J. Expt. Bot.* 34: 915-926.
Maheshwari, P. (1950). *An Introduction to the Embryology of Angiosperms.* New York: McGraw-Hill Book Co.
Maheshwari, P. (ed.). (1963). *Recent Advances in the Embryology of Angiosperms.* Delhi: International Society of Plant Morphologists.
Maheshwari, P., and Chopra, R.N. (1955). The structure and development of the ovule and seed of *Opuntia dillenii* Haw. *Phytomorphology* 5: 112-122.
Maheshwari, P., and Rangaswamy, N.S. (1958). Polyembryony and *in vitro* culture of embryos of *Citrus* and *Mangifera*. *Indian J. Hort.* 15: 275-282.
Maheshwari, P., and Sachar, R.C. (1963). Polyembryony. In: *Recent Advances in*

the *Embryology of Angiosperms,* ed. P. Maheshwari, pp. 265-296. Delhi: International Society of Plant Morphologists.

Maheshwari, S.C., Rashid, A., and Tyagi, A.K. (1982). Haploids from pollen grains – Retrospect and prospect. *Am. J. Bot.* 69: 865-879.

Maheshwari, S.C., Tyagi, A.K., Malhotra, K., and Sopory, S.K. (1980). Induction of haploidy from pollen grains in angiosperms – The current status. *Theor. Appl. Genet.* 58: 193-206.

Maheswaran, G., and Williams, E.G. (1984). Direct somatic embryoid formation on immature embryos of *Trifolium repens, T. pratense* and *Medicago sativa,* and rapid clonal propagation of *T. repens. Ann. Bot.* 54: 201-211.

Mahlberg, P.G. (1960). Embryogeny and histogenesis in *Nerium oleander* L. I. Organization of primary meristematic tissues. *Phytomorphology* 10: 118-131.

Malhotra, K., and Maheshwari, S.C. (1977). Enhancement by cold treatment of pollen embryoid development in *Petunia hybrida. Z. Pflanzenphysiol.* 85: 177-180.

Maliga, P., Lázár, G., Sváb, Z., and Nagy, F. (1976). Transient cycloheximide resistance in a tobacco cell line. *Mol. Gen. Genet.* 149: 267-271.

Maliga, P., Márton, L., and Sz.-Breznovits, A. (1973). 5-Bromodeoxyuridine-resistant cell lines from haploid tobacco. *Plant Sci. Lett.* 1: 119-121.

Maliga, P., Sz.-Breznovits, A., and Márton, L. (1973). Streptomycin-resistant plants from callus cultures of haploid tobacco. *Nature (London), New Biol.* 244: 29-30.

Malik, C.P., Singh, M., and Thapar, N. (1976). Physiology of sexual reproduction. IV. Histochemical characteristics of suspensor of *Brassica campestris. Phytomorphology* 26: 384-389.

Malik, C.P., Vermani, S., and Bhatia, D.S. (1976). Physiology of sexual reproduction. III. Histochemical characteristics of suspensor during embryo development in *Brassica campestris* Linn. var., Sarson. *Acta Histochem.* 57: 178-182.

Manteuffel, R., Müntz, K., Püchel, M., and Scholz, G. (1976). Phase-dependent changes of DNA, RNA and protein accumulation during the ontogenesis of broad bean seeds (*Vicia faba* L. var. minor). *Biochem. Physiol. Pflanzen* 169: 595-605.

Marinos, N.G. (1970a). Embryogenesis of the pea (*Pisum sativum*). I. The cytological environment of the developing embryo. *Protoplasma* 70: 261-279.

Marinos, N.G. (1970b). Embryogenesis of the pea (*Pisum sativum*). II. An unusual type of plastid in the suspensor cell. *Protoplasma* 71: 227-233.

Marsden, M.P.F., and Meinke, D.W. (1984). Abnormal development of the suspensor in an embryo-lethal mutant of *Arabidopsis thaliana. Am. J. Bot.* 71: (No. 5, Part 2), 15 (abstract).

Martin, A.C. (1946). The comparative internal morphology of seeds. *Am. Midl. Naturl.* 36: 513-660.

Masuda, K., Kikuta, Y., and Okazawa, Y. (1984). Embryogenesis and ribosomal DNA in carrot cell suspensions cultured *in vitro. Plant Sci. Lett.* 33: 23-29.

Masuda, K., Koda, Y., and Okazawa, Y. (1977). Callus formation and embryogenesis of endosperm tissues of parsley seed cultured on hormone-free medium. *Physiol. Plant.* 41: 135-138.

Mathan, D.S., and Jenkins, J.A. (1960). Chemically induced phenocopy of a tomato mutant. *Science* 131: 36-37.

Mathan, D.S., and Jenkins, J.A. (1962). A morphogenetic study of lanceolate, a leaf-shape mutant in the tomato. *Am. J. Bot.* 49: 504-514.

Matsubara, S. (1962). Studies on a growth promoting substance, "embryo factor,"

necessary for the culture of young embryos of *Datura tatula in vitro*. *Bot. Mag. Tokyo* 75: 10-18.
Matsubara, S. (1964). Effect of nitrogen compounds on the growth of isolated young embryos of *Datura*. *Bot. Mag. Tokyo* 77: 253-259.
Matsumoto, H., Gregor, D., and Reinert, J. (1975). Changes in chromatin of *Daucus carota* cells during embryogenesis. *Phytochemistry* 14: 41-47.
Maze, J., and Lin, S. (1975). A study of the mature megagametophyte of *Stipa elmeri*. *Can. J. Bot.* 53: 2958-2977.
McComb, J.A., and McComb, A.J. (1977). The cytology of plantlets derived from cultured anthers of *Nicotiana sylvestris*. *New Phytol.* 79: 679-688.
McDaniel, J.K., Conger, B.V., and Graham, E.T. (1982). A histological study of tissue proliferation, embryogenesis, and organogenesis from tissue cultures of *Dactylis glomerata* L. *Protoplasma* 110: 121-128.
McDonnell, R.E., and Conger, B.V. (1984). Callus induction and plantlet formation from mature embryo explants of Kentucky bluegrass. *Crop Sci.* 24: 573-578.
McWilliam, A.A., Smith, S.M., and Street, H.E. (1974). The origin and development of embryoids in suspension cultures of carrot (*Daucus carota*). *Ann. Bot.* 38: 243-250.
Meinke, D.W. (1982). Embryo-lethal mutants of *Arabidopsis thaliana:* Evidence for gametophytic expression of the mutant genes. *Theor. Appl. Genet.* 63: 381-386.
Meinke, D.W., Chen, J., and Beachy, R.N. (1981). Expression of storage-protein genes during soybean seed development. *Planta* 153: 130-139.
Meinke, D.W., and Sussex, I.M. (1979). Isolation and characterization of six embryo-lethal mutants of *Arabidopsis thaliana*. *Develop. Biol.* 72: 62-72.
Mericle, L.W., and Mericle, R.P. (1970). Nuclear DNA complement in young proembryos of barley. *Mutation Res.* 10: 515-518.
Mestre, J.-C. (1967). La signification phylogénétique de l'embryogénie. *Rev. Gen. Bot.* 74: 273-322.
Miao, S., Kuo, C., Kwei, Y., Sun, A., Ku, S., Lu, W., Wang, Y., Chen, M., Wu, M., and Hang, L. (1978). Induction of pollen plants of maize and observations on their progeny. In: *Proceedings of Symposium on Plant Tissue Culture,* pp. 23-33. Peking: Science Press.
Michellon, R., Hugard, J., and Jonard, R. (1974). Sur l'isolement de colonies tissulaires de pêcher (*Prunus persica* Batsch, cultivars Dixired et Nectared IV)'et d'amandier (*Prunus amygdalus* Stokes, cultivar Ai) à partir d'anthères cultivées in vitro. *Compt. Rend. Acad. Sci. Paris* 278D: 1719-1722.
Miflin, B.J. (1969). The inhibitory effects of various amino acids on the growth of barley seedlings. *J. Expt. Bot.* 20: 810-819.
Mii, M. (1976). Relationships betwen anther browning and plantlet formation in anther culture of *Nicotiana tabacum* L. *Z. Pflanzenphysiol.* 80: 206-214.
Mii, M. (1980). Effect of pollen degeneration on the pollen embryogenesis in anther culture of *Nicotiana tabacum* L. *Z. Pflanzenphysiol.* 99: 349-355.
Miller, H.A., and Wetmore, R.H. (1945). Studies in the developmental anatomy of *Phlox drummondii* Hook. I. The embryo. *Am. J. Bot.* 32: 588-599.
Millerd, A., Simon, M., and Stern, H. (1971). Legumin synthesis in developing cotyledons of *Vicia faba* L. *Plant Physiol.* 48: 419-425.
Millerd, A., and Spencer, D. (1974). Changes in RNA-synthesizing activity and template activity in nuclei from cotyledons of developing pea seeds. *Austr. J. Plant Physiol.* 1: 331-341.

Millerd, A., Spencer, D., Dudman, W.F., and Stiller, M. (1975). Growth of immature pea cotyledons in culture. *Austr. J. Plant Physiol.* 2: 51-59.

Millerd, A., and Whitfeld, P.R. (1973). Deoxyribonucleic acid and ribonucleic acid synthesis during the cell expansion phase of cotyledon development in *Vicia faba* L. *Plant Physiol.* 51: 1005-1010.

Misoo, S., Yoshida, K., and Mastubayashi, M. (1979). Studies on the mechanism of pollen embryogenesis. III. Mitotic responses of the pollen to varied sucrose concentrations and the process of embryoid formation in tobacco anther culture. *Sci. Rep. Fac. Agr. Kobe Univ.* 13: 193-202.

Mix, G., Wilson, H.M., and Foroughi-Wehr, B. (1978). The cytological status of plants of *Hordeum vulgare* L. regenerated from microspore callus. *Z. Pflanzenzuchtg.* 80: 89-99.

Mogensen, H.L. (1972). Fine structure and composition of the egg apparatus before and after fertilization in *Quercus gambelii:* The functional ovule. *Am. J. Bot.* 59: 931-941.

Mogensen, H.L., and Suthar, H.K. (1979). Ultrastructure of the egg apparatus of *Nicotiana tabacum* (Solanaceae) before and after fertilization. *Bot. Gaz.* 140: 168-179.

Mok, D.W.S., Mok, M.C., and Rabakoarihanta, A. (1978). Interspecific hybridization of *Phaseolus vulgaris* with *P. lunatus* and *P. acutifolius*. *Theor. Appl. Genet.* 52: 209-215.

Mokhtarzadeh, A., and Constantin, M.J. (1978). Plant regeneration from hypocotyl- and anther-derived callus of berseem clover. *Crop Sci.* 18: 567-572.

Monnier, M. (1975). Action d'un gel de polyacrylamide employé comme support pour la culture de l'embryon immature de *Capsella bursa-pastoris*. *Compt. Rend. Acad. Sci. Paris* 280D: 705-708.

Monnier, M. (1976a). Culture *in vitro* de l'embryon immature de *Capsella bursa-pastoris* Moench. *Rev. Cytol. Biol. Veg.* 39: 1-120.

Monnier, M. (1976b). Action de la pression partielle d'oxygène sur le développement de l'embryon de *Capsella bursa-pastoris* cultivé *in vitro*. *Compt. Rend. Acad. Sci. Paris* 282D: 1009-1012.

Monnier, M. (1982). Culture of mature ecotyledonous embryos of *Phaseolus vulgaris* and the nutritional role of cotyledons. *Am. J. Bot.* 69: 896-903.

Monnier, M. (1984). Survival of young immature *Capsella* embryos cultured *in vitro*. *J. Plant Physiol.* 115: 105-113.

Montague, M.J., Armstrong, T.A., and Jaworski, E.G. (1979). Polyamine metabolism in embryogenic cells of *Daucus carota*. II. Changes in arginine decarboxylase activity. *Plant Physiol.* 63: 341-345.

Montague, M.J., Enns, R.K., Siegel, N.R., and Jaworski, E.G. (1981). A comparison of 2,4-dichlorophenoxyacetic acid metabolism in cultured soybean cells and in embryogenic carrot cells. *Plant Physiol.* 67: 603-607.

Montague, M.J., Koppenbrink, J.W., and Jaworski, E.G. (1978). Polyamine metabolism in embryogenic cells of *Daucus carota*. I. Changes in intracellular content and rates of synthesis. *Plant Physiol.* 62: 430-433.

Morris, D.A. (1978). Germination inhibitors in developing seeds of *Phaseolus vulgaris* L. *Z. Pflanzenphysiol.* 86: 433-441.

Mroginski, L.A., and Fernandez, A. (1979). Cultivo *in vitro* de anteras de especies de *Arachis* (Leguminosae). *Oleagineux* 34: 243-248.

Mujeeb, K.A., Thomas, J.B., Rodriguez R, R., Waters, R.F., and Bates, L.S. (1978). Chromosome instability in hybrids of *Hordeum vulgare* L. with *Triticum turgidum* and *T. aestivum*. *J. Hered.* 69: 179-182.

Mukkada, A.J. (1962). Some observations on the embryology of *Dicraea stylosa* Wight. In: *Plant Embryology. A Symposium*, pp. 139-145. New Delhi: Council of Scientific & Industrial Research.

Müller, A.J. (1963). Embryonentest zum Nachweis rezessiver Letalfaktoren bei *Arabidopsis thaliana. Biol. Zentralbl.* 82: 133-163.

Muniyamma, M. (1977). Triploid embryos from endosperm *in vivo. Ann. Bot.* 41: 1077-1079.

Nabors, M.W., Heyser, J.W., Dykes, T.A., and DeMott, K.J. (1983). Long-duration, high-frequency plant regeneration from cereal tissue cultures. *Planta* 157: 385-391.

Nag, K.K., and Street, H.E. (1973). Carrot embryogenesis from frozen cultured cells. *Nature (London)* 245: 270-272.

Nagato, Y. (1978). Analysis on the growth of embryo in some Gramineae. *Jap. J. Breed.* 28: 97-105.

Nagato, Y. (1979). Incorporation of ^{3}H-uridine and ^{3}H-leucine during early embryogenesis of rice and barley in caryopsis culture. *Plant Cell Physiol.* 20: 765-773.

Nagl, W. (1962). 4096-Ploidie und "Riesenchromosomen" im Suspensor von *Phaseolus coccineus. Naturwissenschaften* 49: 261-262.

Nagl, W. (1965). Die SAT-Riesenchromosomen der Kerne des Suspensors von *Phaseolus coccineus* und ihr Verhalten während der Endomitose. *Chromosoma* 16: 511-520.

Nagl, W. (1967). Die Riesenchromosomen von *Phaseolus coccineus* L.: Baueigentumlichkeiten, Strukturmodifikationen, zusätzliche Nukleolen und Vergleich mit den mitotischen Chromosomen. *Osterr. Bot. Z.* 114: 171-182.

Nagl, W. (1969a). Banded polytene chromosomes in the legume *Phaseolus vulgaris. Nature (London)* 221: 70-71.

Nagl, W. (1969b). Correlation of structure and RNA synthesis in the nucleolus-organizing polytene chromosomes of *Phaseolus vulgaris. Chromosoma* 28: 85-92.

Nagl, W. (1969c). Puffing of polytene chromosomes in a plant (*Phaseolus vulgaris*). *Naturwissenschaften* 56: 221-222.

Nagl, W. (1970a). Temperature-dependent functional structures in the polytene chromosomes of *Phaseolus*, with special reference to the nucleolus organizers. *J. Cell Sci.* 6: 87-107.

Nagl, W. (1970b). Differentielle RNS-Synthese an pflanzlichen Riesenchromosomen. *Ber. Deut. Bot. Ges.* 83: 301-309.

Nagl, W. (1973a). Origin and fate of the micronucleoli in the giant cells of the *Phaseolus* suspensor. *Nucleus* 16: 100-109.

Nagl, W. (1973b). Photoperiodic control of activity of the suspensor polytene chromosomes in *Phaseolus vulgaris. Z. Pflanzenphysiol.* 70: 350-357.

Nagl, W. (1974). The *Phaseolus* suspensor and its polytene chromosomes. *Z. Pflanzenphysiol.* 73: 1-44.

Nagl, W. (1976a). Early embryogenesis in *Tropaeolum majus* L.: Evolution of DNA content and polyteny in the suspensor. *Plant Sci. Lett.* 7: 1-6.

Nagl, W. (1976b). Early embryogenesis in *Tropaeolum majus* L.: Ultrastructure of the embryo-suspensor. *Biochem. Physiol. Pflanzen* 170: 253-260.

Nagl, W. (1976c). Ultrastructural and developmental aspects of autolysis in embryo-suspensors. *Ber. Deut. Bot. Ges.* 89: 301-311.

Nagl, W. (1977). "Plastolysomes" – plastids involved in the autolysis of the embryo-suspensor in *Phaseolus. Z. Pflanzenphysiol.* 85: 45-51.

Nagl, W. (1978). *Endopolyploidy and Polyteny in Differentiation and Evolution.* Amsterdam: North-Holland Publishing Co.

Nagl, W., and Kühner, S. (1976). Early embryogenesis in *Tropaeolum majus* L.: Diversification of plastids. *Planta* 133: 15-19.

Nagl, W., Peschke, C., and van Gyseghem, R. (1976). Heterochromatin underreplication in *Tropaeolum* embryogenesis. *Naturwissenschaften* 63: 198.

Nagmani, R., and Raghavan, V. (1983). Induction of embryogenic divisions in isolated pollen grains of *Hyoscyamus niger* in a single-step method. *Z. Pflanzenphysiol.* 109: 87-90.

Nakajima, T. (1962). Physiological studies of seed development, especially embryonic growth and endosperm development. *Bull. Univ. Osaka Pref. Ser. B* 13: 13-48.

Napier, K.V., and Walton, P.D. (1984). Hybrids between tetraploid *Agropyron spicatum* and *A. intermedium. Z. Pflanzenzuchtg.* 92: 221-228.

Narayanaswami, S. (1959). Experimental studies on growth of excised grass embryos *in vitro*. I. Overgrowth of the scutellum of *Pennisetum* embryos. *Phytomorphology* 9: 358-367.

Nast, C. (1941). The embryogeny and seedling morphology of "*Juglans regia*"L. *Lilloa* 6: 163-205.

Natesh, S., and Rau, M.A. (1984). The embryo. In: *Embryology of Angiosperms*, ed. B.M. Johri, pp. 377-443. Berlin: Springer-Verlag.

Naumova, T.N., and Willemse, M.T.M. (1982). Nucellar polyembryony in *Sarcococca humilis* – Ultrastructural aspects. *Phytomorphology* 32: 94-108.

Negbi, M., and Koller, D. (1962). Homologies in the grass embryo – A re-evaluation. *Phytomorphology* 12: 289-296.

Nesling, F.A.V., and Morris, D.A. (1979). Cytokinin levels and embryo abortion in interspecific *Phaseolus* crosses. *Z. Pflanzenphysiol.* 91: 345-358.

Nessler, C.L. (1982). Somatic embryogenesis in the opium poppy, *Papaver somniferum. Physiol. Plant.* 55: 453-458.

Neuffer, M.G., and Sheridan, W.F. (1980). Defective kernel mutants of maize. I. Genetic and lethality studies. *Genetics* 95: 929-944.

Newcomb, W. (1973a). The development of the embryo sac of sunflower *Helianthus annuus* before fertilization. *Can. J. Bot.* 51: 863-878.

Newcomb, W. (1973b). The development of the embryo sac of sunflower *Helianthus annuus* after fertilization. *Can. J. Bot.* 51: 879-890.

Newcomb, W. (1978). The development of cells in the coenocytic endosperm of the African blood lily *Haemanthus katherinae. Can. J. Bot* 56: 483-501.

Newcomb, W., and Fowke, L.C. (1973). The fine structure of the change from the free-nuclear to cellular conditions in the endosperm of chickweed *Stellaria media. Bot. Gaz.* 134: 236-241.

Newcomb, W., and Fowke, L.C. (1974). *Stellaria media* embryogenesis: The development and ultrastructure of the suspensor. *Can. J. Bot.* 52: 607-614.

Newcomb, W., and Steeves, T.A. (1971). *Helianthus annuus* embryogenesis: Embryo sac wall projections before and after fertilization. *Bot. Gaz.* 132: 367-371.

Newell, C.A., and Hymowitz, T. (1982). Successful wide hybridization between the soybean and a wild perennial relative, *G. tomentella. Crop Sci.* 22: 1062-1065.

Niimi, Y. (1974). Effect of Fe-EDTA on the development of isolated ovules of *Petunia hybrida. J. Jap. Soc. Hort. Sci.* 43: 77-83.

Niizeki, H., and Oono, K. (1968). Induction of haploid rice plant from anther culture. *Proc. Jap. Acad.* 44: 554-557.

Nitsch, C. (1974). La culture de pollen isolé sur milieu synthétique. *Compt. Rend. Acad. Sci. Paris* 278D: 1031-1034.

Nitsch, C., Anderson, S., Godard, M., Neuffer, M.G., and Sheridan, W.F. (1982).

Production of haploid plants of *Zea mays* and *Pennisetum* through androgenesis. In: *Variability in Plants Regenerated from Tissue Culture*, ed. E. D. Earle and Y. Demarly, pp. 69-91. New York: Praeger Publishers.

Nitsch, C., and Norreel, B. (1973). Effet d'un choc thermique sur le pouvoir embryogène du pollen de *Datura innoxia* cultivé dans l'anthère ou isolé de l'anthère. *Compt. Rend. Acad. Sci. Paris* 276D: 303-306.

Nitsch, J.P. (1972). Haploid plants from pollen. *Z. Pflanzenzuchtg.* 67: 3-18.

Nitsch, J.P., and Nitsch, C. (1969). Haploid plants from pollen grains. *Science* 163: 85-87.

Nitsch, J.P., Nitsch, C., and Péreau-Leroy, P. (1969). Obtention de mutants à partir de *Nicotiana* haploides issus de grains de pollen. *Compt. Rend. Acad. Sci. Paris* 269D: 1650-1652.

Nitzsche, W., and Wenzel, G. (1977). *Haploids in Plant Breeding*. Berlin: Verlag Paul Parey.

Noma, M., Huber, J., Ernst, D., and Pharis, R.P. (1982). Quantitation of gibberellins and the metabolism of [^{3}H]gibberellin A_1 during somatic embryogenesis in carrot and anise cell cultures. *Planta* 155: 369-376.

Norreel, B. (1972). Etude comparative de la répartition des acides ribonucléiques au cours de l'embryogenèse zygotique et de l'embryogenèse androgénétique chez le *Nicotiana tabacum* L. *Compt. Rend. Acad. Sci. Paris* 275D: 1219-1222.

Norstog, K. (1961). The growth and differentiation of cultured barley embryos. *Am. J. Bot.* 48: 876-884.

Norstog, K. (1969). Morphology of coleoptile and scutellum in relation to tissue culture responses. *Phytomorphology* 19: 235-241.

Norstog, K. (1970). Induction of embryolike structures by kinetin in cultured barley embryos. *Develop. Biol.* 23: 665-670.

Norstog, K. (1972a). Early development of the barley embryo: Fine structure. *Am. J. Bot.* 59: 123-132.

Norstog, K. (1972b). Factors relating to precocious germination in cultured barley embryos. *Phytomorphology* 22: 134-139.

Norstog, K. (1973). New synthetic medium for the culture of premature barley embryos. *In Vitro* 8: 307-308.

Norstog, K. (1974). Nucellus during early embryogeny in barley: Fine structure. *Bot. Gaz* 135: 97-103.

Norstog, K., and Klein, R.M. (1972). Development of cultured barley embryos. II. Precocious germination and dormancy. *Can. J. Bot.* 50: 1887-1894.

Norstog, K., and Smith, J. (1963). Culture of small barley embryos on defined media. *Science* 142: 1655-1656.

Oettler, G. (1983). Crossability and embryo development in wheat-rye hybrids. *Euphytica* 32: 593-600.

Ono, K., and Harashima, S. (1981). Induction of haploid callus from isolated microspores of peony *in vitro*. *Plant Cell Physiol* 22: 337-341.

Ono, K., and Tsukida, T. (1978). Haploid callus formation from anther cultures in a cultivar of *Paeonia*. *Jap. J. Genet.* 53: 51-54.

Ouyang, J.W., Zhou, S.M., and Jia, S.E. (1983). The response of anther culture to culture temperature in *Triticum aestivum*. *Theor. Appl. Genet.* 66: 101-109.

Ouyang, T., Hu, H., Chuang, C., and Tseng, C. (1973). Induction of pollen plants from anthers of *Triticum aestivum* L. cultured *in vitro*. *Scientia Sinica* 16: 79-95.

Ozias-Akins, P., and Vasil, I.K. (1982). Plant regeneration from cultured immature embryos and inflorescences of *Triticum aestivum* L. (wheat): Evidence for somatic embryogenesis. *Protoplasma* 110: 95-105.

Ozias-Akins, P., and Vasil, I.K. (1983). Improved efficiency and normalization of somatic embryogenesis in *Triticum aestivum* (wheat). *Protoplasma* 117: 40-44.

Ozsan, M., and Cameron, J.W. (1963). Artificial culture of small *Citrus* embryos, and evidence against nucellar embryony in highly zygotic varieties. *Proc. Am. Soc. Hort. Sci.* 82: 210-216.

Pacini, E., Cresti, M., and Sarfatti, G. (1972). Incorporation of integumentary nuclei in *Eranthis hiemalis* endosperm and their disaggregation by the endoplasmic reticulum. *J. Submicros. Cytol.* 4: 19-31.

Pacini, E., Simoncioli, C., and Cresti, M. (1975). Ultrastructure of nucellus and endosperm of *Diplotaxis erucoides* during embryogenesis. *Caryologia* 28: 525-538.

Pal, A. (1983). Isolated microspore culture of the winged bean, *Psophocarpus tetragonolobus* L. (DC.) – Growth, development and chromosomal status. *Indian J. Expt. Biol.* 21: 597-599.

Pan, C., and Kao, K. (1978). The production of wheat pollen embryo and the influence of some factors on its frequency of induction. In: *Proceedings of Symposium on Plant Tissue Culture*, pp. 133-142. Peking: Science Press.

Pan, J., Gao, G., and Ban, H. (1983). Initial patterns of androgenesis in wheat anther culture. *Acta Bot. Sinica* 25: 34-39.

Pareek, L.K., and Chandra, N. (1978). Somatic embryogenesis in leaf callus from cauliflower (*Brassica oleracea* var. Botrytis). *Plant Sci. Lett.* 11: 311-316.

Paris, D., Rietsema, J., Satina, S., and Blakeslee, A.F. (1953). Effect of amino acids, especially aspartic and glutamic acid and their amides, on the growth of *Datura stramonium* embryos *in vitro*. *Proc. Natl. Acad. Sci. U.S.A.* 39: 1205-1212.

Pecket, R.D., and Selim, A.R.A.A. (1965). Embryo-culture in *Lathyrus*. *J. Expt. Bot.* 16: 325-328.

Pedersen, K., Bloom, K.S., Andersen, J.N., Glover, D.V., and Larkins, B.A. (1980). Analysis of the complexity and frequency of zein genes in the maize genome. *Biochemistry* 19: 1644-1650.

Pelletier, G., and Durran, V. (1972). Recherche de tissus nourriciers pour la réalisation de l'androgenèse *in vitro* chez *Nicotiana tabacum*. *Compt. Rend. Acad. Sci. Paris* 275D: 35-37.

Pelletier, G., and Ilami, M. (1972). Les facteurs de l'androgénèse *in vitro* chez *Nicotiana tabacum*. *Z. Pflanzenphysiol.* 68: 97-114.

Pelletier, G., Raquin, C., and Simon, G. (1972). La culture *in vitro* d'anthères d'asperge (*Asparagus officinalis*). *Compt. Rend. Acad. Sci. Paris* 274D: 848-851.

Pence, V.C., Hasegawa, P.M., and Janick, J. (1980). Initiation and development of asexual embryos of *Theobroma cacao* L. *in vitro*. *Z. Pflanzenphysiol.* 98: 1-14.

Pence, V.C., Hasegawa, P.M., and Janick, J. (1981a). Sucrose-mediated regulation of fatty acid composition in asexual embryos of *Theobroma cacao*. *Physiol. Plant.* 53: 378-384.

Pence, V.C., Hasegawa, P.M., and Janick, J. (1981b). *In vitro* cotyledonary development and anthocyanin synthesis in zygotic and asexual embryos of *Theobroma cacao*. *J. Am. Soc. Hort. Sci.* 106: 381-385.

Periasamy, K. (1977). A new approach to the classification of angiosperm embryos. *Proc. Indian Acad. Sci.* 86B: 1-13.

Peters, J.E., Crocomo, O.J., Sharp, W.R., Paddock, E.F., Tegenkamp, I., and Tegenkamp, T. (1977). Haploid callus cells from anthers of *Phaseolus vulgaris*. *Phytomorphology* 27: 79-85.

Phillips, G.C., and Collins, G.B. (1981). Induction and development of somatic embryos from cell suspension cultures of soybean. *Plant Cell Tissue Organ Culture* 1: 123-129.

Phillips, G.C., Collins, G.B., and Taylor, N.L. (1982). Interspecific hybridization of red clover (*Trifolium pratense* L.) with *T. sarosiense* Hazsl. using *in vitro* embryo rescue. *Theor. Appl. Genet.* 62: 17-24.

Picard, E., and de Buyser, J. (1973). Obtention de plantules haploides de *Triticum aestivum* L. à partir de cultures d'anthères *in vitro*. *Compt. Rend. Acad. Sci. Paris* 277D: 1463-1466.

Picard, E., and de Buyser, J. (1977). High production of embryoids in anther culture of pollen derived homozygous spring wheats. *Ann. Amelior. Plant.* 27: 483-488.

Picciarelli, P., Alpi, A., Pistelli, L., and Scalet, M. (1984). Gibberellin-like activity in suspensors of *Tropaeolum majus* L. and *Cytisus laburnum* L. *Planta* 162: 566-568.

Pickering, R.A., and Thomas, H.M. (1979). Crosses between tetraploid barley and diploid rye. *Plant Sci. Lett.* 16: 291-296.

Pierson, E.S., van Lammeren, A.A.M., Schel, J.H.N., and Staritsky, G. (1983). *In vitro* development of embryoids from punched leaf discs of *Coffea canephora*. *Protoplasma* 115: 208-216.

Pissarev, W.E., and Vinogradova, N.M. (1944). Hybrids between wheat and *Elymus*. *Compt. Rend. (Doklady) Acad. Sci. URSS* 45: 129-132.

Pollock, E.G., and Jensen, W.A. (1964). Cell development during early embryogenesis in *Capsella* and *Gossypium*. *Am. J. Bot.* 51: 915-921.

Ponzi, R., and Pizzolongo, P. (1972). The ultrastructure of suspensor cells of *Ipomoea purpurea* Roth. *J. Submicros. Cytol.* 4: 199-204.

Ponzi, R., and Pizzolongo, P. (1973). Ultrastructure of plastids in the suspensor cells of *Ipomoea purpurea* Roth. *J. Submicros. Cytol.* 5: 257-263.

Poulson, R., and Beevers, L. (1973). RNA metabolism during the development of cotyledons of *Pisum sativum* L. *Biochim. Biophys. Acta* 308: 381-389.

Prabhakar, K., and Vijayaraghavan, M.R. (1983). Histochemistry and ultrastructure of suspensor cells in *Alyssum maritimum*. *Cytologia* 48: 389-402.

Preil, W., Huhnke, W., Engelhardt, M., and Hoffmann, M. (1977). Haploide bei *Gerbera jamesonii* aus *in vitro*-Kulturen von Blütenköpfchen. *Z. Pflanzenzuchtg.* 79: 167-171.

Preťová, A. (1974). The influence of the osmotic potential of the cultivation medium on the development of excised flax embryos. *Biol. Plant.* 16: 14-20.

Price, H.J., and Smith, R.H. (1979). Somatic embryogenesis in suspension cultures of *Gossypium klotzschianum* Anderss. *Planta* 145: 305-307.

Pritchard, H.N. (1964). A cytochemical study of embryo development in *Stellaria media*. *Am. J. Bot.* 51: 472-479.

Pritchard, H.N., and Bergstresser, K.A. (1969). The cytochemistry of some enzyme activities in *Stellaria media* embryos. *Experientia* 25: 1116-1117.

Przybyllok, T., and Nagl, W. (1977). Auxin concentration in the embryo and suspensors of *Tropaeolum majus*, as determined by mass fragmentation (single ion detection). *Z. Pflanzenphysiol.* 84: 463-465.

Puri, P. (1964). *In vitro* culture of floral organs of an apomict, *Aerva javanica* (Brum. F.) Spreng. *Phytomorphology* 14: 564-573.

Quatrano, R.S., Ballo, B.L., Williamson, J.D., Hamblin, M.T., and Mansfield, M. (1983). ABA controlled expression of embryo-specific genes during wheat grain development. In: *Plant Molecular Biology*, ed. R. B. Goldberg, pp. 343-353. New York: Alan R. Liss.

Quebedeaux, B., Sweetser, P.B., and Rowell, J.C. (1976). Abscisic acid levels in soybean reproductive structures during development. *Plant Physiol.* 58: 363-366.

Rabakoarihanta, A., Mok, D.W.S., and Mok, M.C. (1979). Fertilization and early

embryo development in reciprocal interspecific crosses of *Phaseolus. Theor. Appl. Genet.* 54: 55-59.
Racchi, M.L., Gavazzi, G., Monti, D., and Manitto, P. (1978). An analysis of the nutritional requirements of the *pro* mutant in *Zea mays. Plant Sci. Lett.* 13: 357-364.
Radojević, L. (1978). In vitro induction of androgenic plantlets in *Aesculus hippocastanum. Protoplasma* 96: 369-374.
Radojević, L. (1979). Somatic embryos and plantlets of *Paulownia tomentosa* Steud. *Z. Pflanzenphysiol.* 91: 57-62.
Radojević, L., Vujičić, R., and Nešković, M. (1975). Embryogenesis in tissue culture of *Corylus avellana.* L. *Z. Pflanzenphysiol.* 77: 33-41.
Radojević, L., Zylberberg, L., and Kovoor, J. (1980). Etude ultrastructurale des embryons androgenetiques d'*Aesculus hippocastanum* L. *Z. Pflanzenphysiol.* 98: 255-261.
Raghavan, P., and Philip, V.J. (1982). Morphological and histochemical changes in the egg and zygote of *Lagerstroemia speciosa*. I. Cell size, vacuole and insoluble polysaccharides. *Proc. Indian Acad. Sci. (Plant Sci.).* 91: 465-472.
Raghavan, V. (1976a). Role of the generative cell in androgenesis in henbane. *Science* 191: 388-389.
Raghavan, V. (1976b). *Experimental Embryogenesis in Vascular Plants.* London: Academic Press.
Raghavan, V. (1977a). Patterns of DNA synthesis during pollen embryogenesis in henbane. *J. Cell Biol.* 73: 521-526.
Raghavan, V. (1977b). Applied aspects of embryo culture. In: *Applied and Fundamental Aspects of Plant Cell, Tissue, and Organ Culture,* ed. J. Reinert and Y.P.S. Bajaj, pp. 375-397. Berlin: Springer-Verlag.
Raghavan, V. (1978). Origin and development of pollen embryoids and pollen calluses in cultured anther segments of *Hyoscyamus niger* (henbane). *Am. J. Bot.* 65: 984-1002.
Raghavan, V. (1979a). Embryogenic determination and ribonucleic acid synthesis in pollen grains of *Hyoscyamus niger* (henbane). *Am. J. Bot.* 66: 36-39.
Raghavan, V. (1979b). An autoradiographic study of RNA synthesis during pollen embryogenesis in *Hyoscyamus niger* (henbane). *Am. J. Bot.* 66: 784-795.
Raghavan, V. (1980). Embryo culture. *Int. Rev. Cytol.Suppl.* 11B: 209-240.
Raghavan, V. (1981a). Pollen embryogenesis in *Hyoscyamus niger:* A review. In: *Tissue Culture of Economically Important Plants,* ed. A.N. Rao, pp. 262-268. Singapore: Costed.
Raghavan, V. (1981b). Distribution of poly(A)-containing RNA during normal pollen development and during induced pollen embryogenesis in *Hyoscyamus niger. J. Cell Biol.* 89: 593-606.
Raghavan, V. (1983). Biochemistry of somatic embryogenesis. In: *Handbook of Plant Cell Culture,* Vol. 1, ed. D.A. Evans, W.R. Sharp, P.V. Ammirato and Y. Yamada, pp. 654-671. New York: Macmillan Publishing Co.
Raghavan, V. (1984). Protein synthetic activity during normal pollen development and during induced pollen embryogenesis in *Hyoscyamus niger Can. J. Bot.* 62: 2493-2513.
Raghavan, V., and Nagmani, R. (1983). Morphogenesis of pollen callus cultures of *Hyoscyamus niger. Am. J. Bot.* 70: 524-531.
Raghavan, V., and Srivastava, P.S. (1982). Embryo culture. In: *Experimental Embryology of Vascular Plants,* ed. B.M. Johri, pp. 195-230. Berlin: Springer-Verlag.

Raghavan, V., and Torrey, J.G. (1963). Growth and morphogenesis of globular and older embryos of *Capsella* in culture. *Am. J. Bot.* 50: 540-551.
Raghavan, V., and Torrey, J.G. (1964). Effects of certain growth substances on the growth and morphogenesis of immature embryos of *Capsella* in culture. *Plant Physiol.* 39: 691-699.
Rajasekaran, K., and Mullins, M.G. (1979). Embryo and plantlets from cultured anthers of hybrid grapevines. *J. Expt. Bot.* 30: 399-407.
Rajasekaran, K., Vine, J., and Mullins, M.G. (1982). Dormancy in somatic embryos and seeds of *Vitis:* Changes in endogenous abscisic acid during embryogeny and germination. *Planta* 154: 139-144.
Raju, C.R., Kumar, P.P., Chandramohan, M., and Iyer, R.D. (1984). Coconut plantlets from leaf tissue cultures. *J. Plantation Crops* 12: 75-78.
Ram, M. (1956). Floral morphology and embryology of *Trapa bispinosa* Roxb. with a discussion on the systematic position of the genus. *Phytomorphology* 6: 312-323.
Ramanna, M.S. (1974). The origin and *in vivo* development of embryoids in the anthers of *Solanum* hybrids. *Euphytica* 23: 623-632.
Ramanna, M.S., and Hermsen, J.G.T. (1974). Embryoid formation in the anthers of some interspecific hybrids in *Solanum. Euphytica* 23: 423-427.
Ramji, M.V. (1975). Histology of growth with regard to embryos and apical meristems in some angiosperms. I. Embryogeny of *Stellaria media. Phytomorphology* 25: 131-145.
Ramming, D. (1983). Embryo culture. In: *Methods in Fruit Breeding*, ed. J. N. Moore and J. Janick, pp. 136-144. West Lafayette: Purdue University Press.
Randolph, L.F. (1936). Developmental morphology of the caryopsis in maize. *J. Agric. Res.* 53: 881-916.
Rangan, T.S., Murashige, T., and Bitters, W.P. (1968). *In vitro* initiation of nucellar embryos in monoembryonic *Citrus. HortSci.* 3: 226-227.
Rangan, T.S., and Vasil, I.K. (1983a). Somatic embryogenesis and plant regeneration in tissue cultures of *Panicum miliaceum* L. and *Panicum miliare* Lamk. *Z. Pflanzenphysiol.* 109: 49-53.
Rangan, T.S., and Vasil, I.K. (1983b). Sodium chloride tolerant embryogenic cell lines of *Pennisetum americanum* (L.) K. Schum. *Ann. Bot.* 52: 59-64.
Rangaswamy, N.S. (1961). Experimental studies on female reproductive structures of *Citrus microcarpa* Bunge. *Phytomorphology* 11: 109-127.
Rangaswamy, N.S., and Rangan, T.S. (1971). Morphogenic investigations on parasitic angiosperms. IV. Morphogenesis in decotylated embryos of *Cassytha filiformis* L. Lauraceae. *Bot. Gaz.* 132: 113-119.
Rangaswamy, N.S., Sethi, M., and Shrotria, A. (1980). Nucellar and somatic embryony—An experimental interpretation. In: *Current Trends in Botanical Research*, ed. M. Nagaraj and C. P. Malik, pp. 35-50. Delhi: Kalyani Publishers.
Rao, P.S., and Ozias-Akins, P. (1985). Plant regeneration through somatic embryogenesis in protoplast cultures of sandalwood (*Santalum album* L.). *Protoplasma* 124: 80-86.
Raquin, C., Amssa, M., Henry, Y., de Buyser, J., and Essad, S. (1982). Origine des plantes polyploides obtenues par culture d'anthères. Analyse cytophotométrique *in situ* et *in vitro* des microspores de *Petunia* et blé tendre. *Z. Pflanzenzuchtg.* 89: 265-277.
Rashid, A. (1982). Induction of embryos in pollen cultures of *Nicotiana sylvestris. Physiol. Plant.* 56: 223-224.
Rashid, A., and Reinert, J. (1981). *In vitro* differentiation of embryogenic pollen,

control by cold treatment and embryo formation in *ab initio* pollen cultures of *Nicotiana tabacum* var. Badischer Burley. *Protoplasma* 109: 285-294.
Rashid, A., and Reinert, J. (1983). Factors affecting high-frequency embryo formation in *ab initio* pollen cultures of *Nicotiana*. *Protoplasma* 116: 155-160.
Rashid, A., Siddiqui, A.W., and Reinert, J. (1981). Ultrastructure of embryogenic pollen of *Nicotiana tabacum* var. Badischer Burley. *Protoplasma* 107: 375-385.
Rashid, A., Siddiqui, A.W., and Reinert, J. (1982). Subcellular aspects of origin and structure of pollen embryos of *Nicotiana*. *Protoplasma* 113: 202-208.
Rashid, A., and Street, H.E. (1974a). Growth, embryogenic potential and stability of a haploid cell culture of *Atropa belladonna* L. *Plant Sci. Lett.* 2: 89-94.
Rashid, A., and Street, H.E. (1974b). Segmentations in microspores of *Nicotiana sylvestris* and *Nicotiana tabacum* which lead to embryoid formation in anther cultures. *Protoplasma* 80: 323-334.
Redenbaugh, M.K., Westfall, R.D., and Karnosky, D.F. (1981). Dihaploid callus production from *Ulmus americana* anthers. *Bot. Gaz.* 142: 19-26.
Reed, S.M., and Collins, G.B. (1978). Interspecific hybrids in *Nicotiana* through *in vitro* culture of fertized ovules. *N. stocktonii* × *N. tabacum; N. nesophila* × *N. tabacum; N. repanda* × *N. tabacum*. *J. Hered.* 69: 311-315.
Reeve, R.M. (1948). Late embryogeny and histogenesis in *Pisum*. *Am. J. Bot.* 35: 591-602.
Reinert, J. (1959). Uber die Kontrolle der Morphogenese und die Induktion von Adventivembryonen an Gewebekulturen aus Karotten. *Planta* 53: 318-333.
Reinert, J., and Bajaj, Y.P.S. (1977). Anther culture: Haploid production and its significance. In: *Applied and Fundamental Aspects of Plant Cell, Tissue, and Organ Culture*, ed. J. Reinert and Y.P.S. Bajaj, pp. 251-267. Berlin: Springer-Verlag.
Reinert, J., Bajaj, Y.P.S., and Heberle, E. (1975). Induction of haploid tobacco plants from isolated pollen. *Protoplasma* 84: 191-196.
Reinert, J., Tazawa, M., and Semenoff, S. (1967). Nitrogen compounds as factors of embryogenesis *in vitro*. *Nature (London)* 216: 1215-1216.
Reuther, G. (1977). Embryoide Differenzierungsmuster im Kallus der Gattungen *Iris* und *Asparagus*. *Ber. Deut. Bot. Ges.* 90: 417-437.
Reynolds, J.F., and Murashige, T. (1979). Asexual embryogenesis in callus cultures of palms. *In Vitro* 15: 383-387.
Reynolds, T.L. (1984a). An ultrastructural and stereological analysis of pollen grains of *Hyoscyamus niger* during normal ontogeny and induced embryogenic development. *Am. J. Bot.* 71: 490-504.
Reynolds, T.L. (1984b). Callus formation and organogenesis in anther cultures of *Solanum carolinense* L. *J. Plant Physiol.* 117: 157-161.
Reynolds, T.L., and Raghavan, V. (1982). An autoradiographic study of RNA synthesis during maturation and germination of pollen grains of *Hyoscyamus niger*. *Protoplasma* 111: 177-188.
Rietsema, J., Satina, S., and Blakeslee, A.F. (1953). The effect of sucrose on the growth of *Datura stramonium* embryos *in vitro*. *Am. J. Bot* 40: 538-545.
Rijven, A.H.G.C. (1952). *In vitro* studies on the embryo of *Capsella bursa-pastoris*. *Acta Bot. Neerl.* 1: 157-200.
Rijven, A.H.G.C. (1955). Effects of glutamine, asparagine and other related compounds on the *in vitro* growth of embryos of *Capsella bursa-pastoris*. *Koninkl. Nederl. Akad. Wetensch. Proc.Ser.C* 58: 368-376.
Rijven, A.H.G.C. (1956). Glutamine and asparagine as nitrogen sources for the growth of plant embryos *in vitro*: A comparative study of 12 species. *Austr. J. Biol. Sci.* 9: 511-527.

Rijven, A.H.G.C. (1960). On the utilization of γ-aminobutyric acid by wheat seedlings. *Austr. J. Biol. Sci.* 13: 132-141.
Rines, H.W. (1983). Oat anther culture: Genotype effects on callus initiation and the production of a haploid plant. *Crop Sci.* 23: 268-272.
Roberts, B.E., and Osborne, D.J. (1973). Protein synthesis and loss of viability in rye embryos. The lability of transferase enzymes during senescence. *Biochem. J.* 135: 405-410.
Roberts, B.E., Payne, P.I., and Osborne, D.J. (1973). Protein synthesis and the viability of rye grains. Loss of activity of protein-synthesizing systems *in vitro* associated with a loss of viability. *Biochem. J.* 131: 275-286.
Robichaud, C.S., Wong, J., and Sussex, I.M. (1980). Control of *in vitro* growth of viviparous embryo mutants of maize by abscisic acid. *Develop. Genet.* 1: 325-330.
Rogers, S.O., and Quatrano, R.S. (1983). Morphological staging of wheat caryopsis development. *Am. J. Bot.* 70: 308-311.
Rondet, P. (1962). L'organogenèse au cours de l'embryogenèse chez l'*Alyssum maritimum* Lamk. *Compt. Rend. Acad. Sci. Paris* 255: 2278-2280.
Rosati, P., Devreux, M., and Laneri, U. (1975). Anther culture of strawberry. *HortSci.* 10: 119-120.
Rudnicki, R., Kamiński, W., and Pieniażek, J. (1971). The interaction of abscisic acid with growth stimulators in germination of partially after-ripened apple embryos. *Biol. Plant.* 13: 122-127.
Russell, S.D. (1982). Fertilization in *Plumbago zeylanica:* Entry and discharge of the pollen tube in the embryo sac. *Can. J. Bot.* 60: 2219-2230.
Ryczkowski, M. (1960). Changes of osmotic value during the development of the ovule. *Planta* 55: 343-356.
Sakai, A., and Noshiro, M. (1975). Some factors contributing to the survival of crop seeds cooled to the temperature of liquid nitrogen. In: *Crop Genetic Resources for Today and Tomorrow,* ed. O. H. Frankel and J. G. Hawkes, pp. 317-326. Cambridge: Cambridge University Press.
Salem, S., Linstedt, D., and Reinert, J. (1979). The cytokinins of cultured carrot cells. *Protoplasma* 101: 103-109.
Sanders, M.E., and Burkholder, P.R. (1948). Influence of amino acids on growth of *Datura* embryos in culture. *Proc. Natl. Acad. Sci. U.S.A.* 34: 516-526.
Sangduen, N., Kreitner, G.L. and Sorensen, E.L. (1983a). Light and electron microscopy of embryo development in perennial and annual *Medicago* species. *Can. J. Bot.* 61: 837-849.
Sangduen, N., Kreitner, G.L. and Sorensen, E.L. (1983b). Light and electron microscopy of embryo development in an annual × perennial *Medicago* species cross. *Can. J. Bot.* 61: 1241-1257.
Sangwan, R.S. (1978). Amino acid metabolism in cultured anthers of *Datura metel.* *Biochem. Physiol. Pflanzen* 173: 355-364.
Sangwan, R.S. (1983). Effects of exogenous amino acids on *in vitro* androgenesis of *Datura. Biochem. Physiol. Pflanzen* 178: 415-422.
Sangwan, R.S., and Camefort, H. (1982). Ribosomal bodies specific to both pollen and zygotic embryogenesis in *Datura. Experientia* 38: 395-397.
Sangwan, R.S., and Camefort, H. (1983). The tonoplast, a specific marker of embryogenic microspores of *Datura* cultured *in vitro. Histochemistry* 78: 473-480.
Sangwan, R.S., and Norreel, B. (1975a). Pollen embryogenesis in *Pharbitis nil* L. *Naturwissenschaften* 62:440.
Sangwan, R.S., and Norreel, B. (1975b). Induction of plants from pollen grains of *Petunia* cultured *in vitro. Nature (London)* 257: 222-224.

Sangwan-Norreel, B.S. (1977). Androgenic stimulating factors in the anther and isolated pollen grain culture of *Datura innoxia* Mill. *J. Expt. Bot.* 28: 843-852.

Sangwan-Norreel, B.S. (1978). Cytochemical and ultrastructural peculiarities of embryogenic pollen grains and of young androgenic embryos in *Datura innoxia*. *Can. J. Bot.* 56: 805-817.

Sangwan-Norreel, B.S. (1979). Evolution du contenu en DNA nucléaire dans les gamétophytes mâles du *Datura innoxia* au cours de la période favorable à androgenèse. *Can. J. Bot.* 57: 450-457.

Sangwan-Norreel, B.S. (1981). Evolution *in vitro* du contenu en ADN nucléaire et de la ploidie des embryons polliniques du *Datura innoxia*. *Can. J. Bot.* 59: 508-517.

Sangwan-Norreel, B.S. (1983). Male gametophyte nuclear DNA content evolution during androgenic induction in *Datura innoxia* Mill. *Z. Pflanzenphysiol.* 111: 47-54.

San Noeum, L.H. (1976). Haploides d'*Hordeum vulgare* L. par culture *in vitro* d'ovaries non fécondés. *Ann. Amelior. Plant.* 26: 751-754.

Sastri, D.C., and Moss, J.P. (1982). Effects of growth regulators on incompatible crosses in the genus *Arachis* L. *J. Expt. Bot.* 33: 1293-1301.

Sauter, J.J. (1969). Autoradiographische Untersuchungen zur RNS- und Proteinsynthese in Pollenmutterzellen, jungen Pollen und Tapetumzellen während der Mikrosporogenese von *Paeonia tenuifolia* L. *Z. Pflanzenphysiol.* 61: 1-19.

Schaeffer, G.W. (1982). Recovery of heritable variability in anther-derived doubled-haploid rice. *Crop Sci.* 22: 1160-1164.

Schaeffer, G.W., Baenziger, P.S., Worley, J. (1979). Haploid plant development from anthers and *in vitro* embryo culture of wheat. *Crop Sci.* 19: 697-702.

Schaeffer, G.W., Sharpe, F.T., Jr., and Cregan, P.B. (1984). Variation for improved protein and yield from rice anther culture. *Theor. Appl. Genet.* 67: 383-389.

Scharpé, A., and van Parijs, R. (1973). The formation of polyploid cells in ripening cotyledons of *Pisum sativum* L. in relation to ribosome and protein synthesis. *J. Expt. Bot.* 24:216-222.

Schlosser-Szigat, G. (1962). Artbastardierung mit Hilfe der Embryokultur bei Steinklee (*Melilotus*). *Naturwissenschaften* 49: 452-453.

Schnepf, E. and Nagl, W. (1970). Uber einige Strukturbesonderheiten der Suspensorzellen von *Phaseolus vulgaris*. *Protoplasma* 69: 133-143.

Schuchmann, R., and Wellmann, E. (1983). Somatic embryogenesis of tissue culture of *Papaver somniferum* and *Papaver orientale* and its relationship to alkaloid and lipid metabolism. *Plant Cell Rep.* 2: 88-91.

Schulz, P., and Jensen W.A. (1969). *Capsella* embryogenesis: The suspensor and the basal cell. *Protoplasma* 67: 139-163.

Schulz, P., and Jensen, W.A. (1974). *Capsella* embryogenesis: The development of the free nuclear endosperm. *Protoplasma* 80: 183-205.

Schulz, P., and Jensen, W.A. (1977). Cotton embryogenesis: The early development of the free nuclear endosperm. *Am. J. Bot.* 64: 384-394.

Schulz, R., and Jensen, W.A. (1968a). *Capsella* embryogenesis: The egg, zygote, and young embryo. *Am. J. Bot.* 55: 807-819.

Schulz, R., and Jensen, W. (1968b). *Capsella* embryogenesis: The early embryo. *J. Ultrastr. Res.* 22: 376-392.

Sehgal, C.B., Arora, S., and Narang, K. (1982). Production of pollen embryoids in anther cultures of *Sambucus nigra* L. *Curr. Sci.* 51: 104-105.

Sen, S., and Osborne, D.J. (1977). Decline in ribonucleic acid and protein synthesis with loss of viability during the early hours of imbibition of rye (*Secale cereale* L.) embryos. *Biochem. J.* 166: 33-38.

Sengupta, C., and Deluca, V., Bailey, D.S., and Verma, D.P.S. (1981). Posttranslational processing of 7S and 11S components of soybean storage proteins. *Plant Mol. Biol.* 1: 19-34.

Sengupta, C. and Raghavan, V. (1980a). Somatic embryogenesis in carrot cell suspension. I. Pattern of protein and nucleic acid synthesis. *J. Expt. Bot.* 31: 247-258.

Sengupta, C. and Raghavan, V. (1980b). Somatic embryogenesis in carrot cell suspension. II. Synthesis of ribosomal RNA and poly $(A)^+$ RNA. *J. Expt. Bot.* 31: 259-268.

Shah, C.K., and Pandey, S.N. (1978). Histochemical studies during embryogenesis in *Limnophyton obtusifolium*. *Phytomorphology* 28: 31-42.

Shanthamma, C.K., and Narayan, K.N. (1977). Formation of nucellar embryos with total absence of embryo sacs in two species of Gramineae. *Ann. Bot.* 41: 469-470.

Sharma, D.R., Chowdhury, J.B., Ahuja, U. and Dhankhar, B.S. (1980). Interspecific hybridization in genus *Solanum*. A cross between *S. melongena* and *S. khasianum* through embryo culture. *Z. Pflanzenzuchtg.* 85: 248-253.

Sharp, W.R., Caldas, L.S., Crocomo, O.J., Monaco, L.C., and Carvalho, A. (1973). Production of *Coffea arabica* callus of three ploidy levels and subsequent morphogenesis. *Phyton* 31: 67-74.

Sharp, W.R., Raskin, R.S., and Sommer, H.E. (1972). The use of nurse culture in the development of haploid clones in tomato. *Planta* 104: 357-361.

Sharp, W.R., Söndahl, M.R., Caldas, L.S., and Maraffa, S.B. (1980). The physiology of *in vitro* asexual embryogenesis. *Hort. Rev.* 2: 268-310.

Shekhawat, N.S., and Galston, A.W. (1983). Isolation, culture, and regeneration of moth bean *Vigna aconitifolia* leaf protoplasts. *Plant Sci. Lett.* 32: 43-51.

Sheridan, W.F., and Neuffer, M.G. (1980). Defective kernel mutants of maize. II. Morphological and embryo culture studies. *Genetics* 95: 945-960.

Sheridan, W.F., and Neuffer, M.G. (1981). Maize mutants altered in embryo development. In: *Levels of Genetic Control in Development*, ed. S. Subtelny and U. K. Abbott, pp. 137-156. New York: Alan R. Liss.

Sheridan, W.F., and Neuffer, M.G. (1982). Maize developmental mutants. Embryos unable to form leaf primordia. *J. Hered.* 73: 318-329.

Shigenobu, T., and Sakamoto, S. (1977). Production of a polyhaploid plant of *Aegilops crassa* (6x) pollinated by *Hordeum bulbosum*. *Jap. J. Genet.* 52: 397-401.

Shii, C.T., Rabakoarihanta, A., Mok, M.C. and Mok, D.W.S. (1982). Embryo development in reciprocal crosses of *Phaseolus vulgaris* L. and *P. coccineus* Lam. *Theor. Appl. Genet.* 62: 59-64.

Simoncioli, C. (1974). Ultrastructural characteristics of "*Diplotaxis erucoides* (L.) DC." suspensor. *Gior. Bot. Ital.* 108: 175-189.

Singh, A.P., and Mogensen, H.L. (1975). Fine structure of the zygote and early embryo in *Quercus gambelii*. *Am. J. Bot.* 62: 105-115.

Singh, M., Bhalla, P.L., and Malik, C.P. (1979). Peroxidase localization during embryogenesis in *Tropaeolum majus*. *Phytomorphology* 29: 306-309.

Singh, M.B., Bhalla, P.L., and Malik, C.P. (1980a). Physiology of sexual reproduction VII. Localization and activity of aminopeptidases during early embryogenesis in *Tropaeolum majus* L. *Biochem. Physiol. Pflanzen* 175: 389-395.

Singh, M.B., Bhalla, P.L., and Malik, C.P. (1980b). Activity of some hydrolytic

enzymes in the autolysis of the embryo suspensor in *Tropaeolum majus* L. *Ann. Bot.* 45: 523-527.
Sinha, S., Jha, K.K., and Roy, R.P. (1978). Segmentation pattern of pollen in anther culture of *Solanum surattense*, *Luffa cylindrica* and *Luffa echinata*. *Phytomorphology* 28: 43-49.
Siriwardana, S., and Nabors, M.W. (1983). Tryptophan enhancement of somatic embryogenesis in rice. *Plant Physiol.* 73: 142-146.
Sita, G.L., Ram, N.V.R., and Vaidyanathan, C.S. (1980). Triploid plants from endosperm cultures of sandalwood by experimental embryogenesis. *Plant Sci. Lett.* 20: 63-69.
Sita, G.L., Shoba, J., and Vaidyanathan, C.S. (1980). Regeneration of whole plants by embryogenesis from cell suspension cultures of sandalwood. *Curr. Sci.* 49: 96-98.
Sivaramakrishna, D. (1978). Size relationships of apical cell and basal cell in two-celled embryos in angiosperms. *Can. J. Bot.* 56: 1434-1438.
Skene, K.G.M., and Barlass, M. (1983). *In vitro* culture of abscissed immature avocado embryos. *Ann. Bot.* 52: 667-672.
Smart, M.G., and O'Brien, T.P. (1983). The development of the wheat embryo in relation to the neighbouring tissues. *Protoplasma* 114: 1-13.
Smith, D.L. (1973). Nucleic acid, protein, and starch synthesis in developing cotyledons of *Pisum arvense* L. *Ann. Bot.* 37: 795-804.
Smith, J.G. (1973). Embryo development in *Phaseolus vulgaris*. II. Analysis of selected inorganic ions, ammonia, organic acids, amino acids, and sugar in the endosperm liquid. *Plant Physiol.* 51: 454-458.
Smith, S.M., and Street, H.E. (1974). The decline of embryogenic potential as callus and suspension cultures of carrot (*Daucus carota* L.) are serially subcultured. *Ann. Bot.* 38: 223-241.
Söndahl, M.R., and Sharp, W.R. (1977). High frequency induction of somatic embryos in cultured leaf explants of *Coffea arabica*. *Z. Pflanzenphysiol.* 81: 395-408.
Sopory, S.K. (1977). Development of embryoids in isolated pollen culture of dihaploid *Solanum tuberosum*. *Z. Pflanzenphysiol.* 84: 453-457.
Sopory, S.K. (1979). Effect of sucrose, hormones, and metabolic inhibitors on the development of pollen embryoids in anther cultures of dihaploid *Solanum tuberosum*. *Can. J. Bot.* 57: 2691-2694.
Sopory, S.K., Jacobsen, E., and Wenzel, G. (1978). Production of monohaploid embryoids and plantlets in cultured anthers of *Solanum tuberosum*. *Plant Sci. Lett.* 12: 47-54.
Sopory, S.K., and Maheshwari, S.C. (1976). Development of pollen embryoids in anther cultures of *Datura innoxia*. II. Effects of growth hormones. *J. Expt. Bot.* 27: 58-68.
Sopory, S.K., and Rogan, P.G. (1976). Induction of pollen divisions and embryoid formation in anther cultures of some dihaploid clones of *Solanum tuberosum*. *Z. Pflanzenphysiol.* 80: 77-80.
Souèges, R. (1919). Les premières divisions de l'oeuf et les différenciation du suspenseur chez le *Capsella bursa-pastoris* Moench (1). *Ann. Sci. Naturl. X Bot.* 1: 1-28.
Souèges, R. (1923). Embryogénie des Joncacées. Développement de l'embryon chez le *Luzula forsteri* DC. *Compt. Rend. Acad. Sci. Paris* 177: 705-708.
Souèges, R. (1932). Recherches sur l'embryogénie des Liliacées (*Muscari comosum* L.). *Bull. Soc. Bot. France* 79: 11-23.

Souèges, R. (1936). *La différenciation. III. La différenciation organique.* Paris: Hermann.
Souèges, R. (1937). Les lois du développement. Paris: Hermann.
Souèges, R. (1950). Embryogénie des Papilionacées. Développement de l'embryon chez le *Phaseolus vulgaris* L. *Compt. Rend. Acad. Sci. Paris* 231: 937-940.
Souèges, R. (1954). L'origine du cône végétatif de la tige et la question de la "Terminalite" du cotylédon des monocotylédones. *Ann. Sci. Naturl. XI Bot.* 15: 1-20.
Sree Ramulu, K., Devreux, M., Ancora, G., and Laneri, U. (1976). Chimerism in *Lycopersicum peruvianum* plants regenerated from *in vitro* cultures of anthers and stem internodes. *Z. Pflanzenzuchtg.* 76: 299-319.
Srinivasan, C., and Mullins, M.G. (1980). High-frequency somatic embryo production from unfertilized ovules of grapes. *Scientia Hort.* 13: 245-252.
Srivastava, P.S. (1982). Endosperm culture. In: *Experimental Embryology of Vascular Plants,* ed. B.M. Johri, pp. 175-193. Berlin: Springer-Verlag.
Stafford, A., and Davies, D.R. (1979). The culture of immature pea embryos. *Ann. Bot.* 44: 315-321.
Stanwood, P.C., and Bass, L.N. (1978). Ultracold preservation of seed germplasm. In: *Plant Cold Hardiness and Freezing Stress,* ed. P. H. Li and A. Sakai, pp. 361-371. New York: Academic Press.
Sterling, C. (1955). Embryogeny in the lima bean. *Bull. Torrey Bot. Cl.* 82: 325-338.
Steward, F.C. (1963). The control of growth in plant cells. *Sci. Am.* 209 (No. 4): 104-113.
Steward, F.C., Ammirato, P.V., and Mapes, M.O. (1970). Growth and development of totipotent cells. Some problems, procedures, and perspectives. *Ann. Bot.* 34: 761-787.
Steward, F.C., and Mapes, M.O. (1971). Morphogenesis and plant propagation in aseptic cultures of *Asparagus. Bot. Gaz.* 132: 70-79.
Steward, F.C., with Mapes, M.O., Kent, A.E., and Holsten, R.D. (1964). Growth and development of cultured plant cells. *Science* 143: 20-27.
Steward, F.C., Mapes, M.O., and Mears, K. (1958). Growth and organized development of cultured cells. II. Organization in cultures grown from freely suspended cells. *Am. J. Bot.* 45: 705-708.
Steward, F.C., Mapes, M.O., and Smith, J. (1958). Growth and organized development of cultured cells. I. Growth and division of freely suspended cells. *Am. J. Bot.* 45: 693-703.
Stewart, J.M., and Hsu, C.L. (1977). *In-ovulo* embryo culture and seedling development of cotton (*Gossypium hirsutum* L.). *Planta* 137: 113-117.
Stewart, J.M., and Hsu, C.L. (1978). Hybridization of diploid and tetraploid cottons through *in-ovulo* embryo culture. *J. Hered.* 69: 404-408.
Stinissen, H.M., Peumans, W.J., and de Langhe, E. (1984). Abscisic acid promotes lectin biosynthesis in developing and germinating rice embryos. *Plant Cell Rep.* 3: 55-59.
Stokes, P. (1953). The stimulation of growth by low temperature in embryos of *Heracleum sphondylium* L. *J. Expt. Bot.* 4: 222-234.
Strauss, A., Bucher, F., and King, P.J. (1981). Isolation of biochemical mutants using haploid mesophyll protoplasts of *Hyoscyamus muticus.* I. A NO_3 non-utilizing clone. *Planta* 153: 75-80.
Street, H.E., and Withers, L.A. (1974). The anatomy of embryogenesis. In: *Tissue Culture and Plant Science 1974,* ed. H.E. Street, pp. 71-100. London: Academic Press.

Stuart, D.A., and Strickland, S.G. (1984a). Somatic embryogenesis from cell cultures of *Medicago sativa* L. I. The role of amino acid additions to the regeneration medium. *Plant Sci. Lett.* 34: 165-174.
Stuart, D.A., and Strickland, S.G. (1984b). Somatic embryogenesis from cell cultures of *Medicago sativa* L. II. The interaction of amino acids with ammonium. *Plant Sci. Lett.* 34: 175-181.
Subrahmanyam, N.C. (1977). Haploidy from *Hordeum* interspecific crosses. I. Polyhaploids of *H. parodii* and *H. procerum*. *Theor. Appl. Genet.* 49: 209-217.
Subrahmanyam, N.C. (1979). Haploidy from *Hordeum* interspecific crosses. Part 2: Dihaploids of *H. brachyantherum* and *H. depressum*. *Theor. Appl. Genet.* 55: 139-144.
Subrahmanyam, N.C. (1980). Haploidy from *Hordeum* interspecific crosses. Part 3: Trihaploids of *H. arizonicum* and *H. lechleri*. *Theor. Appl. Genet.* 56: 257-263.
Sun, C. (1978). Androgenesis of cereal crops. In: *Proceedings of Symposium on Plant Tissue Culture*, pp. 117-123. Peking: Science Press.
Sun, C., Wang, C., and Chu, C. (1974). Cell division and differentiation of pollen grains in *Triticale* anthers cultured *in vitro*. *Scientia Sinica* 17: 47-54.
Sun, S.M., Buchbinder, B.U., and Hall, T.C. (1975). Cell-free synthesis of the major storage protein of the bean, *Phaseolus vulgaris* L. *Plant Physiol.* 56: 780-785.
Sun, S.M., Mutschler, M.A., Bliss, F.A., and Hall, T.C. (1978). Protein synthesis and accumulation in bean cotyledons during growth. *Plant Physiol.* 61: 918-923.
Sunderland, N. (1971). Anther culture: A progress report. *Sci. Prog. (Oxford)* 59: 527-549.
Sunderland, N. (1973). Pollen and anther culture. In: *Plant Tissue and Cell Culture*, Ist ed., ed. H.E. Street, pp. 205-239. Berkeley: University of California Press.
Sunderland, N. (1982). Induction of growth in the culture of pollen. In: *Differentiation* in vitro. British Society for Cell Biology Symposium 4, ed. M.M. Yeoman and D.E.S. Truman, pp. 1-24. Cambridge: Cambridge University Press.
Sunderland, N., Collins, G.B., and Dunwell, J.M. (1974). The role of nuclear fusion in pollen embryogenesis of *Datura innoxia* Mill. *Planta* 117: 227-241.
Sunderland, N., and Dunwell, J.M. (1977). Anther and pollen culture. In: *Plant Tissue and Cell Culture*, IInd ed., ed. H.E. Street, pp. 223-265. Berkeley: University of California Press.
Sunderland, N., and Evans, L.J. (1980). Multicellular pollen formation in cultured barley anthers. II. The A, B, and C pathways. *J. Expt. Bot.* 31: 501-514.
Sunderland, N., Huang, B., and Hills, G.J. (1984). Disposition of pollen *in situ* and its relevance to anther/pollen culture. *J. Expt. Bot.* 35: 521-530.
Sunderland, N., and Roberts, M. (1977). New approach to pollen culture. *Nature (London)* 270: 236-238.
Sunderland, N., and Roberts, M. (1979). Cold-pretreatment of excised flower buds in float culture of tobacco anthers. *Ann. Bot.* 43: 405-414.
Sunderland, N., Roberts, M., Evans, L.J., and Wildon, D.C. (1979). Multicellular pollen formation in cultured barley anthers. I. Independent division of the generative and vegetative cells. *J. Expt. Bot.* 30: 1133-1144.
Sunderland, N., and Wicks, F. (1971). Embryoid formation in pollen grains of *Nicotiana tabacum*. *J. Expt. Bot.* 22: 213-226.
Sunderland, N., and Wildon, D.C. (1979). A note on the pretreatment of excised flower buds in float culture of *Hyoscyamus* anthers. *Plant Sci. Lett.* 15: 169-175.
Sunderland, N., and Xu, Z.H. (1982). Shed pollen culture in *Hordeum vulgare*. *J. Expt. Bot.* 33: 1086-1095.

Sung, Z.R. (1979). Relationship of indole-3-acetic acid and tryptophan concentrations in normal and 5-methyltryptophan-resistant cell lines of wild carrots. *Planta* 145: 339-345.
Sung, Z.R., Lazar, G.B., and Dudits, D. (1981). Cycloheximide resistance in carrot culture: A differentiated function. *Plant Physiol.* 68: 261-264.
Sung, Z.R., and Okimoto, R. (1981). Embryonic proteins in somatic embryos of carrot. *Proc. Natl. Acad. Sci. U.S.A.* 78: 3683-3687.
Sung, Z.R., and Okimoto, R. (1983). Coordinate gene expression during somatic embryogenesis in carrot. *Proc. Natl. Acad. Sci. U.S.A.* 80: 2661-2665.
Sung, Z.R., Smith, R., and Horowitz, J. (1979). Quantitative studies of embryogenesis in normal and 5-methyltryptophan-resistant cell lines of wild carrot. The effects of growth regulators. *Planta* 147: 236-240.
Sussex, I. (1975). Growth and metabolism of the embryo and attached seedling of the viviparous mangrove, *Rhizophora mangle*. *Am. J. Bot.* 62: 948-953.
Sussex, I., Clutter, M., Walbot, V., and Brady, T. (1973). Biosynthetic activity of the suspensor of *Phaseolus coccineus*. *Caryologia Suppl.* 25: 261-272.
Sussex, I.M., and Dale, R.M.K. (1979). Hormonal control of storage protein synthesis in *Phaseolus vulgaris*. In: *Plant Seed: Development, Preservation and Germination*, ed. I. Rubenstein, R.L. Phillips, C.E. Green and B.G. Gengenbach, pp. 129-141. New York: Academic Press.
Sussex, I.M., and Frei, K.A. (1968). Embryoid development in long-term tissue cultures of carrot. *Phytomorphology* 18: 339-349.
Swamy, B.G.L. (1943). Gametogenesis and embryogeny of *Eulophea epidendraea* Fischer. *Proc. Natl. Inst. Sci. India* 9: 59-65.
Swamy, B.G.L. (1949). Embryological studies in the Orchidaceae. II. Embryogeny. *Am. Midl. Natur.* 41: 202-232.
Swamy, B.G.L. (1962). The embryo of monocotyledons: A working hypothesis from a new approach. In: *Plant Embryology. A Symposium*, pp. 113-123. New Delhi: Council of Scientific & Industrial Research.
Swamy, B.G.L. (1979). Embryogenesis in *Cheirostylis flabellata*. *Phytomorphology* 29: 199-203.
Swamy, B.G.L. (1980). Embryogenesis in *Sagittaria sagittaefolia*. *Phytomorphology* 30: 204-212.
Swamy, B.G.L., and Lakshmanan, K.K. (1962a). The origin of the epicotylary meristem and cotyledon in *Halophila ovata* Gaudich. *Ann. Bot.* 26: 243-249.
Swamy, B.G.L., and Lakshmanan, K.K. (1962b). Contributions to the embryology of the Najadaceae. *J. Indian Bot. Soc.* 41: 422-435.
Swamy, B.G.L., and Padmanabhan, D. (1961). Embryogenesis in *Sphenoclea zeylanica*. *Proc. Indian Acad. Sci.* 54B: 169-187.
Syamasundar, J., and Panchaksharappa, M.G. (1976). A histochemical study of some post-fertilization developmental stages in *Dipcadi montanum* Dalz. *Cytologia* 41: 123-130.
Tabaeezadeh, Z., and Khosh-Khui, M. (1981). Anther culture of *Rosa*. *Scientia Hort.* 15: 61-66.
Tagliasacchi, A.M., Forino, L.M.C., Cionini, P.G., Cavallini, A., Durante, M., Cremonini, R., and Avanzi, S. (1984). Different structure of polytene chromosome of *Phaseolus coccineus* suspensors during early embryogenesis. 3. Chromosome pair VI. *Protoplasma* 122: 98-107.
Tagliasacchi, A.M., Forino, L.M.C., Frediani, M., and Avanzi, S. (1983). Different structure of polytene chromosomes of *Phaseolus coccineus* suspensors during early embryogenesis. 2. Chromosome pair VII. *Protoplasma* 115: 95-103.

Taira, T., and Larter, E.N. (1977). Effects of ε-amino-*n*-caproic acid and L-lysine on the development of hybrid embryos of *Triticale* (× *Triticosecale*). *Can. J. Bot.* 55: 2330-2334.
Takahashi, A., Sakuragi, Y., Kamada, H., and Ishizuka, K. (1984). Plant regeneration through somatic embryogenesis in barnyardgrass, *Echinochloa oryzicola* Vasing. *Plant Sci. Lett.* 36: 161-163.
Takeno, K., Koshioka, M., Pharis, R.P., Rajasekaran, K., and Mullins, M.G. (1983). Endogenous gibberellin-like substances in somatic embryos of grape (*Vitis vinifera* × *Vitis rupestris*) in relation to embryogenesis and the chilling requirement for subsequent development of mature embryos. *Plant Physiol.* 73: 803-808.
Takeshita, M., Kato, M., and Tokumasu, S. (1980). Application of ovule culture to the production of intergeneric or interspecific hybrids in *Brassica* and *Raphanus*. *Jap. J. Genet.* 55: 373-387.
Tazawa, M., and Reinert, J. (1969). Extracellular and intracellular chemical environments in relation to embryogenesis *in vitro*. *Protoplasma* 68: 157-173.
Thanh-Tuyen, N.T., and de Guzman, E.V. (1983). Formation of pollen embryos in cultured anthers of coconut (*Cocos nucifera* L.). *Plant Sci. Lett.* 29: 81-88.
Thévenot, C., and Côme, D. (1971). Influence de la température et du mode d'imbibition sur la germination des embryons de pommier (*Pirus malus* L.) non dormants. *Compt. Rend. Acad. Sci. Paris* 273D: 2515-2517.
Thévenot, C., and Côme, D. (1973). Inhibition de la germination de l'axe embryonnaire par les cotylédons chez le pommier (*Pirus malus* L.). *Compt. Rend. Acad. Sci. Paris* 277D: 1873-1876.
Thomas, B.R., and Pratt, D. (1981). Efficient hybridization between *Lycopersicon esculentum* × *L. peruvianum* via embryo callus. *Theor. Appl. Genet.* 59: 215-219.
Thomas, E., Hoffmann, F., Potrykus, I., and Wenzel, G. (1976). Protoplast regeneration and stem embryogenesis of haploid androgenetic rape. *Mol. Gen. Genet.* 145: 245-247.
Thomas, E., Konar, R.N., and Street, H.E. (1972). The fine structure of the embryogenic callus of *Ranunculus sceleratus* L. *J. Cell Sci.* 11: 95-109.
Thomas, E., and Wenzel, G. (1975). Embryogenesis from microspores of *Brassica napus*. *Z. Pflanzenzuchtg.* 74: 77-81.
Tian, H., and Yang, H. (1983). Synergid apogamy and egg cell anomalous division in cultured ovaries of *Oryza sativa* L. *Acta Bot. Sinica* 25: 403-408.
Tilton, V.R. (1981a). Ovule development in *Ornithogalum caudatum* (Liliaceae) with a review of selected papers on angiosperm reproduction. IV. Egg apparatus structure and function. *New Phytol.* 88: 505-531.
Tilton, V.R. (1981b). The influence of individual embryonic tissues on the morphology and developemnt of *Zea mays* (Poaceae) germlings. *Am. J. Bot.* 68: 980-993.
Tilton, V.R., and Mogensen, H.L. (1979). Ultrastructural aspects of the ovule of *Agave parryi* before fertilization. *Phytomorphology* 29: 338-350.
Ting, Y.C., Yu, M., and Zheng, W. (1981). Improved anther culture of maize. *Plant Sci. Lett.* 23: 139-145.
Tisserat, B. (1979). Propagation of date palm (*Phoenix dactylifera* L.) in vitro. *J. Expt. Bot.* 30: 1275-1283.
Tisserat, B., and Murashige, T. (1977a). Effects of ethephon, ethylene, and 2,4-dichlorophenoxyacetic acid on asexual embryogenesis *in vitro*. *Plant Physiol.* 60: 437-439.
Tisserat, B., and Murashige, T. (1977b). Probable identity of substances in *Citrus* that repress asexual embryogenesis. *In Vitro* 13: 785-789.

Tisserat, B., and Murashige, T. (1977c). Repression of asexual embryogenesis *in vitro* by some plant growth regulators. *In Vitro* 13: 799-805.
Triplett, B.A., and Quatrano, R. S. (1982). Timing, localization, and control of wheat germ agglutinin synthesis in developing wheat embryos. *Develop. Biol.* 91: 491-496.
Tsay, H., and Tseng, M. (1979). Embryoid formation and plantlet regeneration from anther callus of sweet potato. *Bot. Bull. Acad. Sinica* 20: 117-122.
Tsay, H.S., and Su, C.Y. (1985). Anther culture of papaya (*Carica papaya* L.). *Plant Cell Rep.* 4: 28-30.
Tyagi, A.K., Rashid, A., and Maheshwari, S.C. (1979). High frequency production of embryos in *Datura innoxia* from isolated pollen grains by combined cold treatment and serial culture of anthers in liquid medium. *Protoplasma* 99: 11-17.
Tyagi, A.K., Rashid, A., and Maheshwari, S.C. (1981a). Sodium chloride resistant cell line from haploid *Datura innoxia* Mill. A resistance trait carried from cell to plantlet and vice versa *in vitro*. *Protoplasma* 105: 327-332.
Tyagi, A.K., Rashid, A., and Maheshwari, S.C. (1981b). Promotive effect of polyvinylpolypyrrolidone on pollen embryogenesis in *Datura innoxia*. *Physiol. Plant.* 53: 405-406.
Tykarska, T. (1979). Rape embryogenesis. II. Development of embryo proper. *Acta Soc. Bot. Poloniae* 48: 391-421.
Tykarska, T. (1980). Rape embryogenesis. III. Embryo development in time. *Acta Soc. Bot. Poloniae* 49: 369-385.
Ulrich, J.M., Finkle, B.J., and Tisserat, B.H. (1982). Effects of cryogenic treatment on plantlet production from frozen and unfrozen date palm callus. *Plant Physiol.* 69: 624-627.
Umbeck, P.F., and Norstog, K. (1979). Effects of abscisic acid and ammonium ion on morphogenesis of cultured barley embryos. *Bull. Torrey Bot. Cl.* 106: 110-116.
Vagera, J., and Havránek, P. (1982). *In vitro* regulation of androgenesis by iron ions and chelate: A common property of two androgenic species (*Nicotiana tabacum* L. and *Datura innoxia* Mill.). *Biol. Plant.* 24: 282-289.
Vallade, J. (1970). Développement embryonnaire chez un *Petunia hybrida* hort. *Compt. Rend. Acad. Sci. Paris* 270D: 1893-1896.
Vallade, J. (1972). Structure et fonctionnement du meristémè lors de la formation de la jeune racine primaire chez un *Petunia hybrida* hort. *Compt. Rend. Acad. Sci. Paris* 274D: 1027-1030.
Vallade, J., and Cornu, A. (1979). Blocage embryonnaire d'origine maternelle chez deux mutants de *Petunia hybrida*. *Bull. Soc. Bot. France Actual. Bot.* 126: 39-52.
Vallade, J., Cornu, A., Essad, S., and Alabouvette, J. (1978). Niveaux de DNA dans les noyaux zygotiques chez le *Petunia hybrida* hort. *Bull. Soc. Bot. France Actual. Bot.* 125: 253-258.
van der Pluijm, J.E. (1964). An electron microscopic investigation of the filiform apparatus in the embryo sac of *Torenia fournieri*. In: *Pollen Physiology and Fertilization*, ed. H.F. Linskens, pp. 8-16. Amsterdam: North-Holland Publishing Co.
van Overbeek, J., Conklin, M.E., and Blakeslee, A.F. (1942). Cultivation *in vitro* of small *Datura* embryos. *Am. J. Bot.* 29: 472-477.
van Overbeek, J., Siu, R., and Haagen-Smit, A.J. (1944). Factors affecting the growth of *Datura* embryos *in vitro*. *Am. J. Bot.* 31: 219-224.
van Staden, J., Gilliland, M.G., and Brown, N.A.C. (1975). Ultrastructure of dry viable and non-viable *Protea compacta* embryos. *Z. Pflanzenphysiol.* 76: 28-35.

van Went, J.L. (1970). The ultrastructure of the egg and central cell of *Petunia*. *Acta Bot. Neerl.* 19: 313-322.
Vardi, A., Spiegel-Roy, P., and Galun, E. (1975). *Citrus* cell culture: Isolation of protoplasts, plating densities, effect of mutagens and regeneration of embryos. *Plant Sci. Lett.* 4: 231-236.
Vasil, I.K. (1980). Androgenetic haploids. *Int. Rev. Cytol. Suppl.* 11A: 195-223.
Vasil, V., and Hildebrandt, A.C. (1965). Growth and tissue formation from single, isolated tobacco cells in microculture. *Science* 147: 1454-1455.
Vasil, V., and Vasil, I.K. (1980). Isolation and culture of cereal protoplasts. Part 2: Embryogenesis and plantlet formation from protoplasts of *Pennisetum americanum*. *Theor. Appl. Genet.* 56: 97-99.
Vasil, V., and Vasil, I.K. (1981a). Somatic embryogenesis and plant regeneration from suspension cultures of pearl millet (*Pennisetum americanum*). *Ann. Bot.* 47: 669-678.
Vasil, V., and Vasil, I.K. (1981b). Somatic embryogenesis and plant regeneration from tissue cultures of *Pennisetum americanum*, and *P. americanum* × *P. purpureum* hybrid. *Am. J. Bot.* 68: 864-872.
Vasil, V., and Vasil, I.K. (1982a). The ontogeny of somatic embryos of *Pennisetum americanum* (L.) K. Schum. I. In cultured immature embryos. *Bot. Gaz.* 143: 454-465.
Vasil, V., and Vasil, I.K. (1982b). Characterization of an embryogenic cell suspension culture derived from cultured inflorescences of *Pennisetum americanum* (Pearl millet, Gramineae). *Am. J. Bot.* 69: 1441-1449.
Vasil, V., Wang, D., and Vasil, I.K. (1983). Plant regeneration from protoplasts of napier grass (*Pennisetum purpureum* Schum.). *Z. Pflanzenphysiol.* 111: 233-239.
Vazart, J. (1969). Organisation et ultrastructure du sac embryonnaire du lin (*Linum usitatissimum* L.). *Rev. Cytol. Biol. Veg.* 32: 227-240.
Veen, H. (1963). The effect of various growth-regulators on embryos of *Capsella bursa-pastoris* growing *in vitro*. *Acta Bot. Neerl.* 12: 129-171.
Verma, D.C., and Dougall, D.K. (1977). Influence of carbohydrates on quantitative aspects of growth and embryo formation in wild carrot suspension cultures. *Plant Physiol.* 59: 81-85.
Verma, D.C., and Dougall, D.K. (1979a). Biosynthesis of *myo*-inositol and its role as a precursor of cell-wall polysaccharides in suspension cultures of wild-carrot cells. *Planta* 146: 55-62.
Verma, D.C., and Dougall, D.K. (1979b). *Myo*-inositol biosynthesis and galactose utilization by wild carrot (*Daucus carota* L. var. *carota*) suspension cultures. *Ann. Bot.* 43: 259-269.
Viegi, L., Pagni, A.M., Corsi, G., and Renzoni, G.C. (1976). Il sospensore embrionale nelle Cruciferae. I. Morfologia e struttura. *Giorn. Bot. Ital.* 110: 347-357.
Vijayaraghavan, M.R., and Prabhakar, K. (1984). The endosperm. In: *Embryology of Angiosperms*, ed. B.M. Johri, pp. 319-376. Berlin: Springer-Verlag.
Villiers, T.A., and Wareing, P. F. (1965). The growth-substance content of dormant fruits of *Fraxinus excelsior* L. *J. Expt. Bot.* 16: 533-544.
Vishnyakova, I.A., Krasnook, N.P., Povarova, R.I., Morgunova, E.A., and Bukhtoyarova, Z.T. (1976). Ultrastructure of cells of the embryos of viable and unviable rice seeds in the course of swelling. *Soviet Plant Physiol.* 23: 307-311.
Vujičić, R., Radojević, L., and Nešković, M. (1976). Orderly arrangement of ribosomes in the embryogenic callus tissue of *Corylus avellana* L. *J. Cell Biol.* 69: 686-692.

Walbot, V. (1971). RNA metabolism during embryo development and germination of *Phaseolus vulgaris. Develop. Biol.* 26: 369-379.
Walbot, V. (1973). RNA metabolism in developing cotyledons of *Phaseolus vulgaris. New Phytol.* 72: 479-483.
Walbot, V. (1978). Control mechanisms for plant embryogeny. In: *Dormancy and Developmental Arrest. Experimental Analysis in Plants and Animals,* ed. M.E. Clutter, pp. 113-166. New York: Academic Press.
Walbot, V., Brady, T., Clutter, M., and Sussex, I. (1972). Macromolecular synthesis during plant embryogeny: Rates of RNA synthesis in *Phaseolus coccineus* embryos and suspensors. *Develop. Biol.* 29: 104-111.
Walbot, V., Clutter, M., and Sussex, I.M. (1972). Reproductive development and embryogeny in *Phaseolus. Phytomorphology* 22: 59-68.
Walbot, V., and Dure, L.S. III. (1976). Developmental biochemistry of cotton seed embryogenesis and germination. VII. Characterization of the cotton genome. *J. Mol. Biol.* 101: 503-536.
Walbot, V., Harris, B., and Dure, L.S. III. (1975). The regulation of enzyme synthesis in the embryogenesis and germination of cotton. In: *The Developmental Biology of Reproduction,* ed. C.L. Markert, pp. 165-187. New York: Academic Press.
Walker, K.A., and Sato, S.J. (1981). Morphogenesis in callus tissue of *Medicago sativa:* The role of ammonium ion in somatic embryogenesis. *Plant Cell Tissue Organ Culture* 1: 109-121.
Walker, R.I. (1947). Megasporogenesis and embryo development in *Tropaeolum majus* L. *Bull. Torrey Bot. Cl.* 74: 240-249.
Wang, C., Chu, C., Sun, C., Wu, S., Yin, K., and Hsü, C. (1973). The androgenesis in wheat (*Triticum aestivum*) anthers cultured *in vitro. Scientia Sinica* 16: 218-222.
Wang, C., Sun, C., and Chu, Z. (1974). On the conditions for the induction of rice pollen plantlets and certain factors affecting the frequency of induction. *Acta Bot. Sinica* 16: 43-54.
Wang, D., and Vasil, I.K. (1982). Somatic embryogenesis and plant regeneration from inflorescence segments of *Pennisetum purpureum* Schum. (napier or elephant grass). *Plant Sci. Lett.* 25: 147-154.
Wang, D., Wergin, W.P., and Zimmerman, R.H. (1984). Somatic embryogenesis and plant regeneration from immature embryos of strawberry. *HortSci.* 19: 71-72.
Wang, D., and Yan, K. (1984). Somatic embryogenesis in *Echinochloa crusgalli. Plant Cell Rep.* 3: 88-90.
Wang, T., and Chang, C. (1978). Triploid *Citrus* plantlet from endosperm culture. *Scientia Sinica* 21: 823-827.
Wardlaw, C.W. (1955). *Embryogenesis in Plants.* London: Methuen & Co.
Wareing, P.F. (1982). Hormonal regulation of seed dormancy – Past, present and future. In: *The Physiology and Biochemistry of Seed Development, Dormancy and Germination,* IInd ed., ed. A.A. Khan, pp. 185-202. Amsterdam: Elsevier Biomedical Press.
Wareing, P.F., and Saunders, P.F. (1971). Hormones and dormancy. *Annu. Rev. Plant Physiol.* 22: 261-288.
Warren, G.S., and Fowler, M.W. (1977). A physical method for the separation of various stages in the embryogenesis of carrot cell cultures. *Plant Sci. Lett.* 9: 71-76.
Warren, G.S., and Fowler, M.W. (1981). Physiological interactions during the initial stages of embryogenesis in cultures of *Daucus carota* L. *New Phytol.* 87: 481-486.
Weatherhead, M.A., Burdon, J., and Henshaw, G.G. (1978). Some effects of

activated charcoal as an additive to plant tissue culture media. *Z. Pflanzenphysiol.* 89: 141-147.
Weatherhead, M.A., and Henshaw, G.G. (1979). The induction of embryoids in free pollen culture of potatoes. *Z. Pflanzenphysiol.* 94: 441-447.
Webster, G.T. (1955). Interspecific hybridization of *Melilotus alba* × *M. officinalis* using embryo culture. *Agron. J.* 47: 138-142.
Wenzel, G., Hoffmann, F., Potrykus, I., and Thomas, E. (1975). The separation of viable rye microspores from mixed populations and their development in culture. *Mol. Gen. Genet.* 138: 293-297.
Wenzel, G., and Uhrig, H. (1981). Breeding for nematode and virus resistance in potato via anther culture. *Theor. Appl. Genet.* 59: 333-340.
Wernicke, W., and Brettell, R. (1980). Somatic embryogenesis from *Sorghum bicolor* leaves. *Nature (London)* 287: 138-139.
Wernicke, W., Brettell, R., Wakizuka, T., and Potrykus, I. (1981). Adventitious embryoid and root formation from rice leaves. *Z. Pflanzenphysiol.* 103: 361-365.
Wernicke, W., Harms, C.T., Lörz, H., and Thomas, E. (1978). Selective enrichment of embryogenic microspore populations. *Naturwissenschaften* 65: 540.
Wernicke, W., and Kohlenbach, H.W. (1977). Versuche zur Kultur isolierter Mikrosporen von *Nicotiana* und *Hyoscyamus*. *Z. Pflanzenphysiol.* 81: 330-340.
Wernicke, W., Potrykus, I., and Thomas, E. (1982). Morphogenesis from cultured leaf tissue of *Sorghum bicolor* − The morphogenetic pathways. *Protoplasma* 111: 53-62.
Wetherell, D.F. (1984). Enhanced adventive embryogenesis resulting from plasmolysis of cultured wild carrot cells. *Plant Cell Tissue Organ Culture* 3: 221-227.
Wetherell, D.F., and Dougall, D.K. (1976). Sources of nitrogen supporting growth and embryogenesis in cultured wild carrot tissue. *Physiol. Plant.* 37: 97-103.
Wetherell, D.F., and Halperin, W. (1963). Embryos derived from callus tissue cultures of the wild carrot. *Nature (London)* 200: 1336-1337.
Wheeler, C.T., and Boulter, D. (1967). Nucleic acids of developing seeds of *Vicia faba* L. *J. Expt. Bot.* 18: 229-240.
Williams, E., and White, D.W.R. (1976). Early seed development after crossing of *Trifolium ambiguum* and *T. repens*. *New Zealand J. Bot.* 14: 307-314.
Williams, E.G., and de Lautour, G. (1980). The use of embryo culture with transplanted nurse endosperm for the production of interspecific hybrids in pasture legumes. *Bot. Gaz.* 141: 252-257.
Williams, E.G., Knox, R.B., Kaul, V., and Rouse, J.L. (1984). Post-pollination callose development in ovules of *Rhododendron* and *Ledum* (Ericaceae): Zygote special wall. *J. Cell Sci.* 69: 127-135.
Williams, L., and Collin, H.A. (1976). Embryogenesis and plantlet formation in tissue cultures of celery. *Ann. Bot.* 40: 325-332.
Wilmar, C., and Hellendoorn, M. (1968). Growth and morphogenesis of *Asparagus* cells cultured *in vitro*. *Nature (London)* 217: 369-370.
Wilms, H.J. (1981). Ultrastructure of the developing embryo sac of spinach. *Acta Bot. Neerl.* 30: 75-99.
Wilms, H.J., van Went, J.L., Cresti, M., and Ciampolini, F. (1983). Adventive embryogenesis in *Citrus*. *Caryologia* 36: 65-78.
Wilson, H.J., Israel, H.W., and Steward, F.C. (1974). Morphogenesis and the fine structure of cultured carrot cells. *J. Cell Sci.* 15: 57-73.
Wilson, H.M. (1977). Culture of whole barley spikes stimulates high frequencies of pollen calluses in individual anthers. *Plant Sci. Lett.* 9: 233-238.

Wilson, H.M., Mix, G., and Foroughi-Wehr, B. (1978). Early microspore divisions and subsequent formation of microspore calluses at high frequency in anthers of *Hordeum vulgare* L. *J. Expt. Bot.* 29: 227-238.
Wilson, H.M., and Street, H.E. (1975). The growth, anatomy and morphogenetic potential of callus and cell suspension cultures of *Hevea brasiliensis*. *Ann. Bot.* 39: 671-682.
Wilson, K.J., and Mahlberg, P.G. (1977). Investigations of laticifer differentiation in tissue cultures derived from *Asclepias syriaca* L. *Ann. Bot.* 41: 1049-1054.
Withers, L.A. (1979). Freeze preservation of somatic embryos and clonal plantlets of carrot (*Daucus carota* L.). *Plant Physiol.* 63: 460-467.
Wochok, Z.S. (1973). Microtubules and multivesicular bodies in cultured tissues of wild carrot: Changes during transition from the undifferentiated to the embryonic condition. *Cytobios* 7: 87-95.
Wochok, Z.S., and Wetherell, D.F. (1972). Restoration of declining morphogenetic capacity in long term tissue cultures of *Daucus carota* by kinetin. *Experientia* 28: 104-105.
Woodard, J.W. (1956). DNA in gametogenesis and embryogeny in *Tradescantia*. *J. Biophys. Biochem. Cytol.* 2: 765-776.
Woodard, J.W. (1958). Intracellular amounts of nucleic acids and protein during pollen grain growth in *Tradescantia*. *J. Biophys. Biochem. Cytol.* 4: 383-389.
Woodcock, C.L.F., and Bell, P.R. (1968). Features of the ultrastructure of the female gametophyte of *Myosurus minimus*. *J. Ultrastr. Res.* 22: 546-563.
Wright, D.J., and Boulter, D. (1972). The characterisation of vicilin during seed development in *Vicia faba* (L.). *Planta* 105: 60-65.
Wright, J.E., and Srb, A.M. (1950). Inhibition of growth in maize embryos by canavanine and its reversal. *Bot. Gaz.* 112: 52-57.
Xu, Z., and Huang, B. (1984). Anther factor(s) in barley anther culture. *Acta Bot. Sinica* 26: 1-10.
Xu, Z., Wang, D., Yang, L., and Wei, Z. (1984). Somatic embryogenesis and plant regeneration in cultured immature inflorescences of *Setaria italica*. *Plant Cell Rep.* 3: 149-150.
Xu, Z.H., Huang, B., and Sunderland, N. (1981). Culture of barley anthers in conditioned media. *J. Expt. Bot.* 32: 767-778.
Xu, Z.H., and Sunderland, N. (1981). Glutamine, inositol and conditioning factor in the production of barley pollen callus *in vitro*. *Plant Sci. Lett.* 23: 161-168.
Yakovlev, M.S., and Yoffe, M.D. (1957). On some peculiar features in the embryogeny of *Paeonia* L. *Phytomorphology* 7: 74-82.
Yang, C., Pan, N., and Liu, C. (1979). Plantlet formation in *Brassica pekinensis* Rupr. on the anther culture. *Acta Bot. Sinica* 21: 378-379.
Yang, H., and Zhou, C. (1979). Experimental researches on the two pathways of pollen development in *Oryza sativa* L. *Acta Bot. Sinica* 21: 345-351.
Yang, H., and Zhou, C. (1984). Observations on megasporogenesis and megametophyte development in *Paulownia* sp. and *Sesamum indicum* by enzymatic maceration technique. *Acta Bot. Sinica* 26: 355-358.
Yang, H.Y., and Zhou, C. (1982). *In vitro* induction of haploid plants from unpollinated ovaries and ovules. *Theor. Appl. Genet.* 63: 97-104.
Yeung, E.C. (1980). Embryogeny of *Phaseolus:* The role of the suspensor. *Z. Pflanzenphysiol.* 96: 17-28.
Yeung, E.C., and Brown, D.C.W. (1982). The osmotic environment of developing embryos of *Phaseolus vulgaris*. *Z. Pflanzenphysiol.* 106: 149-156.

Yeung, E.C., and Clutter, M.E. (1978). Embryogeny of *Phaseolus coccineus:* Growth and microanatomy. *Protoplasma* 94: 19-40.

Yeung, E.C., and Clutter, M.E. (1979). Embryogeny of *Phaseolus coccineus:* The ultrastructure and development of the suspensor. *Can. J. Bot.* 57: 120-136.

Yeung, E.C., and Sussex, I.M. (1979). Embryogeny of *Phaseolus coccineus:* The suspensor and the growth of the embryo-proper *in vitro. Z. Pflanzenphysiol.* 91: 423-433.

Yoo, B.Y., and Chrispeels, M.J. (1980). The origin of protein bodies in developing soybean cotyledons: A proposal. *Protoplasma* 103: 201-204.

You, R., and Jensen, W.A. (1985). Ultrastructural observations of the mature megagametophyte and the fertilization in wheat (*Triticum aestivum*). *Can. J. Bot.* 63: 163-178.

Zapata, F.J., and Sink, K.C. (1981). Somatic embryogenesis from *Lycopersicon peruvianum* leaf mesophyll protoplasts. *Theor. Appl. Genet.* 59: 265-268.

Zatykó, J.M., Kiss, F., and Szalay, F. (1981). Induction of adventive embryony in cultured ovules of black currant (*Ribes nigrum* L.). *Hort. Res.* 21: 99-101.

Zatykó, J.M., Simon, I., and Szabó, C. (1975). Induction of polyembryony in cultivated ovules of red currant. *Plant Sci. Lett.* 4: 281-283.

Zee, S.-Y., Wu, S.C., and Yue, S.B. (1979). Morphological and SDS-polyacrylamide gel electrophoretic studies of pro-embryoid formation in petiole explants of Chinese celery. *Z. Pflanzenphysiol.* 95: 397-403.

Zeng, J. (1983). Application of anther culture technique to crop improvement in China. In: *Plant Cell Culture in Crop Improvement,* ed. S. K. Sen and K. L. Giles, pp. 351-363. New York: Plenum Press.

Zenkteler, M., Misiura, E., and Ponitka, A. (1975). Induction of androgenetic embryoids in the *in vitro* cultured anthers of several species. *Experientia* 31: 289-291.

Zenkteler, M., and Nitzsche, W. (1984). Wide hybridization experiments in cereals. *Theor. Appl. Genet.* 68: 311-315.

Zenkteler, M., and Stefaniak, B. (1982). Induction of androgenesis in anthers of *Hordeum vulgare* L. cultured *in vitro* on leaves and calluses. *Plant Sci. Lett.* 26: 219-225.

Zhong, Z., Ren, Y., and Dai, W. (1978). Preliminary studies on the anther culture of *Brassica chinensis. Acta Bot. Sinica* 20: 180-181.

Zhou, C., and Yang, H. (1980). Anther culture and androgenesis of *Hordeum vulgare* L. *Acta Bot. Sinica* 22: 211-215.

Zhou, C., and Yang, H. (1982). Enzymatic isolation of embryo sacs in angiosperms: Isolation and microscopical observation on fixed materials. *Acta Bot. Sinica* 24: 403-407.

Zhou, C., and Yang, H. (1984). The enzymatic isolation of embryo sacs from fixed and fresh ovules of *Antirrhinum majus* L. *Acta Biol. Expt. Sinica* 17: 141-147.

Zhou, J. (1980). Pollen dimorphism and its relation to the formation of pollen embryos in anther culture of wheat (*Triticum aestivum*). *Acta Bot. Sinica* 22: 117-121.

Zhou, M.D., and Lee, T.T. (1984). Selectivity of auxin for induction and growth of callus from excised embryo of spring and winter wheat. *Can. J. Bot.* 62: 1393-1397.

Zhu, Z., Shen, R., and Tang, X. (1980a). Studies on the developmental biology of embryogenesis in higher plants. II. Biochemical changes during embryogenesis in *Oryza sativa* L. *Acta Phytophysiol. Sinica* 6: 141-148.

Zhu, Z., Shen, R., and Tang, X. (1980b). Studies on the developmental biology of embryogenesis in higher plants. III. Kinetic changes of nucleic acids and protein during embryogenesis of wheat (*Triticum vulgaris* L.). *Acta Bot. Sinica* 22: 122-126.

Zhu, Z., Sun, J., and Wang, J. (1978). Cytological investigation on androgenesis of *Triticum aestivum*. *Acta Bot. Sinica* 20: 6-12.

Ziebur, N.K., Brink, R.A., Graf, L.H., and Stahmann, M.A. (1950). The effect of casein hydrolysate on the growth *in vitro* of immature *Hordeum* embryos. *Am. J. Bot.* 37: 144-148.

Zou, C., and Li, P. (1981). Induction of pollen plants of grape (*Vitis vinifera* L.). *Acta Bot. Sinica* 23: 79-81.

Author index

Abo El-Nil, M.M., 157
Ackerson, R.C., 107, 110
Adams, W.R. Jr., 75
Ahée, J., 127
Ahloowalia, B.S., 130
Ahn, C.S., 216
Ahuja, U., 217
Alabouvette, J., 52
Al-Abta, S., 124, 145
Alpi, A., 70–1, 102
Alvarez, M.N., 215
Alvarez, M.R., 50
Ammirato, P.V., 120, 146–7, 221
Amos, J.A., 155
Amssa, M., 174
An, H., 161
Anand, V.V., 155, 168
Ancora, G., 188
Andersen, J.N., 41
Anderson, S., 161
Andersson, B., 156, 158
Andrews, C.J., 91
Andronescu, D.I., 96
Anguillesi, M.C., 80
Arditti, J., 94
Arekal, G.D., 155, 168
Armstrong, K.C., 156, 213
Armstrong, T.A., 200
Arnott, H.J., 25
Arora, S., 157
Arthuis, P., 127
Ascher, P.D., 215
Ashihara, H., 198
Ashley, T., 48, 209
Aspart, L., 196
Atal, C.K., 131
Augsten, H., 87
Avanzi, S., 65–7

Ba, L.T., 23
Babbar, S.B., 155, 179
Backs-Hüsemann, D., 119
Baenziger, P.S., 159, 187
Bailey, D.S., 74

Bajaj, Y.P.S., 133, 153, 157, 162, 181, 187, 212–14, 216–17, 226–8
Baker, J.C., 79
Balfour, E., 27
Ballo, B.L., 109
Ban, H., 166, 168
Banerjee, N., 124
Banerjee, S., 124
Bapat, V.A., 126, 155
Barclay, I., 36
Barclay, I.R., 213
Barlass, M., 217
Barrow, J., 157
Barthe, P., 92
Barton, K.A., 76
Bass, L.N., 227
Bates, L.S., 213
Battaglia, E., 41, 192
Batygina, T.B., 24, 86
Bayliss, M.W., 34, 36, 89
Beachy, R.N., 34, 74, 76, 78
Bedford, I.D., 80
Beevers, L., 56
Bell, P.R., 47
Bell, S.L., 157
Benbadis, A., 157
Ben-Hayyim, G., 140, 148
Bennett, M.D., 34, 36, 52, 186
Bennici, A., 70–1, 102, 127
Beranger-Novat, N., 91
Bergstresser, K.A., 68
Berjak, P., 80
Bernard, S., 168
Beversdorf, W.D., 138
Beyl, C.A., 127
Bhalla, P.L., 52, 69, 70–1
Bhatnagar, S.P., 17, 37, 42
Bhatia, D.S., 68
Bhojwani, S.S., 17, 175
Bi, P., 158
Bianco, J., 91
Bigot, C., 133
Binding, H., 128, 224
Binding, K., 224
Bingham, E.T., 138

Bitters, W.P., 131
Blakeslee, A.F., 87, 98, 100
Bliss, F.A., 75–6
Bloom, K.S., 41
Bohdanowicz, J., 63
Bollini, R., 72
Bolton, J.L., 214
Bornman, C.H., 135
Borthwick, H.A., 119
Botha, C.E.J., 125
Botti, C., 128, 134
Bouharmont, J., 159, 168
Boulter, D., 53–4, 73, 75, 79
Bourgin, J., 225
Bourgin, J.P., 133
Bowes, B.G., 124
Bown, D., 73, 79
Boyes, C.J., 128
Boyle, S.A., 52
Braak, J.P., 215
Bradley, P. M., 122
Brady, T., 55, 62, 65, 67, 69, 70–1
Bray, C.M., 81–2
Breekland, A.E., 109, 111
Breidenbach, R.W., 53, 77, 193–4
Breton, A.M., 150
Brettell, R., 128
Brettell, R.I.S., 128, 161
Brewster, V., 56
Bright, S.W.J., 87–8
Brink, R.A., 99, 106, 210
Brinkhorst-van der Swan, D.L.C., 109, 111
Brocklehurst, P.A., 81
Broekaert, D., 57
Broué, P., 218
Brown, D.C.W., 100
Brown, J.W.S., 41
Brown, N.A.C., 80
Brown, S., 140
Buchbinder, B.U., 76
Bucher, F., 225
Buell, K.M., 25
Buffard-Morel, J., 87
Bui Dang Ha, D., 133, 161
Bukhtoyarova, Z.T., 80
Bulard, C., 91–3
Bullen, M.R., 214
Burdon, J., 182
Burghardtová, K., 86
Burkholder, P.R., 87
Button, J., 125, 135

Cadic, A., 183
Caldas, L.S., 123, 157
Caldiroli, E., 81
Camefort, H., 172–3
Cameron, J.W., 86

Cameron-Mills, V., 86
Cantliffe, D.J., 124, 131
Caponetti, J.D., 157
Carasco, J.F., 75
Carlson, P.S., 138, 224–5
Carniel, K., 25
Carvalho, A., 157
Cas, G., 127
Casey, R., 74
Cass, D.D., 47–8
Cauderon, Y., 217
Cavallini, A., 66
Cavé, G., 23
Cave, M.S., 25
Ceccarelli, N., 71
Chaleff, R.S., 159
Chandler, S.F., 130
Chandra, N., 131
Chandramohan, M., 131
Chang, C., 126
Chang, W., 124
Chang, W.C., 124
Chatterjee, A., 81
Cheah, K.S.E., 82
Chen, C., 158, 166
Chen, J., 34, 74, 78
Chen, M., 161, 166, 168
Chen, X., 128
Chen, Y. 165
Chen, Z., 157
Cheng, J., 126
Cheng, K., 161
Cheng, W., 161
Chien, N., 159
Chlan, C., 75, 109
Chlan, C.A., 79, 195
Choi, J.H., 199, 200
Choinski, J.S. Jr., 109
Chopra, R.N., 43
Chow, T. -Y., 81
Chowdhury, J.B., 217
Chrispeels, M.J., 72
Christianson, M.L., 138
Chu, C., 155, 157–8, 161, 166, 170, 175, 180, 187
Chu, C.C., 128
Chu, Z., 159
Chuang, C., 159
Chueca, M., 217
Chupeau, Y., 133
Ciampolini, F., 43
Cionini, P.G., 65–7, 70–1, 102
Clapham, D., 159, 166
Clapham, D.H., 181
Clowes, F.A.L., 30
Clutter, M., 34, 55, 69, 70, 105
Clutter, M.E., 38, 59, 61, 65, 67

Cockerline, A.W., 89
Cocking, E.C., 133
Cocucci, A., 47–9
Collin, H.A., 124, 145
Collins, G.B., 138, 148, 155, 161, 169, 188, 216, 219–21
Côme, D., 91–3, 97
Conger, B.V., 89, 128, 130–1
Conklin, M.E., 98
Conover, R.A., 126
Constabel, F., 127
Constantin, M.J., 156–7
Cook, S.A., 25
Cooper, D.C., 210
Cooper, K.V., 220
Corduan, G., 157, 182
Cornejo-Martin, M.J., 158, 163
Cornish, M., 156, 179
Cornu, A., 52, 112
Corsi, G., 62–3, 101
Cregan, P.B., 223
Cremonini, R., 65–7
Cresti, M., 39, 43
Crété, P., 17–8
Crocomo, O.J., 156–7
Cronauer, S., 145
Cross, J.W., 75
Crouch, M.L., 34, 75, 79, 105, 107, 109, 120, 156
Croy, R., 75
Croy, R.R.D., 73, 79
Cullis, C.A., 57
Cummings, D.P., 89

Dai, W., 155
D'Alascio-Deschamps, R., 47, 49
Dale, J.E., 34, 211, 220
Dale, P.J., 89, 128, 185
Dale, R.M.K., 80, 105, 107, 109
D'Amato, F., 65–7, 70–1, 102
Danovich, K.N., 75
Dasgupta, J., 82
Das Gupta, P., 81
Dauphiné, A., 28
Davey, M.R., 133
Davidson, E.H., 194
Davies, D.R., 56, 74, 80, 100
Davis, B.D., 87
Davis, D.W., 215
Davis, G.L., 2, 18
Deambrogio, E., 89
Debata, B.K., 157
Debergh, P., 162
de Buyser, J., 159, 165, 174, 187
de Courcel, A.G.L., 156, 179
de Guzman, E.V., 87, 161, 217
de Langhe, E., 103, 109, 124

de la Roche, A.I., 156
de Lautour, G., 218
de Leo, P., 81
Dell'Aquila, A., 81
del Rosario, A.G., 87
del Rosario, D.A., 217
Delseny, M., 196
Deluca, V., 74
DeMott, K.J., 130
de Nettancourt, D., 214
Dennis, F.G. Jr., 131
Derbyshire, E., 75
Deschamps, R., 49
Deumling, B., 68
Devi, H.M., 44
Devreux, M., 156, 188, 224
Dhanju, M.S., 157, 213
Dhankhar, B.S., 217
Dhillon, S.S., 53
Diboll, A.G., 47–9
Diez, J.L., 67
Ditta, G.S., 53, 77, 193–4
Dix, P.J., 224
Doležel, J., 217
Dollmantel, H.-J., 165
Domoney, C., 74
Donovan, C.M., 87–8
Dorion, N., 133
Dougall, D.K., 137–8, 140, 201, 227
Douglass, J., 218
Drew, R.L.K., 147, 221
Drira, N., 157
Du, R., 175
Du, Z.H., 128
Dublin, P., 131
Dudits, D., 150
Dudman, W.F., 73
Duffus, C., 207
Duffus, C.M., 53–4, 86
Duncan, E.J., 179
Dunn, S.D.M., 89
Dunwell, J.M., 86, 156, 166, 169, 172–3, 175, 177, 179–80, 182, 187–8
Durand, M., 91–2
Durante, M., 65–7
Dure, III., 75, 77, 79, 80, 106–7, 109, 195–6
Dure, L.S., 90
Dure, L.S., III, 34, 53, 106
Durran, V., 182
Duval, Y., 127
Dyer, A.F., 220
Dykes, T.A., 130

Eeuwens, C.J., 40
Egorova, N.A., 158
Eichholtz, D.A., 131

Eid, A.A.H., 103
El-Fiki, F., 122
Emershad, R.L., 220
Emerson, C.P., 7
Engelhardt, M., 157
Engvild, K.C., 188
Enns, R.K., 141
Epstein, E., 144
Erdelská, O., 38
Eriksson, T., 147, 156, 158
Ernst, D., 145–6
Ersland, D.R., 41
Esau, K., 95
Esen, A., 42–3, 211
Essad, S., 52, 174
Eusebio, E.C., 87
Evans, A.M., 215
Evans, I.M., 73, 79
Evans, L.J., 166–70, 188

Facciotti, D., 133
Favre-Duchartre, M., 15
Fedak, G., 213, 217
Feirer, R.P., 200
Ferl, R., 75, 79
Fernandez, A., 157
Fienberg, A.A., 199, 200
Finger, J., 128
Finkelstein, R.R., 75, 105
Finkle, B.J., 227
Fischer, R.L., 54
Fisher, D.B., 52
Fitch, M.M., 161
Fitch, M.M.M., 163
Floris, C., 80–1
Foard, D.E., 96
Folsom, M.W., 48
Fong, F., 110
Forino, L.M.C., 65–6
Forman, M., 51
Foroughi-Wehr, B., 159, 166, 187–8
Forster, B.P., 34, 211, 220
Fowke, L.C., 39, 61
Fowler, M.W., 92, 121, 147
Franckowiak, J.D., 213
Franklin, J., 89
Fraser, R.S.S., 81
Frediani, M., 66
Freed, H.J., 62–3
Frei, K.A., 138
Fridborg, G., 147
Fridriksson, S., 214
Friedt, W., 187
Fujimura, T., 122, 141, 143, 145–6, 197–8, 202
Furuya, M., 89

Galau, G.A., 75, 77, 79, 80, 107, 109, 195–6
Galitz, D.S., 56
Galston, A.W., 133
Galun, E., 133
Gamborg, O.L., 127, 132
Ganapathy, P.S., 120, 126, 131
Ganguly, S.N., 81
Gao, G., 166, 168
Gärtner, P.-J., 68–9
Gatehouse, J.A., 73, 79
Gaul, H., 159
Gavazzi, G., 113
Gayler, K.R., 74
Gazit, S., 126, 131
Gebhardt, C., 225
Gennai, D., 127
Genovesi, A.D., 159, 161
George, L., 155
Gharyal, P.K., 126, 157
Ghosh, P.D., 157
Giles, K.L., 122
Gill, B.S., 217
Gill, K.S., 213
Gill, M.S., 217
Gilliland, M.G., 80
Giovannozzi-Sermanni, G., 81
Giuliano, G., 122
Gleddie, S., 144, 148
Glover, D.V., 41
Godard, M., 161
Godin, B., 133
Godineau, J.-C., 47–8
Goel, S., 155
Goldberg, R.B., 53–4, 77–8, 113, 193–4
Gonzalez-Medina, M., 159
Gosal, S.S., 157, 212, 216
Gosch, G., 133
Grace, J.P., 218
Graf, L.H., 106
Graham, E.T., 128
Graham, T.A., 73
Grambow, H.J., 132
Granatek, C.H., 89
Granström, H., 67
Grant, W.F., 62–3, 214
Gray, D., 34
Gray, D.J., 130–1
Green, C.E., 87–9
Greenway, S.C., 107, 195
Gregor, D., 202
Grewal, S., 131
Gröbler, A., 133
Grout, B.W.W., 227
Grunewaldt, J., 159
Gu, S., 179
Gu, Z., 161

Guénin, G., 127
Guha, S., 152, 158
Guha-Mukherjee, S., 158
Gui, Y., 126, 179
Guignard, J., 23, 134
Guignard, J.L., 23
Gunning, B.E.S., 73
Gupta, N., 158
Gupta, S., 124, 157
Gupta, S.C., 155, 179

Haagen-Smit, A.J., 86, 98
Haber, A.H., 96
Haccius, B., 44, 115, 131
Hall, T.C., 41, 75–6
Halperin, W., 117, 119, 121, 135, 137, 141, 145
Hamblin, M.T., 109
Hammerschlag, F.A., 156
Hanawa, J., 27
Hang, A., 213
Hang, L., 161, 166, 168
Hannig, E., 2, 85
Hanning, G.E., 128, 130–1
Hanower, J., 127
Hanower, P., 127
Harada, H., 138, 141, 147, 163, 180
Harashima, S., 163
Hardham, A.R., 39
Harms, C.T., 162
Harris, B., 34
Harris, G.P., 87
Harry, E., 134
Hartmann, R.W., 216
Hasegawa, P.M., 120, 131
Hasitschka-Jenschke, G., 62–3
Hausner, G., 131
Havránek, P., 180
Haydu, Z., 128
He, Z. (Ho, C.), 225
Heberle, E., 162, 179
Heberle-Bors, E., 163, 178, 180, 185–7
Heirwegh, K.M.G., 124
Hellendoorn, M., 127
Henry, M., 23
Henry, Y., 165, 174
Henshaw, G.G., 163, 182
Hepler, P.K., 173
Hermsen, J.G.T., 186
Hesemann, C.U., 156
Heyne, E.G., 159
Heyser, J.W., 128, 130
Hidaka, T., 157
Hildebrandt, A.C., 116, 157
Hills, G.J., 183
Hinchee, M.A.W., 163
Ho, W., 130, 134

Hoffmann, F., 131, 133, 157, 162
Hoffmann, M., 157
Holsten, R.D., 117
Hòmes, J.L.A., 122, 130
Honma, S., 86, 214
Horner, M., 162, 182–3, 185
Horowitz, J., 145
Hoschek, G., 53, 77–8, 113, 193–4
Howell, R.W., 56
Hsing, Y., 124
Hsing, Y.I., 124
Hsü, C., 158–9, 166
Hsu, C.L., 103, 219
Hsu, F.C., 110
Hu, C., 94
Hu, C.Y., 94, 124, 131
Hu, H., 159
Hu, S., 38–9, 61
Huang, B., 160, 183
Huang, C., 161
Huang, P.C., 111
Huang, Q., 42
Huber, J., 146
Hugard, J., 156
Hughes, K.W., 157
Hughes, W.G., 186
Huhnke, W., 157
Humphreys, T., 7
Hymowitz, T., 220

Idzikowska, K., 173, 183
Ihle, J.N., 106
Ilami, M., 183–4
Imamura, J., 163, 180
Ingram, D.S., 187
Inomata, N., 220
Irikura, Y., 155
Isaia, A., 91
Ishizuka, K., 128
Israel, H.W., 135
Ivanov, V.N., 75
Iyer, R.D., 131, 158, 168

Jacobsen, E., 155, 187
Jalouzout, M.-F., 47
Janick, J., 120, 126, 131, 146
Jarvis, B.C., 91–3
Jaworski, E.G., 141, 199, 200
Jean, R., 157
Jenkins, J.A., 112
Jensen, C.J., 213
Jensen, W., 50–1
Jensen, W.A., 33, 38–9, 47–52, 59, 61, 90, 135
Jha, K.K., 157, 166
Jia, S.E., 187
Johansen, D.A., 15, 18
Johansson, L., 156, 158, 182

Johri, B.M., 2, 4, 42, 103, 120, 131
Jonard, R., 156
Jones, L.H., 134, 227
Jones, P.A., 30
Jones, R.A., 41
Jones, W.T., 218
Joó, F., 224
Jordan, M., 156
Joshi, P.C., 103
Juncosa, A.M., 27-8

Kallarackal, J., 37
Kamada, H., 128, 138, 141, 147
Kameya, T., 132
Kamiński, W., 92
Kao, K., 159
Kao, K.N., 132-3, 148, 192, 213
Kaplan, D.R., 25, 27-9
Karas, I., 47-8
Karihaloo, J.L., 186
Karnosky, D.F., 157
Karssen, C.M., 109, 111
Kasha, K.J., 130, 192, 213
Kato, H., 117, 122, 138, 143
Kato, M., 220
Katterman, F., 157
Kaul, V., 49
Kavathekar, A.K., 120, 126, 131
Kazimierska, E.M., 209
Keim, W.F., 215
Keller, W., 144, 148
Keller, W.A., 155-6, 180
Kent, A.E., 117
Kent, N., 99, 106
Keyes, G.J., 148
Khan, S.K., 157
Khavkin, E.E., 75
Khosh-Khui, M., 156
Kikuta, Y., 198
King, P.J., 225
King, R.W., 110
Kiss, F., 131
Klein, R.M., 106
Klimaszewska, K., 155
Knight, R.L., 126, 131
Knight, R.J. Jr., 126
Knox, R.B., 49
Kochba, J., 124-5, 135, 140, 143-4, 146, 148, 201
Koda, Y., 126
Koehler, D.E., 110
Kohlenbach, H.W., 123, 133, 162, 182
Koller, D., 96
Komamine, A., 122, 141, 143, 145-6, 197-8, 202
Konar, R.N., 124, 130, 136
Kononowicz, A.K., 126, 146

Kononowicz, H., 126
Kooistra, E., 215
Koornneef, M., 109
Koppenbrink, J.W., 199
Koshioka, M., 146
Kott, L.S., 130
Koul, A.K., 186
Kovoor, J., 173
Kowyama, Y., 52
Krasnook, N.P., 80
Kreitner, G.L., 39, 61, 210-11
Krikorian, A.D., 145, 222
Krul, W.R., 126, 131
Krumbiegel-Schroeren, G., 128
Kruse, A., 213, 218
Ku, M., 161
Ku, S., 161, 166, 168
Kuan, Y., 161
Kubicki, B., 156
Kühner, S., 61
Kumar, P., 214
Kumar, P.P., 131
Kuo, C., 161, 166, 168
Kuo, L., 161
Kwei, Y., 161, 166, 168
Kyo, M., 163

Labana, K.S., 157, 214
Lacadena, J., 192
Lacapra, J., 155, 180
La Cour, L.F., 174
Laibach, F., 211
Lakshmanan, K.K., 22-3
Lanaud, C., 131
Lancaster, V.A., 89
Landström, L., 147
Laneri, U., 156, 188
Lang, H., 123
Larkins, B.A., 41
Laroche-Raynal, M., 196
Larson, D.A., 47-8
Larter, E.N., 213
Lavee, S., 144, 201
Lázár, G., 224
Lazar, G.B., 150
Lazar, M.D., 150, 187
Lee, D.W., 201
Lee, T.T., 89
Leelavathi, S., 155
Lersten, N.R., 31
Li, L., 133
Li, M., 186
Li, P., 157
Liang, G.H., 159
Lichter, R., 163
Lievoux, D., 127
Lima-de-Faria, A., 67

Lin, M., 158
Lin, S., 47
Linde-Laursen, I., 188
Linder, R., 157
Linstedt, D., 145
Lioret, C., 127
Litvay, J.D., 200
Litz, R.E., 126, 131
Liu, C., 155
Liu, H., 175
Liu, J.R., 124, 131
Loh, C.-S., 187
Long, S.R., 80, 105, 107, 109
Loo, S., 225
Lorenzi, R., 70–1, 102
Lörz, H., 133, 162
Lu, C., 128, 130, 133
Lu, D., 165
Lu, D.Y., 133
Lu, W., 161–2, 166, 168
Lubich, W.P., 199, 200
Lundqvist, A., 188
Lupi, C., 127
Lužný, J., 217
Lyne, R.L., 220

Ma, Y., 76
Maddock, S.E., 89
Madison, J.T., 76
Magill, C.W., 159
Maheshwari, P., 2, 13, 15–18, 37, 41–3, 186
Maheshwari, S.C., 126, 152–3, 155, 157, 162, 165, 178–9, 181–2, 184, 225
Maheswaran, G., 131
Mahlberg, P.G., 27–8, 120
Malaurie, B., 127
Malepszy, S., 159
Malhotra, K., 155, 179
Maliga, P., 224–5
Malik, C.P., 52, 68–71
Manitto, P., 113
Mansfield, M., 109
Manteuffel, R., 54
Mapes, M.O., 116–17, 122, 127
Maraffa, S.B., 123
Maretzki, A., 130
Marinos, N.G., 38–9, 61
Marsden, M.P.F., 112
Marshall, D.R., 218
Martin, A.C., 94
Márton, L., 224
Masset, A., 133
Mastubayashi, M., 167
Masuda, K., 126, 198
Mathan, D.S., 112
Matsubara, S., 87, 99, 100

Matsumoto, H., 197, 202
Maze, J., 47
McComb, A.J., 182, 187
McComb, J.A., 182, 187
McDaniel, J.K., 128, 131
McDonnell, R.E., 89
McWilliam, A.A., 119
Mears, K., 116
Meinke, D.W., 34, 74, 78, 112
Meletti, P., 81
Mericle, L.W., 52
Mericle, R.P., 52
Mestre, J., 134
Mestre, J. -C., 17
Miao, S., 161, 166, 168
Michayluk, M.R., 133, 148
Michellon, R., 156
Miflin, B.J., 87–8
Mignon, G., 200
Mii, M., 184
Miksche, J.P., 53
Milewska-Pawliczuk, E., 156
Miller, H.A., 25, 27–8
Miller, R.A., 127, 131
Millerd, A., 53, 56–7, 73
Misharin, S.I., 75
Misiura, E., 156
Misoo, S., 167
Mittal, A. (née Vishnoi), 155
Mix, G., 159, 166, 188
Młodzianowski, F., 183
Mogensen, H.L., 47–51
Mok, D.W.S., 209, 215
Mok, M.C., 209, 215
Mokhtarzadeh, A., 156–7
Monaco, L.C., 157
Monin, J., 91, 93
Monnier, M., 97, 101–2
Montague, M.J., 141, 199, 200
Monti, D., 113
Moore, P.H., 161
Morgunova, E.A., 80
Morris, D.A., 110, 210
Moss, J.P., 212, 214
Mott, R.L., 185
Mroginski, L.A., 157
Mu, X., 126
Mudgal, A.K., 155
Mujeeb, K.A., 213
Mukkada, A.J., 32
Müller, A.J., 112
Mullins, M.G., 120, 125–6, 146
Muniyamma, M., 42
Müntz, K., 54
Murashige, T., 127, 131, 142, 146–7
Mutschler, M.A., 75

Nabors, M.W., 128, 130
Nag, K.K., 227
Nagato, Y., 34, 191
Nagl, W., 61-6, 68-9, 71, 101
Nagmani, R., 163, 182
Nagy, F., 224
Nakajima, T., 99
Nakamura, C., 217
Napier, K.V., 217
Narang, K., 157
Narayan, K.N., 42
Narayanaswami, S., 90
Nast, C., 27-8
Nataraja, K., 130
Natesh, S., 18
Naumova, T.N., 42
Negbi, M., 96
Nešković, M., 126, 135
Nesling, F.A.V., 210
Nessler, C.L., 120
Neuffer, M.G., 111, 113. 161
Neumann, H., 140, 144, 146, 148
Newcomb, W., 39, 47, 51, 61
Newell, C.A., 219
Niimi, Y., 103
Niizeki, H., 153
Nitsch, C., 155, 161-2, 179, 183-4, 224
Nitsch, J.P., 155, 168, 224
Nitzsche, W., 209, 223
Noma, M., 146
Norreel, B., 51, 133, 157-8, 162-3, 179, 183
Norstog, K., 39, 61, 90, 95, 99, 106, 127
Noshiro, M., 227
Novák, F.J., 217

O'Brien, T.P., 38
Oesterhelt, D., 145
Oettler, G., 213
Okabe, E., 163
Okazawa, Y., 126, 198
Okimoto, R., 150, 199
Ono, K., 156, 163
Oono, K., 153
Osborne, D.J., 81-2
Ottma, M., 133
Ouyang, J.W., 187
Ouyang, T., 159
Ozias-Akins, P., 128, 133
Ozsan, M., 86

Pacini, E., 39
Paddock, E.F., 111, 156
Padmanabhan, D., 25, 27
Pagni, A.M., 62-3
Pal, A., 157
Palevitz, B.A., 173

Pan, C., 159
Pan, J., 166, 168
Pan, N., 155
Panchaksharappa, M.G., 51
Pandey, S.N., 50-1
Pannetier, C., 127
Pareek, L.K., 131
Paris, D., 87
Patnaik, S.N., 157
Payne, P.I., 81
Pecket, R.D., 214
Pedersen, K., 41
Pedersén, M., 147
Pelletier, G., 161, 182-4
Pence, V.C., 120, 131
Penon, P., 196
Péreau-Leroy, P., 224
Periasamy, K., 17
Pernes, J., 161
Pero, R., 67
Peschke, C., 68
Peters, J.E., 156
Peterson, C.M., 48
Peumans, W.J., 109
Pharis, R.P., 146
Philip, V.J., 48
Phillips, G.C., 138, 148, 216, 221
Phillips, R.L., 87, 89
Picard, E., 159, 187
Picciarelli, P., 70
Pickering, R.A., 213
Pieniażek, J., 92
Pierson, E.S., 131
Pilet, P.-E., 133
Pissarev, W.E., 218
Pistelli, L., 70
Pizzolongo, P., 61
Pollock, E.G., 33
Ponitka, A., 156, 183
Ponzi, R., 61
Potrykus, I., 128, 131, 133-4, 162
Poulson, R., 56
Povarova, R.I., 80
Prabhakar, K., 37, 61
Pratt, D., 220
Pratt, M.L., 183
Preil, W., 157
Preťová, A., 100
Price, H.J., 139
Primo-Millo, E., 158, 163
Pritchard, H.N., 50-1, 68
Pritchard, H.W., 227
Przybyllok, T., 71
Püchel, M., 54
Pullaiah, T., 44
Puri, P., 192

Pyle, J., 109
Pyle, J.B., 79
Pyne, J.W., 76

Qian, C., 157
Qin, M., 157
Quatrano, R.S., 34, 107, 109
Quebedeaux, B., 110

Rabakoarihanta, A., 209, 215
Racchi, M.L., 113
Radojević, L., 126, 135, 157, 173
Raghavan, P., 48
Raghavan, V., 4, 40, 85, 90, 98, 101–2, 126, 136, 163, 168, 174–7, 182–4, 197–8, 203–5, 212, 217
Raillot, D., 127
Raina, S.K., 168
Rajasekaran, K., 120, 126, 146
Rajhathy, T., 155, 180
Raju, C.R., 131
Ram, A.K., 157
Ram, M., 33
Ram, M.V.R., 126
Ramanna, M.S., 186
Ramji, M.V., 25
Ramming, D., 217
Ramming, D.W., 220
Randolph, L.F., 24
Rangan, T.S., 96, 128, 131, 148
Rangaswamy, N.S., 42–3, 96, 124
Rao, M.K., 34, 36
Rao, P.S., 126, 133, 155
Raquin, C., 161, 174
Rashid, A., 130, 153, 157, 162–3, 165, 168, 173, 178, 181–2, 184, 225
Raskin, R.S., 182
Rau, M.A., 18
Reddy, V.S., 155
Redenbaugh, M.K., 157
Reed, S.M., 219–20
Reeve, R.M., 27
Reichert, H., 44
Reinert, J., 117, 119, 133, 137–9, 145, 162–3, 165, 173, 178, 181, 185, 202, 226
Ren, Y., 155
Renzoni, G.C., 62–3
Reuther, G., 127
Reynolds, J.F., 127
Reynolds, T.L., 155, 174, 176, 203
Reznikova, S.A., 158
Rietsema, J., 87, 100
Rijven, A.H.G.C., 86–7, 100
Rines, H.W., 161
Risiott, R., 89
Rivière, S., 28
Roberts, B.E., 81–2

Roberts, M., 165, 168, 170, 178–9
Robichaud, C.S., 110
Robitaille, H.A., 131
Rodriguez R,R., 213
Rogalski, F., 94
Rogan, P.G., 155
Rogers, S.O., 34
Rondet, P., 29, 51
Rosati, P., 156
Rosellini, D., 122
Rosie, R., 53–4
Ross, M.D., 218
Rouse, J.L., 49
Rowell, J.C., 110
Roy, R.P., 157, 166
Rudnicki, R., 92
Russell, S.D., 47
Ryczkowski, M., 100

Saad, S., 140, 146, 148
Saccardo, F., 224
Sachar, R.C., 41–2
Sachdeva, U., 131
Sagawa, Y., 50
Saha, P.K., 81
Sakai, A., 227
Sakamoto, S., 213
Sakuragi, Y., 128
Salem, S., 145
Sanders, M.E., 87
Sandha, G.S., 213
Sangduen, N., 39, 61, 159, 210–11
Sangwan, R.S., 157–8, 163, 172–3, 184
Sangwan-Norreel, B.S., 172, 174–5, 179, 183
San Noeum, L.H., 192
Sarfatti, G., 39
Sastri, D.C., 212, 214
Satina, S., 87, 100
Sato, S.J., 138
Saunders, P.F., 93
Sauter, J.J., 174
Scalet, M., 70
Schaeffer, G.W., 158–9, 187, 223
Scharpé, A., 53, 56–7
Schel, J.H.N., 131
Schlosser-Szigat, G., 214
Schnebli, V., 225
Schnepf, E., 61
Scholl, R.L., 155
Scholz, G., 54
Schroeren, V., 128
Schuchman, R., 120
Schulz, P., 38–9, 59, 61
Schulz, R., 47–51, 59
Schwabe, W.W., 40
Sears, R.G., 159

Sehgal, C.B., 157
Selim, A.R.A.A., 214
Semenoff, S., 137–8
Sen, S., 82
Sen, S.K., 155
Sengupta, C., 74, 198
Sethi, M., 42
Setterfield, G., 144, 148
Shah, C.K., 50–1
Shannon, P.R.M., 93
Shanthamma, C.K., 42
Sharma, D.R., 217
Sharma, G.C., 127
Sharma, H.C., 217
Sharp, W.R., 123, 156–7, 182
Sharpe, F.T., Jr., 223
Shekhawat, N.S., 133
Shelton, K., 227
Shen, R., 53, 55
Sheridan, W.F., 11, 113, 161
Shichijo, T., 157
Shigenobu, T., 213
Shii, C.T., 209, 215
Shoba, J., 126
Shrotria, A., 42
Siddiqui, A.W., 173
Siegel, N.R., 141
Simon, A.E., 75, 79
Simon, G., 161
Simon, I., 130–1
Simon, M., 73
Simoncioli, C., 38–9, 61
Simpson, G.M., 91
Singh, A.P., 50–1
Singh, H., 157
Singh, M., 68, 71
Singh, M.B., 52, 69, 70
Singh, M.M., 214
Sinha, S., 157, 166
Sink, K.C., 131, 133
Sircar, S.M., 81
Siriwardana, S., 130
Sita, G.L., 126
Siu, R., 86, 98
Sivaramakrishna, D., 16
Skene, K.G.M., 217
Slaughter, C., 207
Smart, M.G., 38
Smith, D.L., 53, 56
Smith, J., 99, 116
Smith, J.B., 34, 36, 52
Smith, J.D., 110
Smith, J.G., 100
Smith, R., 145
Smith, R.H., 139
Smith, S.M., 119, 222
Soma, K., 89

Sommer, H.E., 182
Söndahl, M.R., 123
Soost, R.K., 42–3, 211
Sopory, S.K., 155, 163, 180–1, 187
Sorensen, E.L., 39, 61, 210–11
Souèges, R., 16–7, 19, 22
Spencer, D., 56–7, 73
Spiegel-Roy, P., 124, 133, 140, 143–4, 146, 148, 201
Spix, C., 157
Srb, A.M., 87
Sree Ramulu, K., 188
Srinivasan, C., 125
Srivastava, P.S., 41, 85
Stafford, A., 100
Stahle, U., 67
Stahmann, M.A., 106
Stanwood, P.C., 227
Staritsky, G., 131
Steckel, J.R.A., 34
Steeves, T.A., 39
Stefaniak, B., 182
Sterling, C., 29
Stern, H., 73
Steward, F.C., 116–17, 122, 127, 135, 221
Stewart, J.M., 103, 219
Stiller, M., 73
Stinissen, H.M., 109
Stokes, P., 87–8
Stolarz, A., 159
Straub, J., 224
Strauss, A., 225
Street, H.E., 115, 119, 124, 126, 130, 135–6, 162, 168, 182, 185, 222, 224, 227
Strickland, S.G., 138
Stuart, D.A., 138
Stuthman, D.D., 89
Su, C.Y., 157
Subrahmanyam, N.C., 213
Sun, A., 161, 166, 168
Sun, C., 158–9, 166, 168, 170
Sun, C.S., 128
Sun, J., 166, 170
Sun, S.M., 75–6
Sunderland, N., 155, 160, 165–70, 172–3, 175, 178–9, 183, 185–8
Sung, Z.R., 142, 145, 149–50, 199, 200
Sussex, I., 55, 69, 70, 110
Sussex, I.M., 34, 71, 75, 79, 80, 102, 105, 107, 109–10, 112, 126, 131, 138
Suthar, H.K., 47–9
Sváb, Z., 224
Swaminathan, M.S., 158
Swamy, B.G.L., 22–3, 25, 27, 31, 42, 168
Sweetser, P.B., 110
Syamasundar, J., 51

Sykes, G.E., 74
Szabó, C., 130-1
Szalay, F., 131
Sz.-Breznovits, A., 224

Tabaeezadeh, Z., 156
Tagliasacchi, A.M., 65-6
Taira, T., 213
Takahashi, A., 128
Takeno, K., 146
Takeshita, M., 220
Takeuchi, M., 117, 138
Tam, S.H., 77, 193-4
Tang, X., 53, 55
Taylor, N.L., 148, 216, 221
Tazawa, M., 137-9
Tegenkamp, I., 156
Tegenkamp, T., 156
Telezynska, J., 156
Tempé, J., 217
Tenbarge, K.L., 75, 79
Terzi, M., 122
Thanh-Tuyen, N.T., 161
Thapar, N., 68
Thévenot, C., 91-3, 97
Thomas, B.R., 220
Thomas, E., 124, 128, 131, 133-4, 136, 155, 157, 161-2
Thomas, H.M., 213
Thomas, J.B., 213
Thompson, J.F., 76
Tian, H., 42
Tian, W., 165
Tilton, V.R., 47-8, 96
Ting, Y.C., 161
Tisserat, B., 127, 142, 146-7
Tisserat, B.H., 227
Tognoni, F., 70-1, 102
Tokumasu, S., 220
Torrey, J.G., 90, 101
Trelease, R.N., 109
Triplett, B.A., 107, 109
Tsai, C.Y., 41
Tsay, H., 157-8
Tsay, H.S., 157
Tseng, C., 159
Tseng, M., 157-8
Tsukida, T., 156
Tupý, J., 86
Tyagi, A.K., 153, 155, 162, 165, 178, 181-2, 184, 225
Tykarska, T., 24, 61

Uchimiya, H., 132
Uhrig, H., 187
Ulrich, J.M., 227
Umbeck, P.F., 106

Vagera, J., 180
Vaidyanathan, C.S., 126
Vallade, J., 29, 51-2, 112
van der Pluijm, J.E., 47-8
van Gyseghem, R., 68
van Lammeren, A.A.M., 131
van Nerum, K., 1214
van Overbeek, J., 86, 98
van Parijs, R., 53, 56-7
van Staden, J., 80
van Went, J.L., 43, 47-8
Vardi, A., 133
Varechon, C., 127
Vasil, I.K., 127-8, 130, 133-4, 148, 155, 161, 166, 170
Vasil, V., 116, 127-8, 133-4
Vasilyeva, V.E., 86
Vazart, J., 47-8
Veen, H., 90, 100
Verma, D.C., 138, 140
Verma, D.P.S., 74
Verma, M.M., 213
Vermani, S., 68
Viegi, L., 62-3
Vijayaraghavan, M.R., 37, 61
Villiers, T.A., 80, 93
Vine, J., 120
Vinogradova, N.M., 218
Vishnyakova, I.A., 80
Vodkin, L.O., 78, 113
Vujičić, R., 126, 135

Waines, J.G., 217
Wakizuka, T., 128
Walbot, V., 34, 53-6, 69, 70, 105, 111
Walker, J.T., 220
Walker, K.A., 138
Walker, R.I., 33
Walthall, E.D., 71
Walton, P.D., 217
Wang, C., 158-9, 166, 170
Wang, D., 128, 131, 133
Wang, J., 166, 170
Wang, R., 165
Wang, T., 126
Wang, Y., 161, 166, 168
Ward, C., 94
Ward, J.A., 34
Wardlaw, C.W., 3
Wareing, P.F., 93
Warnick, D.A., 138
Warren, G.S., 121, 147
Waterkeyn, L., 103
Waters, R.F., 213
Weatherhead, M.A., 163, 182
Weber, A., 133
Webster, G.T., 214

Wei, Z., 128
Wellmann, E., 120
Wenzel, G., 131, 133, 155, 157, 162, 187, 223
Wergin, W.P., 131
Wernicke, W., 128, 134, 161-2, 182
Westfall, R.D., 157
Wetherell, D.F., 117, 121, 137-8, 140-1, 145, 227
Wetmore, R.H., 25, 27-8
Wheeler, C.T., 53-4
White, D.W.R., 209-10
Whitfeld, P.R., 53, 56
Wicks, F., 166, 178, 185
Wildon, D.C., 168, 170, 179
Willemse, M.T.M., 42
Williams, D., 157
Williams, E., 209-10
Williams, E.G., 49, 131, 218
Williams, L., 124
Williamson, J.D., 109
Wilmar, C., 127
Wilms, H.J., 43, 47-8
Wilson, D., 91-2
Wilson, D.A., 92
Wilson, H.J., 135
Wilson, H.M., 126, 159, 166, 188
Wilson, K.J., 120
Withers, L.A., 115, 135, 227
Wochok, Z.S., 135, 145
Wong, J., 110
Wood, E.A., 87-8
Woodard, J.W., 52, 174
Woodcock, C.L.F., 47
Worley, J., 159
Worley, J.F., 126, 131
Wright, D.C., 120
Wright, D.J., 73
Wright, J.E., 87
Wu, M., 161, 166, 168
Wu, S., 166
Wu, S.C., 201

Xiao, Y., 157
Xu, S., 225
Xu, T., 126, 179
Xu, X., 157
Xu, Z., 128, 160, 225
Xu, Z.H., 160, 165, 183

Yakovlev, M.S., 24, 26
Yamada, Y., 157
Yan, K., 128
Yang, C., 155
Yang, H., 15, 42, 166, 170
Yang, H.Y., 192
Yang, L., 128
Yeung, E.C., 38, 52, 58, 61, 71, 100, 102
Yin, K., 158-9, 166
Yoffe, M.D., 24, 26
Yoo, B.Y., 72
Yoshida, K., 167
You, R., 47
Yu, M., 161
Yue, S.B., 201

Zapata, F.J., 133
Zatykó, J.M., 130-1
Zee, S.-Y., 38-9, 61, 201
Zeng, J., 223
Zenkteler, M., 156, 182, 209
Zháng, G., 165
Zhang, W.X., 128
Zheng, S., 165
Zheng, W., 161
Zhong, Z., 155
Zhou, C., 15, 42, 166, 170, 192
Zhou, J., 185
Zhou, M.D., 89
Zhou, S.M., 187
Zhu, C., 38-9, 61
Zhu, Z., 53, 55, 166, 170
Ziebur, N.K., 106
Zimmerman, R.H., 131
Zocchi, G., 81
Zou, C., 157
Zuckerman, L., 127
Zuo, Q., 165
Zylberberg, L., 173

Subject index

abscisic acid (ABA), 79, 80, 128, 147
  in embryo dormancy, 92–3
  levels in seeds, 110
  in precocious germination, 106, 107f, 109–11
  restoration of embryoid form by, 146–7
  in storage protein synthesis, 79, 80,
ABA-deficient mutants, 109–11
ABA-induced genes, 109
ab initio pollen cultures, 162–5, 172–3
Actinidia chinensis, 126
actinomycin D
  effects on chromatin fibers, 64–5
  effects on pollen embryogenesis, 203–4
activated charcoal, 147, 155, 161, 182
adenine (adenine sulfate), 90, 98, 101, 125
adventive embryos (asexual embryos), 41–4
Aegilops, 161
  crassa × *Hordeum bulbosum*, 213
  squarrosa × *Triticum boeoticum*, 217
Aerva javanica, 192
Aesculus
  hippocastanum, 157, 173
  woerlitzensis, 40
affinity chromatography, 8
afterripening, 91, 217
Agave, 48
  parryi, 47
Agropyron, 161
  spicatum × *intermedium*, 217
Albizzia lebbeck, 157
alfalfa (*see also Medicago sativa*), 138
Alisma
  lanceolata, 62
  plantago-aquatica, 62–3
Allium cepa × *A. fistulosum*, 217
Alyssum maritimum, 29, 61
Ammi majus, 130
amino acids
  antagonism and synergism between, 87–8
  enhancement of somatic embryogenesis by, 138
  growth induction in cultured embryos by, 87
  morphogenetic effects on embryos of, 88

  promotion of pollen embryogenesis by, 184
ammonium ion, in somatic embryogenesis, 137–9
androgenesis, 152
Androsaemum (=*Hypericum*), 17
Anemone, 156
  canadensis, 182
  coronaria, 169
anther, amino acid changes in, 184
anther culture, 154–61
anther extract, 162, 183
anther wall
  effect on pollen embryogenesis of, 162
  somatic embryos from, 126
antiauxins, in somatic embryogenesis, 142
antipodals, 14, 42
Apium graveolens, 123
apomict, 192
apple (*see also Pyrus malus*), 91, 93, 97, 131, 156
Arabidopsis, 155, 206
  thaliana, 109, 110, 112
Arachis
  correntina, 157
  glabrata, 157
  hypogaea, 157, 187
  hypogaea × *glabrata*, 214t
  hypogaea × *monticola*, 214t
  hypogaea × *villosa*, 214t
  villosa, 157
archesporial cell, 13
Asclepias syriaca, 120
Asparagus, 161
  officinalis, 127, 133
Asterad type of embryo, 17, 49
Atropa, 180
  belladonna, 124, 130, 133, 154, 169, 187, 228
autoradiography, technique of, 5
autotrophic phase, of embryos, 86, 124
auxins (*see also* individual listings)
  in embryo growth, 88–9, 214–16
  in pollen embryogenesis, 155, 157–8, 181–2
  in somatic embryogenesis, 140–5

294

auxotrophs, 113, 148, 206, 224
Avena, 141, 161
  fatua, 91
  sativa, 87

barley (*see also Hordeum vulgare*), 34, 39, 40, 42, 51, 87–90, 95, 99, 106, 127, 130, 159, 160, 165, 168, 169, 183, 187, 188, 209, 211–13
Bellevalia romana, 127
benzyladenine (BA), 124–5, 217
benzylaminopurine (BAP), 155, 159, 181
bisporic embryo sac, 14
Brachiaria setigera, 42
Brassica, 223
  campestris, 68, 155–6, 180, 187, 220
  chinensis, 155
  hirta (Sinapis alba), 155
  juncea, 155
  napus, 61, 75, 79, 105, 107, 109, 131, 133, 155–6, 163, 179
  napus spp. *oleifera*, 156, 187
  nigra, 62
  oleracea, 131, 220
  pekinensis, 155
Bromus inermis, 127
bulbosum method, 213
Bruguiera exaristata, 27, 28

cacao (*see also Theobroma cacao*), 131, 146
Cajanus cajan, 157, 187
callus, from pollen grains, 153, 155–61, 181–2
callus-specific proteins, 150, 206
Capsella, 19, 22, 25, 29, 30, 33, 39, 47–51, 58, 90, 102
  bursa-pastoris, 18, 20f, 21f, 38, 47, 86–7, 89, 100–1
Capsicum
  annuum, 154, 166
  frutescens, 154
caraway (*see also Carum carvi*), 146
carbohydrates
  effects on cultured embryos, 86–7
  in pollen embryogenesis, 155, 161, 180–1
  in somatic embryogenesis, 140
Carica papaya, 125, 157
carrot (*see also Daucus carota*), 6, 34, 116–17, 118f, 119–21, 127, 130, 132, 134–8, 139t, 140–1, 142t, 143, 145–50, 182, 197–9, 200f, 201f, 202, 206, 221, 222, 227
carrot cells
  cycloheximide-resistance in, 149f
  5-methyltryptophan-resistance in, 142, 145, 148
carrot embryo, development of, 119
Carum carvi, 146

Caryophyllad type of embryo, 17, 59
casein hydrolyzate, in precocious germination, 106
Cassia siamea, 157
Cassipourea elliptica, 27–8
Cassytha filiformis, 96
castor bean (*see also Ricinus communis*), 28, 40
celery (*see also Apium graveolens*), 123–4, 145, 201
cell doubling times, in embryos, 34, 211
cell-free protein synthesis (*in vitro* protein synthesis), 8, 41, 76, 79, 195–6
central cell, 14
Chenopodiad type of embryo, 17, 49
Chicorium intybus, 124
chimera, 168, 224
chromosome banding, 64–5
chromosome elimination, 192
Cicer, 31
  arietinum, 157, 187
  soongaricum, 32f
Citrus, 42, 86, 124, 131, 140, 146, 148, 211
  grandis, 126
  limon, 157
  medica, 147
  mitis (microcarpa), 124
  reticulata, 42, 43f, 147
  sinensis, 42–3, 124, 133
Clematis, 156
clonal multiplication, 221–2
Cochlearia, 85
coconut (*see also Cocos nucifera*), 40, 87, 131, 217
coconut milk, embryo factor from, 98
Cocos, 161
  nucifera, 40
Coelogyne, 31
Coffea
  arabica, 123, 157
  canephora, 131
coffee (*see also Coffea arabica*), 123, 131
Coix, 161
cold stress
  of anthers and pollen grains, 160, 162–3, 173, 183
  of tillers, spikes, and flower buds, 160–1, 163, 165, 178–80
coleoptile, 23, 94, 106
coleorhiza, 23, 94–6
complementary DNA (cDNA), 9, 10, 73, 76–7, 79, 109, 193–4, 196
  cloning of, 10–11, 77
conditioned medium, 160, 162, 183
conglycinin, 74, 76, 77–8
connective, somatic embryos from, 126
convicilin, 73

Subject index

cordycepin, 198
Coriandrum sativum, 158
Corylus avellana, 91, 135
Cot value, 54
cotton (*see also Gossypium hirsutum*), 33, 39, 47–8, 50–3, 75–80, 90, 103, 104f, 106–7, 108f, 109, 195f, 196, 217, 219
cotyledons
  DNA synthesis in, 52–4
  effects on growth of embryos of, 93, 96–7
  initiation of , 19
  RNA and protein metabolism of, 56–8
  storage proteins of, 73–80, 106
cowpea (*see also Vigna unguiculata*), 75
Crambe maritima, 124
Crepis, 48
  tectorum, 47
Crotalaria pallida, 157
Crucifer type of embryo, 16, 19, 22, 119
cruciferin, 75, 79, 105, 156
cryopreservation, 226–8
Cucubalus baccifer, 62
cucumber (*see also Cucumis sativus*), 99
Cucumis sativus, 99
Cucurbita
  maxima, 99
  moschata, 99
C values for DNA, 52, 63t
Cyananchum vincetoxicum, 131
Cymbidium, 31
Cypripedium, 31
Cytisus laburnum, 70
cytochemical analysis
  of pollen development, 174–5
  of pollen embryogenesis, 174–5
cytokinins (*see also* individual listings)
  effects on embryo growth, 90, 214–16
  in embryo dormancy, 92, 93
  in pollen embryogenesis, 152–5, 157–8, 181–2
  promotion of somatic embryogenesis by, 145

Dactylis glomerata, 128, 130–1, 132f
Dactylorhiza maculata, 86
Datura, 154, 180, 187
  innoxia, 87, 152–3, 155, 162, 165–6, 168–9, 172, 174–5, 178–9, 181–4, 186, 188
  metel, 166, 172, 179, 184
  stramonium, 86–7, 98, 99f, 100
  tatula, 100
Daucus carota, 6
decline in embryogenic potential, 145, 147
deoxyribonucleic acid (DNA)
  amplification, 53
  content of egg and zygote, 52, 66–8
  content of suspensor cells, 62, 63t
  in nonviable embryos, 82
  in pollen grains and pollen embryoids, 174–5
  underreplication of, 67–8
Dianthus chinensis, 25, 62
2, 4-dichlorophenoxyacetic acid (2,4-D)
  callus induction by, 89
  in pollen embryogenesis, 153, 155, 159, 181–2
  in somatic embryogenesis, 117, 121, 124–7, 141–4, 146–7, 150, 158, 199–201
Dicraea stylosa, 31
Digitalis purpurea, 157, 187
α-difluoromethylarginine, 200, 201f
dimorphic pollen grains, 185–6
Diplotaxis erucoides, 38–9, 61
DNA–DNA hybridization, 9, 10, 53–4, 55f
dormancy
  embryo culture and, 91
  role of hormones in, 91–4
  in somatic embryos, 120
double fertilization, 15
Downingia, 29
  bacigalupii, 25
  pulchella, 25
Drosera, 17
Drusa type of embryo sac, 14

Echinocloa
  crusgalli, 128
  oryzicola, 128
Echinodorus tenellus, 62
egg
  formation of, 14
  polarity of, 48
  structure of, 47
Elaeis guineensis, 87
electrophoresis
  sodium dodecyl sulfate (SDS) gel, 8
  two-dimensional gel, 8
Elymus canadensis × *Secale cereale*, 213
embryo
  cryopreservation of, 227
  culture of, 84–102
  measurement of growth of, 34
  ontogenesis, 15–25, 119
  origin of quiescent center in, 29, 30f
  origin of shoot and root apices in, 21–2, 28–9
  tissue organization in, 25, 27–8
  viability of, 80–2
embryo abortion, 208–10, 213
embryo axis, storage proteins of, 74-5, 105
embryo implantation (transplantation) 80–1, 218
embryo lethal mutants, 111–3

embryo rescue, 211–21
embryo sac
  enzymatic isolation of, 15
  wall outgrowths on, 39
embryo-specific proteins, 150, 206
embryogenesis, historical aspects of, 2–4
embryogenic cells
  cryopreservation of, 227–8
  protein and RNA synthesis in, 197–8
  poly(A)RNA synthesis in, 198
embryogenic competence, 134–5, 137, 143, 184–7, 227
embryogenic pollen grains
  biochemical cytology of, 174–7, 203–5
  enrichment of fraction of, 163
  ultrastructural cytology of, 171–4
embryoid
  cryopreservation of, 227
  division sequences of, 119
  flowering of, 124
  habituation of, 125
  of pollen grain origin, 153–61
  of somatic cell origin, 120–32
endogenous hormones
  in pollen embryogenesis, 164–5
  in somatic embryogenesis, 141–2, 149
endopolyploidy, 158, 188, 223, 228
endoreduplication (endoreplication), 53, 57, 62, 67
endosperm
  constituents of, 40–1
  in embryo nutrition, 38–9
  embryoid origin in, 126
  haustoria, 37, 210
  in inviable crosses, 210
  nuclear divisions in, 36
  ontogenesis of, 36–7
  storage proteins of, 75
environmental factors, in somatic embryogenesis, 146–7
epiblast, 23, 90, 94–6
Epidendrum, 31, 48–9
  scutella, 47
epidermal strips, embryoid origin in, 122, 143
Epipactis, 31
Eranthis hiemalis, 39, 44
Eruca sativa, 62–3, 101
Erythronium, 42
Eschscholzia californica, 120, 131
ethylene, 142, 147, 182
Eulophia, 31
  epidendraea, 42
Euonymus europaeus, 91
Exocarpus, 42

F1 hybrids, pollen embryogenesis in anthers of, 187
feedback inhibition, 149
feedback sensitive mutants, 88
female gametophyte (*see also* embryo sac)
  development of, 13
  organization of, 14
Festuca, 161
Feulgen DNA measurements, 52, 63, 68, 174–5
flax (*see also Linum usitatissimum*), 100
fluridone, 110
Fragaria × *Ananassa*, 131
Fraxinus excelsior, 93
free nuclear divisions
  in endosperm, 36
  in pollen cells, 166, 168, 170
  in proembryo, 24
French bean (*see also Phaseolus vulgaris*), 22, 38, 74, 76, 97
Fritillaria type of embryo sac, 14
fusion nucleus, 14

Gagea lutea, 62–3
Garrya, 42
Gasteria verrucosa, 127
gene activity
  in embryogenesis, 193–7
  in pollen embryos, 203–5
  in somatic embryos, 148–50, 197–202
Gastrodia, 31
generative cell
  embyoid formation from, 167–8
  nucleic acid and protein metabolism of, 174–7, 203–5
  ultrastructural cytology of, 173–4
Genista monosperma, 31, 32f
genotypic effects
  on pollen embryogenesis, 158–61, 186
  on somatic embryogenesis, 148
Geodorum, 31
Geranium phaeum, 62
Gerbera jamesonii, 157
germ plasm, preservation of, 225–8
gibberellic acid (GA), gibberellins
  effects on embryo growth, 89–90
  in embryo dormancy, 91–3
  in somatic embryogenesis, 124, 126, 145–6
ginseng (*see also Panax ginseng*), 124
globulins, in storage proteins, 75
glutamine
  enhancement of somatic embryogenesis by, 138–9
  induction of embryo growth by, 87
Glycine, 218
  max, 34, 48, 107
  tomentella, 219

Subject index

glycinin, 54, 74, 77–8
glycoproteins, 74, 80
Gossypium, 217
  arboreum, 219
  barbadense, 219
  herbaceum, 219
  hirsutum, 33, 157, 219
  klotzschianum, 139
grape (*see also Vitis vinifera × rupestris*), 120, 125f, 146

Habenaria, 31
Haemanthus katherinae, 39
Halophila ovata, 22
haploid cells, 224–5
haploid embryogenesis, 152
haploid embryoids, 152
haploid mutants, 224
Haworthia fasciata, 127
hazel (*see also Corylus avellana*), 91–3
Helianthus annuus, 28
Helleborus foetidus 156
Heracleum sphondylium, 87–8
heterotrophic phase, of embryos, 38, 97, 124
Hevea brasiliensis, 157
Hibiscus
  costatus, 48, 209
  costatus × furcellatus, 209
  furcellatus, 48, 209
high frequency embryogenesis
  from pollen grains, 163, 165
  from somatic cells, 123, 125f, 138, 145, 148
histochemical methods, 5
histones, 67, 201–3, 205
Hordeum, 192, 218
  bulbosum, 36, 192, 213
  distichum, 52, 54, 86
  distichum var. Hannchen, 52
  distichum × Secale cereale, 220
  vulgare, 34, 36, 86, 159, 166, 170, 173, 180, 188, 192, 213
  vulgare × S. cereale, 213, 220
  vulgare × Triticum aestivum, 128, 213
  vulgare × T. dicoccum, 213
  vulgare × T. monococcum, 213
  vulgare × T. turgidum, 213
  (*vulgare × T. aestivum*) *× S. cereale*, 213
  vulgare var. Akka, 185
  vulgare var. Dissa, 159
  vulgare var. Sabarlis, 159, 160f, 166, 170
Hordeum × Agropyron, 218
Hordeum × Secale, 218
Hordeum × Triticale, 218
horse chestnut (*see also Aesculus woerlitzensis*), 40

hybrid arrested translation, 11, 109
hybrid embryos
  callus formation on, 220
  culture of, 212–18, 214t
  development of, 209
  somatic embryos from, 222
hybrid selected translation, 11, 109
Hyoscyamus, 154, 180
  muticus, 225
  niger, 153, 154f, 162–3, 168, 169f, 173, 175, 176f, 177, 179, 181, 183–4, 186–7, 203, 204f, 205
hypocotyl, initiation of, 21
hypophysis, 19, 21–2, 29, 50–1

Iberis amara, 155
Ilex, 94
  aquifolium, 131
  opaca, 95f
indoleacetic acid (IAA)
  in pollen embryogenesis, 152–3, 155, 158–9, 181
  in somatic embryogenesis, 117, 127, 141
indolebutyric acid (IBA), 141, 181
inhibitors of embryogenesis, 141, 147, 182
in ovulo embryo culture, 102–3, 104f, 219–20
in situ hybridization, 9, 11, 65, 68, 203, 204f
inviable crosses
  embryo development in, 209
  endosperm development in, 210
Ipomoea
  batatas, 124, 131, 157–8
  purpurea, 61
Iris, 127
iron, in culture medium, 103, 180
N^6 (Δ^2-isopentenyl) adenine (2iP), 121, 124, 145
isopentenyladenosine, 145
Isotoma, 42
isozymes, 52, 201, 221

Juglans regia, 27, 28

kinetin, 2, 6, 90, 95, 101, 103, 106, 121, 123–4, 127, 145, 152–3, 155, 158–9, 181, 214, 216

Lagerstroemia speciosa, 48
Lathyrus clymenum × articulatus, 214t
laticifers, 120
lectin, 74, 78, 109, 113
Ledum groenlandicum, 49
legumin, 73, 74f, 78–9
Leguminosae, pollen embryogenesis in, 156–7
Lilium, 161
lima bean (*see also Phaseolus lunatus*)

Limnophyton obtusifolium, 50
Linum, 48
  catharticum, 47, 49
  perenne × *austriacum*, 211
  usitatissimum, 38, 47, 49
Listera, 31
Lobelia, 42
Lobularia maritima, 155
Lolium, 161
  multiflorum, 128
lotus (*see also Nelumbo nucifera*), 86
Lotus, 31, 63, 218
  carmeli, 62
  corniculatus, 32f
  japonicus × *alpinus*, 214t
  japonicus × *filicaulis*, 214t
  japonicus × *frondosus*, 214t
  japonicus × *schoelleri*, 214t
  pedunculatus, 62
  purshianus, 62
Luffa
  cylindrica, 157, 166
  echinata, 157
lupine (*see also Lupinus albus*), 28
Lupinus, 31, 62
  albus, 28
  luteus, 99
  pilosus, 32f
Luzula forsteri, 22
Lycium, 179
  barbarum, 155
  halimifolium, 155
Lycopersicon, 154
  esculentum, 111, 162, 220
  peruvianum, 133, 188, 220
  pimpinellifolium, 162

Macleaya cordata, 122
maize (*see also Zea mays*), 24, 30, 40–1, 47–9, 75, 86–9, 110–3, 160–1, 168, 206
male gametophyte (*see* pollen grains)
Mangifera, 42
  indica, 43, 126, 131
maternal mRNA, 191–2, 197
Medicago
  sativa, 39, 61, 131, 133, 148, 211, 214t
  sativa × *scutellata*, 209, 211
  scutellata, 61
megaspore, 14
megaspore mother cell (megasporocyte), 13, 192
Melandrium rubrum, 62
Melilotus officinalis × *alba*, 214t
mesophyll
  cells, somatic embryogenesis of, 122–3
  protoplasts, somatic embryogenesis of, 133–4

messenger RNA (mRNA)
  changes during embryogenesis, 78–9
  changes in sequence complexity of, 193–5, 196–7
  changes in subsets of, 195f, 196–7
micropropagation, 221–2
microspore, 15
microspore mother cell (microsporocyte), 15
Mirabilis jalapa, 28
monosporic embryo sac, 14
Musa, 144
Muscari comosum, 22
myo-inositol
  biosynthesis of, 140
  effects on pollen embryogenesis, 160
Myosurus minimus, 47

Najas lacerata, 22
naphthaleneacetic acid (NAA), 44, 89, 123, 127, 141, 143–4, 155, 159, 181
naphthoxyacetic acid (NOA), 125, 141
napin, 75, 79
Narcissus biflorus, 186
Nelumbo nucifera, 86
Neottia, 31
Nerium oleander, 27–8
Nicotiana, 154–5, 166, 180, 187, 211, 219
  glutinosa, 183
  nesophila, 219
  nesophila × *tabacum*, 220
  repanda, 219
  rustica, 163
  rustica × *glutinosa*, 210
  rustica × *tabacum*, 210
  stocktonii, 219
  stocktonii × *tabacum*, 220
  sylvestris, 133, 163, 169, 187
  tabacum, 47, 133, 155, 166, 169, 172–3, 178–9, 183, 185, 219, 228
  tabacum var. Badischer Burley, 163, 164f
  tabacum var. Coulo, 162
  tabacum var. Samsun, 163
  tabacum var. White Burley, 166, 167f
Nigella sativa, 124
nitrates
  in embryo growth, 87
  in somatic embryogenesis, 137–8
nonembryogenic cells
  poly(A)RNA synthesis in, 198
  protein and RNA synthesis in, 197–8
nonhistone nuclear proteins, 202–3, 205
nucellar callus, somatic embryogenesis in, 124–5
nucellus
  adventive embryogenesis from, 42, 43f
  hyperplastic growth of, 211

300   Subject index

nuclear fusion, in pollen embryogenesis, 167–8, 170, 174, 188
oat (see also Avena sativa), 87, 89
Oenothera
    coronifera, 157
    lamarckiana, 47
Oenothera type of embryo sac, 14
oil palm (see also Elaeis guineensis), 87, 227
Ononis, 31
Ophrys, 31
Opuntia dillenii, 43
orchids (Orchidaceae), 31, 33, 42, 94
Orchis, 31
Ornithogalum, 48
Ornithopus, 218
Oryza
    sativa, 34, 128, 166, 170
    sativa var. indica, 158
osmolality
    of culture medium, 79–80, 101, 106
    of embryo sac, 100
ovary culture, 220
ovule culture, 102–3, 104f, 219–20

Paeonia, 25, 166, 186
    anomala, 24, 26f, 86
    lactiflora, 156, 163
    lutea, 156
    moutan, 24
    suffruticosa, 156
    wittmanniana, 24, 26f
Panax ginseng, 124
Panicum
    maximum, 128, 130, 133
    miliaceum, 128
    miliare, 128
Papaver
    radicatum, 158
    setigerum, 158
    somniferum, 120
papaya (see also Carica papaya), 126
parsley (see also Petroselinum hortense), 126
parthenogentic activation, 191–2
pathways of pollen embryogenesis, 165–70, 171f
pea (see also Pisum sativum), 27, 30, 38–40, 73, 74f, 78, 82
Pelargonium hortorum, 157
Penaea type of embryo sac, 14
Pennisetum, 161
    americanum, 127, 128, 129f, 133–4
    americanum × purpureum, 128
    purpureum, 128, 133
    typhoideum, 90
peony (see also Paeonia), 24

Peperomia type of embryo sac, 14
Persea americana, 217
Petroselinum hortense, 126
Petunia, 48, 154, 174, 187
    hybrida, 30, 47, 52, 102, 112, 163, 179, 183, 224, 228
Pharbitis nil, 157–8
phaseolin, 75, 105
Phaseolus, 31, 59, 62, 64–5, 209
    acutifolius × vulgaris, 215t
    aureus, 157
    coccineus, 38, 55, 59, 60f, 61–2, 63t, 65–71, 101–2, 105
    coccineus × acutifolius, 215t
    coccineus × vulgaris, 209, 215t
    hysterinus, 62
    lunatus, 29
    multiflorus, 62
    vulgaris, 22, 34, 35f, 51, 54–5, 61–2, 64–6, 68, 80, 88, 105, 107, 109–10, 156, 210
    vulgaris × acutifolius, 86, 209–10, 214t, 215t
    vulgaris × coccineus, 209, 215t
    (vulgaris × coccineus) × acutifolius, 215t
    vulgaris × lunatus, 209, 215t
    vulgaris × ritensis, 215t
Phlaenopsis, 31
Phlox drummondii, 25, 28
Phoenix dactylifera, 227
Physalis
    ixocarpa, 155
    minima, 155
Pimpinella anisum, 145
Piperad type of embryo, 18
Pisum
    arvense, 53, 54f, 82
    sativum, 27, 30f, 53, 61, 73, 157
placenta, somatic embryogenesis from, 126
plant extracts
    effects on embryo growth, 98–9, 106
    effects on pollen embryogenesis, 152, 155, 182
Plumbagella type of embryo sac, 14
Plumbago, 48
    zeylanica, 47
Plumbago type of embryo sac, 14
Poa pratensis, 89
pollen callus, 153, 155–9, 160f, 161
pollen embryoids, cryopreservation of, 228
pollen grains
    culture of, 162–5
    cytochemical observations on, 174
pollen sterility, 185–6
polyadenylic acid-containing RNA [poly(A)RNA]
    isolation of, 7

localization of, 10
polyamines, 199
Polygonum divaricatum, 14
Polygonum type of embryo sac, 14
polyteny, 62-8, 70
polyvinylpolypyrrolidone, 182
Poncirus trifoliata, 157
Populus, 157
potassium ion, role in somatic embryogenesis, 139-40
potato (*see also Solanum tuberosum*), 187
Potomogeton
   densus, 62
   lucens, 23
precocious germination, 79, 80, 89, 103-11
   molecular aspects of, 107-9
Primula, 42
   obconica, 157, 228
proembryo
   culture of, 98-102
   definition of, 15
proembryogenic mass, 134-5, 143, 145
proline requiring mutant, 113, 206
protein synthesis
   developmental regulation of, 73-80, 193-7
   in nonviable embryos, 81
   in pollen embryoids, 178
   in pollen grains, 177
   regulation by ABA, 79-80, 196
   in unfertilized egg, 191
protoplasts
   embryogenic division of, 132-3
   isolation and culture of, 6
   isolation of mutants from, 225
Prunus
   amygdalus, 156
   avium, 156
   persica, 156
Pseudomonas tabaci, 225
Psophocarpus tetragonolobus, 157
pumpkin (*see also Cucurbita maxima* and *C. moschata*), 99
Pyrus malus, 91

Quercus, 48-51
   gambelii, 47

radiation, effects on somatic embryogenesis, 143, 144f
Ranunculus sceleratus, 130, 133, 136
Raphanus, 85
   sativus, 196
respiratory enzymes, in embryogenesis, 51-2
Rhizophora mangle, 27, 110
Rhododendron, 49

Ribes
   nigrum, 131
   rubrum, 130, 131
ribonucleic acid (RNA)
   lesion in nonviable embryos, 81-2
   localization in embryos, 50-1
   metabolism in embryos, 54-8
   metabolism in pollen embryoids, 175-7
   metabolism in pollen grains, 174-5, 203
   monitoring synthesis of, 7
ribosomal RNA (rRNA), genes for, 65, 67, 198
rice (*see also Oryza sativa*), 34, 42, 52, 55, 81, 87, 109, 153, 158, 163, 165, 168, 187, 212, 223, 227-8
Ricinus communis, 28
RNA-DNA hybridization, 9, 67, 73, 77, 79, 109, 193-4, 196, 198
root apex, 21, 28-9
Rosa
   damascena, 156
   hybrida, 156
rye (*see also Secale cereale*), 34, 81-2, 87, 162, 168, 187, 209, 211, 213, 218

Saccharum, 161
   officinarum, 128, 180
   spontaneum, 163
Sagittaria sagittifolia, 22
Saintpaulia, 180
   ionantha, 157
Sambucus nigra, 157
sandalwood (*see also Santalum album*), 126
Santalum album 126, 133
Saxifraga, 17
Scopolia, 154
scutellum, 23, 75, 89, 90, 94-6, 106, 127-8
Secale, 161
   cereale, 34, 128, 130, 166
Sechium edule, 99
seed embryo
   carbohydrate nutrition of, 86-7
   hormonal requirements for growth of, 88-91
   nitrogen nutrition of, 87-8
seeds, with rudimentary embryos, 44, 94
serine
   in anthers, 183-4
   in culture medium, 162
Sesamum indicum, 27
Setaria, 161
   italica, 128
Shamouti orange (*see also Citrus sinensis*), 124, 135, 143, 144f, 201
shed pollen culture, 165
shoot apex, 21-3, 28-9
single copy DNA, 10, 77, 193-4

Solanad type of embryo, 17, 119
Solanum, 154–5, 180, 186
  melongena, 144, 148
  melongena × khasianum, 217
  nigrum, 187
  surattense, 166
  tuberosum, 163, 180–1
somatic embryos (somatic embryogenesis), 118f, 125f
  auxin specificity in, 144–5
  in cereals and grasses, 127–30
  origin from callus, 121–30, 221
  origin from explants, 130–2
  synchrony in yield of, 122
  ultrastructure of, 135–6
somatoplastic sterility, 211
Sophora flavescens, 62
Sorbus aucuparia, 91
Sorghum
  arundinaceum, 128
  bicolor, 128
  vulgare, 87
soybean (see also Glycine max), 34, 53, 55f, 74, 76–7, 78f, 110, 113, 138, 148, 193–4, 219
Sphenoclea zeylanica, 25, 27
spinach (see also Spinacia oleracea), 47–8
Spinacia oleracea, 47
Spiranthes, 31
Stellaria media, 25, 39, 50, 59, 68
Stipa elmeri, 47
storage proteins, 53–4, 72–80
  composition of, 73–5
  expression of genes for, 76–80, 78f
  synthesis of, 73, 109
strawberry (see also Fragaria × Ananassa), 131, 156
structural genes, 192–4
sucrose
  as osmotic stabilizer, 100, 106
  in pollen embryogenesis, 155, 158–9, 161, 180–1
sugarcane (see also Saccharum officinarum), 130
sunflower (see also Helianthus annuus), 28, 39, 47, 51
suspensor
  cytology of, 61–8
  in embryo growth, 71, 101–2
  of embryo lethal mutants, 112
  enzyme activity in, 68–9
  haustoria, 31–3, 69
  hormone synthesis in, 70–1
  in inviable crosses, 210
  morphology of, 30–3

  RNA and protein metabolism of, 55, 69–70
  ultrastructure of, 38, 58–61
synergids, 14, 42

tapetum, 162, 182–3
temperature-sensitive variants, 150
tetrasporic embryo sac, 14
Theobroma cacao, 131
tissue culture, 6
tobacco (see also Nicotiana tabacum), 47–9, 116, 162, 165, 167–8, 173–5, 179–80, 184–5, 187, 223–4
tomato (see also Lycopersicon esculentum), 111, 217
tonoplast, of embryogenic pollen, 172
Torenia, 48
  fournieri, 47
totipotency, 116
transfer cells, 38–9, 61
Trapa
  bispinosa, 32
  natans, 62
Trifolium, 218
  alexandrinum, 156–7
  ambiguum, 209–10
  ambiguum × hybridum, 215t
  ambiguum × repens, 210
  hybridum × ambiguum, 215t
  nigrescens × repens, 215t
  pratense, 131, 148, 221
  repens, 131, 209
  repens × medium, 209
  repens × nigrescens, 215t
  repens × uniflorum, 215t
  sarosiense, 221
  sarosiense × pratense, 216t
  uniflorum × repens, 215t
Triticale, 36, 161, 168, 170, 180, 213
Triticum, 36, 217
  aestivum, 23, 24f, 36, 86, 107f, 159, 166, 170, 180
  aestivum × Hordeum bulbosum, 213
  crassum × H. vulgare, 217
  durum, 80–1
  sativum, 86
  sativum × Aegilops speltoides, 217
  ventricosum × H. bulbosum, 213
  vulgare, 159
  vulgaris, 52, 55
Tropaeolum, 69
  majus, 32f, 33, 61–3, 67, 69–71
Tulipa, 42
Tunica saxifraga, 62

Ulmus americana, 157
unfertilized ovule, origin of embryos in, 192

Vanda, 31, 50
variant cell lines, 143, 148–50, 200, 206, 224–5
vegetative cell
  embryoid formation from, 166–8
  nucleic acid and protein metabolism of, 174–7, 203–5
  ultrastructural cytology of, 172–3
Vicia faba, 53–4, 73, 86, 156
vicilin, 73, 74f, 78–9
Vigna
  aconitifolia, 133
  mungo × *radiata,* 216t
  radiata × *angularis,* 216t
  sinensis, 38–9, 61
  umbellata × *angularis,* 216t
  unguiculata, 74

Vitis
  vinifera, 131, 157
  vinifera × *rupestris,* 120
vivipary, 110–11, 147

wheat (*see also Triticum aestivum, T. vulgare*), 23, 34, 38, 40, 47, 87–9, 95, 109–10, 128, 159, 165, 168, 174, 185–7, 209, 213, 218, 223, 227–8
wheat-germ agglutinin, synthesis of, 109
wild carrot, 117, 121, 137, 140, 142
wildfire toxin, 225
Withania somnifera, 155

Zea mays, 24, 96, 128, 166, 180
zeatin, 40, 102, 121, 145–6, 181
zeatin riboside, 102
zein, 41
zygote, 15, 48
  in inviable hybrids, 209